Explanatory Models, Unit Standards, and Personalized Learning in Educational Measurement

William P. Fisher Jr. · Paula J. Massengill
Editors

Explanatory Models, Unit Standards, and Personalized Learning in Educational Measurement

Selected Papers by A. Jackson Stenner

 Springer

Editors
William P. Fisher Jr.
Living Capital Metrics LLC
Sausalito, CA, USA

Graduate School of Education, BEAR
Center
University of California, Berkeley
Berkeley, CA, USA

Research Institute of Sweden (RISE)
Gothenburg, Sweden

Paula J. Massengill
University of Maryland
College Park, MD, USA

A. Jackson Stenner, 12715 Morehead, Chapel Hill, NC 27517, USA

ISBN 978-981-19-3749-1   ISBN 978-981-19-3747-7 (eBook)
https://doi.org/10.1007/978-981-19-3747-7

This Springer imprint is published by the registered company Springer Nature Singapore Pte Ltd.
The registered company address is: 152 Beach Road, #21-01/04 Gateway East, Singapore 189721, Singapore

This collection of reprinted articles celebrates the collegial collaborations of an extensive network of contributors whose creativity and passion combined with Jack's in over four decades of remarkable advances in educational research and practice.

Preface

This selection of articles from the many dozens Jack Stenner authored and co-authored from the 1970s through the 2010s tracks the development of statistical, measurement, methodological, and theoretical themes in his research and applications. Jack's works document the science enabling alignments of measured human and social values with financial values. Regarding human and social value, Jack shows how educational assessment and instruction can be simultaneously custom-tailored to individual student needs, abstractly represented in common metrics, and formally explained by theoretical models. Jack saw that the consistent and broad-scale reproduction of this value over time and across readers, texts, and assessments sets the stage for coordinating educators', publishers', testing agencies', and students' and their families' shared interests in improved instructional outcomes.

What Jack and his networks of collaborators accomplished provides a model of applied science and commercialized research that ought to be emulated across all fields employing test-, assessment-, and survey-based measurements. Much remains to be done to follow through on what Jack and his colleagues began, not only in emulating the model they demonstrated but also in extending it in further expansions of the relational networks they created.

Jack holds a Ph.D. from Duke University and a Bachelor's degree in Psychology and Education from the University of Missouri. He served as President of the Institute for Objective Measurement in Chicago, Illinois, and as a board member for The National Institute for Statistical Sciences (NISS).

Over the course of his career, Jack Stenner was involved in over 20 business startups ranging from a popular restaurant (Anotherthyme, in Durham, North Carolina) to a furniture factory. His legacy as a serial entrepreneur includes having been Chairman and co-founder of National Technology Group, a 700-person firm specializing in computer networking and systems integration that was sold to VanStar Corporation in December, 1996. He was also co-founder and Chairman of MetaMetrics, Inc., a privately held educational research and development corporation. Jack was the Principal Investigator on five Small Business Innovation Research grants from the National Institute of Health, (1984–1996) dealing with the measurement of literacy.

A little-known fact about Jack is that for many years he competed in billiards tournaments. The 1993 edition of the Professional Billiard Tour Association's Billiards: Official Rules, Records, & Player Profiles shows Jack Stenner as Vice President and Acting President. Jack often played a round during lunch breaks.

And there's more: according to the men's basketball record book published by the University of Missouri, St. Louis, Jack Stenner is "Among the most celebrated players in UMSL men's basketball history and UMSL's first all-American selection" (UMSL, 2021, p. 13). He "earned all-American status as a senior during the 1968–1969 season, after averaging 24.3 points per game and leading his team to a 19–7 record…and a trip to the NAIA National Tournament" (UMSL, 2021, p. 13). Jack "ranks eighth in UMSL history with 1,267 career points and second best career scoring average at 20.1 points per game. He was drafted by the Carolina Cougars in the ABA Draft after his career at UMSL" (UMSL, 2021, p. 12).

Jack's accomplishments in competitive sports, in business, and in science share a common theme that might be summed up as his grasp of the fact that "the play is the thing." Where Hamlet focused on catching the conscience of the king, Jack intuitively understood how to captivate educators' imaginations. Jack saw how to maximize the potential for learners to experience the creative flow that follows from appropriately targeted reading challenges and timely feedback, as was so aptly described by Csikszentmihalyi (1990). Jack furthermore also tapped intuitively but directly into Gadamer's (1989, pp. 101–134) insight that play is a key clue to authenticity in the method. Jack saw mathematical measurement models as tools for telling absorbing stories about the development of reading ability. These stories, like all great literature and science, are true of us all in their broad, abstract generality even though they never happen in their specific details to anyone in particular.

In providing a common language through which developmental processes of growth in reading ability can express themselves, Jack Stenner gave humanity gifts of models and methods informing a new range of artful sciences and scientific arts. His accomplishments will directly and indirectly influence the relations of science, technology, education, and business for decades to come.

Sausalito, California, USA   William P. Fisher Jr.
April 2022

References

Csikszentmihalyi, M. (1990). *Flow: The psychology of optimal experience*. New York: Harper & Row.

Gadamer, H.-G. (1989). *Truth and method* (J. Weinsheimer & D. G. Marshall, Trans.) (Rev. ed.). New York: Crossroad (Original work published 1960 in German).

Meurin, D. (1993). *Billiards: Official Rules, Records, & Player Profiles*. New York: SPI Books.

UMSL. (2021). *UMSL Triton Men's Basketball Record Book*. St. Louis: University of Missouri, St. Louis Sports. https://s3.amazonaws.com/sidearm.sites/umsl.sidearmsports.com/documents/2020/6/17/20_21_MBB_recordbook.pdf.

The original version of this book was revised. In the first paragraph on page xxiii, the spelling of "Robert Millikan's" has been changed to "Robert Mullikan's".

Acknowledgements

The editors would like to extend their appreciation and thanks

- to the co-authors of the articles included here for their permission to reprint their works,
- to all of Jack's friends and colleagues at MetaMetrics for their years of active involvement in shaping and advancing the theory and practice of educational measurement,
- to Jennifer Stenner for initiating the book project, and
- to Jack himself for providing an inspiring model of intelligence, imagination, and compassion in conceiving and nurturing an artful science and scientific art of educational research and practice.

Introduction: Koans, Semiotics, and Metrology in Stenner's Approach to Measurement-Informed Science and Commerce

Abstract Over the course of the last 40 years and more, Alfred Jackson Stenner, signed as A. Jackson Stenner and known by all as Jack, in collaboration with an extensive network of colleagues, has led the introduction of a new model of integrated science and commerce. Papers authored and co-authored by Jack on measurement science, reading comprehension, and the creation of more systematic states of affairs in education policy and practice are collected together in this book. Jack notably operationalized technical issues involved in measuring psychological and social constructs, especially reading comprehension. In the background, behind these explicit themes, playful Zen koans, metrological network alliances, and commercial business agreements are all implicated. Most details of the events that transpired will have to be left to others with the needed expertise in contracts, accounting, pricing, marketing, and sales, and who have access to the relevant documents and personal stories to be told. That broader biography and detailed history of Jack's work seems likely to become of considerable interest, given the innovations incorporated into the model of integrated psychological science and commerce he and his colleagues enacted. But the commercial implementations of the technical processes reported in Jack's publications are of less relevance in the context of the papers collected in this book than the ideas informing those processes, and their still-unfolding implications. What is of interest here is how Jack and his allies demonstrated the creation of shared social realities that co-evolve in dialogue with objectively repeatable and reproducible measurements. Finally, I concentrate here on Jack's work in reading comprehension as it provides the most salient focus for considering the development of his ideas on measurement.

Introduction

Working closely with key allies possessing complementary skills and resources, Jack Stenner demonstrated new ways of blending science and commerce. He showed how to model and represent objective facts so they could be put into circulation as common currencies for the exchange of actionable ideas and observations. He intuitively grasped the value obtained when a phenomenon persistently shows itself in characteristic ways over time and space, and across samples of persons and test items. He moreover saw how that value could be made fungible and pragmatically actionable. He understood, seemingly innately, that real value could be translated into a viable business model only by linking together networks of actors sharing a common frame of reference. Jack understood how measurement models parameterize these frames of reference in ways that make them exportable from the laboratory into the real world. He saw that statistically sufficient estimators for the parameters in identified models lay the foundations for efficient markets matching supply and demand. Economists like Frisch, Koopmans, and Reiersøl and V. Smith would have been proud to see the ways that Jack implicitly proved the truth of their ideas (Fisher, 2002, 2015, 2021b).

As was so clearly stated by Bernstein (2004) and De Soto (2000) after Jack was well underway, Jack saw that the science had to be complemented with arrays of social, legal, and economic relationships if it was to be put into general circulation. Those relationships have to be formatted so as to incorporate the language by which the thing of interest is recognized for what it is. Absent those connections, Jack saw, science can only ever exist as an area of activities isolated and alienated from the broader spheres of human interests. And though Jack and his colleagues have accomplished a great deal, they have only scratched the surface of what is needed. Bringing together some of the key articles documenting high points in Jack's long career may increase the probability that others will extend and improve on what he has done.

So, of particular interest here are Jack's intensive investments in practical implementations of measurement research products. He saw that technical processes for systematically integrating concrete, abstract, and formal levels of complexity opened up opportunities for re-organizing social relationships. From the time he and I met at an International Objective Measurement Workshop held in the mid-1980s in Ben Wright's Judd Hall classroom at the University of Chicago, Jack and I shared a passion for engaging with these topics. Though we conversed endlessly on Zen koans as much as measurement, I will here focus primarily on the measurement side of things, reserving a fuller consideration of Jack's love of Zen koans for another occasion.

To what degree was Jack's profound understanding of how to proceed in his work gleaned from careful and prolonged study, and to what degree was it a product of highly developed intuitions? Plainly, his expertise in mathematical modeling, statistics, and experimental science required years of careful development. But it is also the case that, being widely read across multiple fields, including history and philosophy, Jack's imagination was captivated by multiple intriguing transdisciplinary connections.

My focus in this foreword is an expansion on some of the most pointed themes recurring in conversations with Jack, themes that also are implicated in the background of his writings. What I have to say is, then, a quite personal combination of Jack's and my own perspectives. Jack and I both often remarked that we could not recall who had thought of this or that idea first. We also often could not recall if or when our ideas were drawn from the works of one author in particular, or another, of the many we both engaged with in our thinking.

Our relationship was characterized by a particular kind of dialogue. On the one hand, Jack's intuitions as to how to proceed with commercializing the value produced in reading measurement research were complemented by my deep familiarity with philosophy and the history of science and technology. I could connect Jack's experience with examples from the past providing pertinent lessons and analogies relevant to the problems he faced. On the other hand, conversely, my interests in generalizing the lessons of history and philosophy were tested and challenged by Jack's successes and failures in creating practical measurement implementations in a business environment intolerant of pointless academic wool gathering.

That Jack and I both had extensive experience in the application of Rasch's probabilistic models of measurement, in instrument design and data analysis, and in the testing of predictive construct theories provided a background of shared values. That common background was further informed by Jack's expertise in language learning and my own in-depth studies of how metaphor and analogy infuse cognitive processes. A matter of key concern to be taken up below falls under the heading of the ecological fallacy and the importance of distinguishing and separating levels of meaning in language. Even though Jack did not foreground these issues in his writing, an early emphasis (Hayman et al., 1979) on keeping levels of data aggregation in close correspondence with the organizational levels in which they are put to use became an enduring theme.

That theme found highly salient expression in O. D. Duncan's (1984, pp. 38–39) suggestion "that social measurement should be brought within the scope of historical metrology, while that discipline learns to take advantage of sociological perspectives." Duncan took an historical perspective on the role of measurement in the economy, seeing lessons for sociology in the counterproductive confusions and opportunities for deception associated with false weights and diverse measures. His authoritative grasp of Rasch's probabilistic models for measurement complemented his understanding that "All measurement is...social measurement. Physical measures are made for social purposes" (Duncan, 1984, p. 35). Duncan speculated, "it may be that...the measurement of time and the temporal framework of social organization will prove to be the entering wedge for a sociological metrology" (Duncan, 1984, p. 27). Jack and I thought that the temporal framework of social organization in education—focusing on cognitive development and reading ability, in particular—would provide an actionable point of entry for psychological metrology. Similar opportunities have been opened for measured constructs involving learning progressions, healing trajectories, and developmental sequences in general.

In that context, this foreword can be neither a perspective on Jack's organizational and business accomplishments nor a precise exposition of philosophical and historical influences on Jack's accomplishments. My intention is not to document exact or fully consistent connections or motivations, but rather to convey something of how ideas in the background of the technical work provide a context for a more intimate understanding of what Jack presents in his formal arguments. I recount in a series of meditations the broad strokes of Jack's thinking on the topic of Zen koans and draw out his intuitive pursuit of a thoroughly unmodern (or amodern, terms that will be clarified in what follows) approach to integrating science and commerce. This foreword closes with a brief synopsis of the papers reprinted in the following chapters.

Koans

Over the years we worked together, Jack often spoke of his love for Zen koans. Koans are paradoxical statements traditionally used in meditation by Buddhists as a way of freeing thinking from rigid preconceptions. The basic idea as Jack and I explored is that when words in a grammatically correct sentence do not immediately cohere into a meaningful statement, the resulting perplexity may lead one to reflect on the arbitrariness of idea-word-thing connections.

This foreword is not the time or place for a thorough treatment of even the informal understanding Jack and I shared of the potential relevance Zen koans might have regarding the role of language in understanding and in the measurement of reading comprehension. A brief summary of the basic idea will suffice to give some background to the papers collected together here.

A couple of famous koans include "What is the sound of one hand clapping?" and "If a tree in a distant wood falls and no one hears it, does it make a sound?" Focusing attention on these seeming nonsensical questions may aid in releasing thinking from the bonds of inflexible habits of mind. Encounters with the randomness of word forms and syntactic structures can lead to new ways of seeing the world. Becoming aware of the arbitrariness of any particular language's alphabet, grammar, and pronunciation can be a factor encouraging creative insights, and compassionate understanding, as to how alternatives to existing concepts and things might be formulated.

Jack was fascinated by the different paths forward in thinking opened up when the limits of language are viewed in this way. Being a student of the cognitive processes involved in reading comprehension led Jack to a particularly acute perspective on the value of koans. Meditatively allowing word forms to dissolve into the arbitrary social constructs that they are open perception to previously unnoticed ways of seeing. Now things are experienced in a different ways and are less automatically categorized as something in particular. The effect is in some respects akin to learning a second language, which also can reveal the capricious randomness and historical contingency of the metaphors incorporated in concepts, and in linguistic standards themselves.

Meditating on language in this way can then lead to a pointed awareness of formlessness in relation to the multiplicities of forms, with interesting mathematical implications (Szymanski, 1995). On the one hand, idealizations persist as invariant structures and invisible forces asserting their objective existence independent of particular instances and words. Though meanings are shaped by cultural circumstances in particular ways, things and processes (mountains, trees, rivers, the sun, running, walking, etc.) must exhibit broad consistencies to sustain ongoing ostensive references. On the other hand, the availability of a common language, arbitrary as it is, makes a shared reality possible, one in which meanings within specific and unique circumstances can be negotiated via contrasts with ideal concepts.

Jack came to see the value that could be obtained by systematically separating and balancing formal conceptual ideals, abstract instrumental standards, and concrete local circumstances. Seeing that there are infinite ways of imagining, conceiving, and representing things, one becomes more sensitive to the multiplicity of possibilities implicated by the way language focuses attention. Language itself serves as a model for how representing concrete things systematically requires clear distinctions between unrealistic conceptual ideals and abstract standardized words and semantics. This semiotic triangle of ideas-words-things opens onto an infinite array of possible representations at the same time that it does so by restricting attention to salient matters of interest.

The norms of everyday communication set up opportunities, then, in the context of limits. The limits are not completely inviolable, as humor, irony, and poetry all open onto creative configurations of circumstances that can provide insights into how things might be different from the way they are. But no matter how powerful one's experience of infinite formlessness may have been, articulating it inevitably confronts the constraints imposed by the forms of communication in the language. The idea of sharing a fulfilling meditative experience of a seamless whole immediately confronts a conundrum: whatever is said will have to be shaped in the particular forms of a linguistic expression, a reduction of the infinite to a text of finite length, which, of course, immediately contradicts the point of trying to communicate the entirety of the experience. Thus, raucous laughter is a typical response of those experienced in meditation when asked questions about the meaning of life, or the essence of Zen.

Modern, Postmodern, and Unmodern Issues in Science and Measurement

A valuable lesson Jack took away from these meditations was seeing that the implicit and informal modeling function performed by everyday language could be made explicit and formal. In the same way that linguistic standards (alphabets, phonemes, grammars, etc.) are abstractions mediating unrealistic ideals and concrete circumstances, so, too, measurement standards (quantity values, unit definitions, instrument design principles, estimation methods, etc.) are abstractions mediating the relations of theoretical axioms, laws, and models with locally unique concrete events.

A relevant personal connection with a figure from Jack's childhood in St. Louis illuminates the point of departure from which Jack and I began our conversations. The philosopher Robert J. Ackermann was a young Assistant Professor at Washington University in the 1960s when Jack's father was also on the faculty of the philosophy department there. When I mentioned Ackermann's 1985 book, *Data, Instruments, and Theory: A Dialectical Approach to Understanding Science*, Jack said he was acquainted with Ackermann and as a teenager had earned some spending money babysitting his children. In addition, Jack's father's immersion in philosophical issues was such that a 1992 Washington University dissertation (Campbell, 1992) includes an acknowledgment of the Senior Professor Stenner, thanking him for introducing the author (Campbell) to the theory of metaphor. My (Fisher, 1988) dissertation brought Ackermann's ideas to bear on metaphor in a way that framed my thinking on much of the subsequent work Jack and I did together.

A bit of background is needed to bring out the relevance of Ackermann's ideas. When Jack and I first met in Ben Wright's Chicago classroom, I had been pursuing my interest in situating Rasch's measurement models in the context of science and technology studies, though it was not referred to by that name at the time. Latour and Woolgar's (1979) book, *Laboratory Life: The Social Construction of Scientific Facts*, had opened the door to a new domain of ethnographic, historical, social, and philosophical studies of science and technology. This body of work seemed to me to point in a productive direction in which to take psychological measurement theory and practice.

A key distinguishing feature of science and technology studies that sets it apart from postmodern deconstruction and hermeneutics is its emphasis on the essential roles played by linguistic and scientific standards and technologies. Postmodernism tends to become lost in endless futile debates with modernism, as advocates of transcendental universals spar with advocates of temporally and spatially situated meanings. The infinite spiraling of this contentious opposition ought, itself, to indicate something is amiss. A question that should be asked more often is what makes the arguments unproductive and impossible to resolve?

The answer to that question has, perhaps, a lot to do with hidden assumptions. As Latour (1991, p. 17) puts it, "Postmodernism is a disappointed form of modernism. It shares with its enemy all its features but hope." Modernist vs. postmodernist arguments often proceed with little or no cognizance on the parts of the participants that both sides are making use of a system of standardized signs. Those signs are human-made tools entirely imbued with cultural and historical significance. Ignoring the use being made of language results in internally inconsistent positions. Modern transcendentalists argue in a specific and unique technologically dependent local circumstance for universal objectivity, while postmodernists argue on behalf of relativized subjectivity using formal conceptual ideals represented by globally accessible language.

Ignoring the role of language as a medium leads to mainstream takes on postmodernism that accordingly then range from condemnations of its alleged "suppression of reason" (Bloom, 1987, p. 387) to inconsistently accepting the limits of measurements and interpretations "not assumed to be consistent or similar across time, contexts, or individuals" (Delandshere & Petrosky, 1994; also see Cupples, 2019, p. 15). The irony of saying, in effect, "this statement cannot be interpreted as consistent," seems lost on many writers, though the form of the statement is exactly that of the liar's paradox ("this statement is false"), long recognized as a limit in recursive, self-referential systems (Sethy, 2021). Both rejecting and accepting postmodern ideas necessitate applications of shared systems of symbolic standards, though these are ignored by both sides in the debate.

In contrast to the self-contradictory tensions characterizing the modern-postmodern debates, there are philosophers associated with postmodernism who "leave room for a structural science" that "must in effect liberate the mathematization of language" (Derrida, 1981, pp. 28, 34). Gasché (1987) and Spivak (1990), for instance, explicitly recognize the need for strategic pauses breaking the potentially endless cycle of deconstructions,and applying new understandings in practice. "Sadly, though," in Tasić's (2001, p. 99) opinion, "the mathematical roots of 'high' postmodern theory seem to be well beyond the reach of most of its representatives (with the possible exception of Derrida)."

Though Tasić's endorsement of Derrida's awareness of mathematical implications is tentative, further evidence is provided in Derrida's (2003, p. 62) explicit statement that,

...many people who read me...don't know what I know.... When I take liberties, it's always by measuring the distance from the standards I know or that I've been rigorously trained in.

Derrida's reputation as a Socratic gadfly afflicting the comfortable seems diametrically opposite to Gadamer's image as a Socratic midwife comforting the afflicted (Bernasconi, 1989; Risser, 1989), but they both acknowledged the role of language as the medium of thought.

So Gadamer (1989, p. 412; original emphasis), a leading hermeneutic philosopher, understood that "it is *not word but number* that is the real paradigm of the noetic," the domain of being and existence. Gadamer (1989, p. 413) continued, recognizing that:

The more univocally a sign-thing signifies, the more the sign is a pure sign—i.e., it is exhausted in the co-ordination. Thus, for example, written signs are co-ordinated with particular sounds, numerical signs with particular numbers, and they are the most ideal signs because their position in the order completely exhausts them.

These reflections on language as a medium (Gadamer, 1989, p. 351) effectively crystallize the value of the deconstructive moment in the phenomenological, or onto-logical, method (Heidegger, 1982, pp. 19–23, 320–330; Gadamer, 1989; Nuzzo, 1999, 2011; Fisher, 2004, 2010; Fisher & Stenner, 2011a). All method (a) reduces complex experiences to simplified expressions in language, and then proceeds by (b) applying linguistic expressions in communication, which evolve as (c) inconsistencies are apprehended via critical engagement, and (d) a return to a new reduction begins a new cycle.

Where Gadamer tends to be read as a traditional formalist methodically moving from (a) to (b), focused on idealized ways of appropriating meaning, Derrida, as Tasić (2001, p. 58) points out, seems more aligned with Weyl's intuitionism, methodically moving from (c) to (d) (Gasché, 1987, 2014). Furthermore, where Tasić (2001, p. 106) is "searching for common patterns that could serve as the ground for developing analogies and comparisons as to the 'fate' of both structural linguistics and mathematics," Jack and I aimed to address the need for analogies and comparisons of language and measurement encompassing the full range of qualitative to quantitative mathematical expressions.

The culminating unmodern, nonmodern, or amodern (Dewey, 2012; Latour, 1990, 1993a, 1995b; Fisher, 2018; Fisher & Cavanagh, 2016; Fisher & Stenner, 2018b) insight of postmodern deconstructions of modern language is an expanded semiotic understanding of the inescapable truth that:

- "When I think in language, there aren't 'meanings' going through my mind in addition to the verbal expressions: the language is itself the vehicle of thought" (Wittgenstein, 1958, p. 107).
- "Meanings run in channels formed by instrumentalities of which, in the end, language, the vehicle of thought as well as of communication, is the most important" (Dewey, 1954, p. 210).
- "We have no power of thinking without signs" (Peirce, 1868/1992, p. 30).
- "From the moment that there is meaning there are nothing but signs. *We think only in signs*" (Derrida, 1976, p. 50; original emphasis).
- "...the function of substituting signs for things and of representing things by the means of signs, appears to be more than a mere effect in social life. It is its very foundation. We should have to say, according to this generalized function of the semiotic, not only that the symbolic function is social, but that social reality is fundamentally symbolic" (Ricoeur, 1981, p. 219).

The idea that there is no thought without expression in language goes back at least to the late eighteenth-century works of Hamann and Herder: "They did not use the word 'structure,' but romanticists like Schleiermacher and Humboldt did" (Tasić, 2001, p. 104; also see Surber, 2011, pp. 245–247). Philosophy has increasingly engaged with language as a semiotic model of reason via developments quite independent of the emergence of semiotics as a field of its own.

The contemporary semiotic perspective in the philosophy of science emphasizes instruments, models, and technologies as media embodying relational connections between theory and data (Blok et al., 2020; Bowker et al., 2015; Bud & Cozzens, 1992; Haraway, 1996; Ihde, 1991; Latour, 1987, 2005; Wise, 1995). This point is emphasized in the way Princeton University Press marketed Ackermann (1985) with a description saying the book documents how instruments mediate theory and data, creating a reasoned basis for scientific objectivity that breaks the dependency of observation on theory. Re-expressed in the semiotic context, thing-word-idea and data-instrument-theory connections are embedded in the external environment as material artifacts (phonemic, alphabetical, grammatical, spelling, product definition, unit, etc. standards) that serve as common points of reference.

The unproductive, yet still ongoing, tension between modern and postmodern worldviews is defused when both have to confront the facts that (a) transcendent universals cannot be demonstrated in the absence of technical media (linguistic signs and measuring instruments), just as (b) locally situated unique circumstances cannot be described or shared without communications standards. Unmodern philosophical investigations and studies overthrow contentious modernist and postmodernist debates by means of fundamentally radical critiques that "follow *the exact same paths* through which the extension of standards, templates, or metrological chains occurs" (Latour, 2014, p. 10; original emphasis). Deconstructive, philological, and hermeneutic analyses can follow those paths just as much as transcendental idealizations do. All are involved in taking apart the historical, social, and economic interests informing taken-for-granted concepts embedded in institutionalized linguistic routines.

But deconstructively retracing the paths by which conceptions became instituted as standards for shared social realities is not an end in itself:

- "Deconstructive doubt is not a doubt about things but about the unrevisability of established linguistic formulas" (Staten, 1984, p. 156).
- In a 1981 interview, Derrida remarked that "it is totally false to suggest that deconstruction is a suspension of reference.... I never cease to be surprised by critics who see my work as a declaration that there is nothing beyond language" (Kearney, 1984, p. 123).
- "Derrida is not trying to bury the idea of 'objectivity' ... [since] it is not that texts and languages have no 'referents' or 'objectivity' but that the referent and objectivity are not what they pass themselves off to be, a pure transcendental signified" (Caputo, 1997, p. 80).
- "Derrida's concern with foregrounding the constitutive power of inscription...- parallels our interest in mediating machines. Applied to the domain of science studies, Derrida's manner of proceeding suggests that we should attend to the empirical, material character of the experimental system as bound up with the production of a graphic trace" (Lenoir, 1998, p. 6).

Revising—better, transforming—languages serving as vehicles of thought requires new ways of apprehending the semiotic relations of ideas, words, and things. By tracing the routes taken by established linguistic formulas as they circulate in societies and economies, we can come to learn how improved, more complex formulas, ones more sensitive to differences that make a difference, might be created.

Measurement Modeling as Establishing Semiotic Correspondences between Things and Thought

Jack and I felt that Ackermann's 1985 book provided vitally important pointers as to how to find and follow the paths along which traceable standards and metrological chains are extended. Ackermann's (1985) inclusion of instruments in relation to data and theory was independently and simultaneously asserted by a number of additional writers, such as Heelan (1983a, 1983b), Hacking (1983), Latour (1987), and others characterized by Ihde (1991) as an emerging school of instrumental realists.[1] The

---

[1] My pursuit of these connections began about a year before Jack and I met in the mid-1980s. After establishing my philosophical and scientific interest in language as the vehicle of thought, I mentioned Ackermann's book to David Tracy, a philosopher and theologian who supervised my

basic theme on which Jack and I focused concerns what Latour and Woolgar (1979, pp. 151–186) called the "microprocessing of facts." This was described by Ackermann as an achievement establishing connections between real things in the world and representations in language:

> Once clear statements of fact have been achieved through instrumental investigation, the reference of fact seems fixed and objective, and indeed it is. The world has been discovered to show a fixed and repeatable response in certain interactions as described in the language, and this response is an objective consequence of these interactions. … This process of achieving or constructing reference for language by development of a domain we will call the microprocessing of fact, after discussion of this phenomenon by Latour and Woolgar. When the process is complete, the evidence of microprocessing disappears, and mere correspondence, the very correspondence that has been slowly and carefully constructed, is all that remains. (Ackermann, 1985, pp. 143–144)

It was evident to Jack and me that applications of Rasch's (1960, 1977) probabilistic measurement models microprocess facts with the aim of setting up correspondences between linguistic representations and reality. The point of constructing models is plainly focused on identifying repeatedly observable patterns in response data, doing so in ways that learn from the past to predict the future, as Wright (1997a) puts it. The potential for the removal of the evidence of microprocessing after establishing correspondences with real-world phenomena was recognized by 1985 across multiple fields (Andersen, 1977; Andrich, 1978; Andrich & Douglas, 1982; Fischer, 1973; Wright, 1968b, 1977, 1984; Wright & Masters, 1982; Wright & Stone, 1979). Anchoring item estimates adaptively administered from pre-calibrated banks and linking classroom tests with each other as well as with high stakes examinations (Choppin, 1968, 1974; Masters, 1984; Wright, 1977; Wright & Bell, 1984) both make explicit use of carefully constructed correspondences between models and real-life situations.

Philosophically, the problem being addressed here is exactly that taken up in hermeneutic phenomenology (Fisher, 1988, 1992, 2003, 2004, 2010; Fisher & Stenner, 2011a). In Husserl's (1965, p. 108) conception of philosophy as rigorous science, he calls for phenomenologists to return "to the things themselves!" Husserl's, Heidegger's (1962, pp. 50, 195; 1967, p. 105), and Gadamer's (Gadamer, 1991, pp. 14–15; 1994, p. 171) responses to that call focus on

> things as they show themselves before the work of abstraction and theorizing has carved out a language of fixed essences for them removed from human praxis, history and culture (Heelan, 1994, p. 369).

Modernism ignores the erasure of the evidence of microprocessing, so it can focus productively on making use of the practical correspondence with representations and the language of fixed essences that have been established. Unfortunately, the success

---

master's thesis and later co-chaired my Ph.D. dissertation with Ben Wright. Tracy then pointed me toward Heelan (1983a) and Toulmin (1982). Heelan and Toulmin exchanged views at a 1984 conference held at the University of Chicago (Heelan, 1984). I was able to ask both of them about their dialogue when I met the former at a meeting of the Society for Phenomenology and Existential Philosophy and when I enrolled in a University of Chicago seminar with Toulmin.

of modern science has altered the planetary environment to a degree that makes it increasingly impossible to continue assuming there is little or no value to be found in the evidence of microprocessing. Postmodernism, within which most hermeneutic phenomenology remains trapped, invests all of its interest in the infinite variety of ways in which human praxis, history, and culture carve out those correspondences. It then often refuses to let any of the evidence be erased, which productively maintains a memory of what might have been lost, but also does not offer any generalized, methodical ways forward.

Unmodern philosophy, in contrast to both modernism and postmodernism, foregrounds the role of linguistic and measurement standards in ways that aid in tracing out the good implicit hermeneutic phenomenologies that must be concealed within any effective, productive science (as suggested by Ricoeur, 1967b, p. 219). Unmodern semiotics articulates the work of abstraction and theorizing that makes generalized representations possible. Latour's (1993b) study of Pasteur on lactic acid yeast and Nersessian's (2002) study of Maxwell's method of analogy (utilized by Rasch; see Fisher & Stenner, 2013) provide pertinent examples from the history of science.

Now, consider what Ackermann said in the context of Ben Wright's approach to measurement. A few years before Jack and I met in Wright's classroom, Wright had written that:

> Because we are born into a world full of well-established variables it can seem that they have always existed as part of an external reality which our ancestors have somehow discovered. But science is more than discovery. It is also an expanding and ever-changing network of practical inventions. Progress in science depends on the creation of new variables constructed out of imaginative selections and organizations of experience. (Wright & Masters, 1982, p. 1)

Wright grasped the crux of the problem described by Ackermann, seeing science as an "expanding and ever-changing network of practical inventions" that can make the existing world—"full of well-established variables"—seem to "have always existed as part of an external reality." Wright implicitly augments the microprocessing of facts documented by Latour and Woolgar by setting up new models and methods for slowly and carefully constructing correspondences between words and things. He was suggesting that improved measurement research in psychology and the social sciences could contribute to expanding and evolving networks of new practical inventions.

With no familiarity at all with the work of Ackermann or any of the other "instrumental realists," Wright focused directly on idealized models preserving the simplicity of fixed and repeatable responses to interactions that could stand forth independent of the particular data and theory employed in experimental tests validating the correspondences. What is more, Wright also appreciated Rasch's (1960, pp. 34–35; also see Box, 1979) point that data never fit models, devising individual student reports mapping response patterns in ways useful to teachers (Wright, Mead, & Ludlow, 1980). Though neither Wright nor Stenner made explicit use of "unmodern" philosophy or "metrology" in their writing, their intuitions are well characterized by writers who do, largely in the domain of science and technology studies.

Unmodern Semiotics Across the Sciences

An additional avenue of support for an unmodern perspective that Jack and I discussed deserves mention here. This involves the semiotic implications of twentieth-century advances in physics. The capacity to take language as a model for meaningful communication across all fields of human endeavor requires firm reasons establishing exactly how "The whole of science is nothing more than a refinement of everyday thinking," as Einstein (1954, p. 290) puts it. Recent proposals by teams of collaborating metrology engineers, physicists, and psychometricians (Cano et al., 2019; Mari & Wilson, 2014; Mari et al., 2021; Pendrill, 2019; Pendrill & Fisher, 2015; Wilson & Fisher, 2016, 2019) of a common frame of reference for measurement across the sciences would seem to be viable only via this kind of grounding in a common root source, or seed. Modernist efforts at imitating the methods of the natural sciences in the social sciences have plainly produced disappointing results (de Marchi, 1993; Gould, 1996). An unmodern semiotic perspective may offer ways forward innovative enough to transform what needs to change while leaving intact what works.

Neils Bohr's essentially semiotic focus on the role of the instrument in quantum physics is, then, of considerable interest. As Bohr once stressed in a discussion, his principle of complementarity implies scientific observers are suspended in language to the point that words can no longer be seen as grounded on an independent reality. Instead, Bohr said, our thinking and communications are suspended in language to the point that we cannot tell up from down (Petersen, 1968, pp. 187–188; French & Kennedy, 1985, p. 302).

The physics Nobelist Wheeler (1980, p. 153; 1982, p. 201; 1984) follows through from Bohr's emphasis on the arrangement of the experimental apparatus to the essential role played by the reading of instruments in science. Wheeler was reportedly influenced by Peirce's (1868/1992) semiotics (Merrell, 2011, p. 263), seeing it as a way of situating humanity as participants in an evolving, meaningful universe. Sebeok (1991, pp. 143, 153), in turn, drew from Wheeler and Peirce, then, in developing his own perspective on how sign systems inform a new transdisciplinary research paradigm of semiotic modeling (Anderson et al., 1984; Brier, 2013; Maran, 2007; Merrell, 2011; Nöth, 1990, 2018; Sebeok, 2001; Xu & Feng, 2004). Others arrive via their own independent analyses at similar conclusions, as with Cartwright's (1983, p. 129; quoted in Burdick et al., 2006) statement that "Fundamental equations do not govern objects in reality; they govern only objects in models."

Bohr's theme of being suspended in language echoes a similar strain voiced by another noted physicist, Ernst Mach, who noted that language provides a labor-saving service allowing us to take up and use sign systems already in circulation. Mach referred to this removal of the need for us all to make up our own sign systems as an economy of thought (Banks, 2004; Mach, 1919, pp. 481–495; Franck, 2002, 2019). In this economy, language communities are the original efficient markets, where the common currency of a shared symbol system reduces transaction costs to negligible levels.

Another noted physicist mining this same vein was John Platt, Ben Wright's supervisor in Nobelist Robert Mullikan's physics lab at the University of Chicago in the 1950s. Platt wrote extensively on how "a social chain reaction with positive feedback" could "amplify the delicate and uncertain powers of our highest intelligence into directed and effective acts that will reduce our collective tensions and meet our collective needs" (Platt, 1961, pp. 39–40). Communicable thing-word-idea correspondences can "go viral" in efficient linguistic markets leveraging the economy of thought, giving rise to social contagions (Goyal, Heidari, & Kearns, 2019; Hodas & Lerman, 2014; Latour & Lépinay, 2010; Pastor-Satorras, Castellano, Van Miegham, et al., 2015). Platt was well known for his views in this area and is quoted repeatedly on the power of decentralized networks in a best-selling book I obtained in 1981 during a sabbatical year of reflection in Alaska (Ferguson, 1980). This book and the follow-up reading I did in the next couple of years were preparatory to the events that then unfolded in Chicago.

A philosophical motivation for semiotic modeling was indirectly provided by Ricoeur (1981), with whom I was enrolled in seminars at the time Jack and I first met in the mid-1980s. Ricoeur had formulated a "model of the text" grounded in a textual

> 'objectivity' [from which] derives a possibility of explaining which is not derived in any way from another field, that of natural events, but which is congenial to this kind of objectivity. Therefore there is no transfer from one region of reality to another—let us say, from the sphere of facts to the sphere of signs. It is within the same sphere of signs that the process of objectification takes place and gives rise to explanatory procedures. And it is within the same sphere of signs that explanation and comprehension are confronted. (Ricoeur, 1981, p. 210; see Fisher & Stenner, 2005, 2011a; Fisher, 1992, 2003, 2004, 2010)

This theme of language as the vehicle of thought and the environment in which science operates motivated Peirce's pragmatist semiotics, as he felt there was a surer basis in it than in traditional epistemology (Houser, 1992, p. xxxvi). Latour (1987, 1995a, 2005) similarly recommends tracing out the propagation of representations across media as superior to epistemology. Documenting science in action in that way, via its inscriptions, we can be thoroughly positivistic about the factuality of what is said and done, as that evidence can only work to inform, substantiate, and test the tracing of communicative functions through ecological economies of thought.

Hamann advanced a variation on this kind of argument in direct response to Kant in the eighteenth century (Surber, 2011), saying that "language is reason." My own work on metaphor (Fisher, 1988, 1990, 1994, 1995a/b, 2012a) led to much the same conclusion. Taking up the direction indicated by Black (Ben Wright's philosophy professor at Cornell in the 1940s) in his work on models and metaphors (Black, 1962; also see Ricoeur, 1977, 1979, 1981, pp. 165–181; Cartwright, 1983; Lakoff and Johnson, 1980) led to a mathematical way of setting up models as heuristic fictions, and vice versa. Seeing science and art as overlapping in this way does not lead to a negative conclusion or mere historical or cultural relativization, as would be the case in a postmodernist context. The unmodern conclusion more productively sees in the artful craft of making useful things like texts, textiles, techniques, and technologies (all from the ancient Greek, *techne*) a semiotic basis for a pragmatically idealized (complex, multilevel, partly disunified, and partly unified) integration of the arts and sciences.

Counting, for instance, is inherently metaphorical in the way it ignores what is different about unique individual entities, events, or persons, as has been recognized since Plato (see the *Phaedo*, 96b; Ballard, 1978, pp. 186–190; Fisher, 1994). Just as irrational numbers involved in geometry cannot be exactly specified numerically or drawn as lines or arcs, so, too, do laws like Newton's inertial law also describe unobservable ideals, such as an object left entirely unto itself with no forces acting on it. These imaginary frames of reference captivate thinking in the play of signifiers in highly productive ways (Gadamer, 1989, pp. 101–134; Ricoeur, 1981, p. 185).

Plato resolved the crisis of irrationality in Pythagorean geometry by redefining the elements of geometry, making a point an indivisible line, a line an indivisible plane, etc. (Cajori, 1985, p. 26; Gadamer, 1980, pp. 35, 100–101; Ricoeur, 1965, p. 202). In so doing, Plato made irrational numbers equivalent to rational ones at a higher order formal level of complexity. This distinction between names and concepts effectively founded philosophy. Plato then required familiarity with mathematics for entrance to his Academy because that experience orients students to the fact that numeric and geometric figures are not the mathematical relationships themselves (Gadamer, 1980, pp. 35, 100–101).

The theory of heuristic fiction (Ricoeur, 1981, p. 187), and understanding that mathematical relationships are primarily qualitative, led to the idea that metaphors are models just as much as models are metaphors. In addition, understanding that quantification requires a linear geometric metaphor, it was plain that variation in the pertinence or relevance of what Black called the system of associated commonplaces had to encompass the full range from the positive to the negative. Laying out the implications of metaphors as statements on a survey that people could agree and disagree with provided fairly clear and simple tests of the hypothesis that the ratio of rationality is embedded in the metaphoric roots of all concepts.

The predictability of the statements was consistent enough that the theory-informed blueprint mapping the metaphoric constructs sufficed to calibrate parallel forms on which statements took expected positions on the scales in clearly distinguished ranges (Fisher, 1988, 1990, 1995a/b, 2012a). The metaphoric process effected a virtual calibration of language as an instrument providing a measure of meaning. Here, in the metaphor of measurement and in the measurement of metaphor, we transition in the phenomenological method from the deconstruction of language back to a new reduction (Ricoeur, 1977, p. 22), achieving the mathematization of language foreseen by Derrida (1981, p. 34).

Hamann was, unfortunately, largely forgotten, and Peirce left much of his work incomplete, was notorious for inventing new words, sometimes confused readers by changing the meanings of those words, and was systematic to a fault with his profusion of categorizations (Houser, 1992, pp. xxx-xxxix).

Peirce nonetheless established the U.S. as a reputable metrological partner in the global geodetic survey, is widely recognized alongside Frege as a leading early logician, proposed a measurement model today recognized as possessing all the features included by Rasch (Linacre, 2000; Peirce, 1878/1992), initiated the economic study of scientific research (though without connecting semiotics with metrology) (Peirce, 1879/2002), and took up the study of method, meaning, and interpretation as a hermeneutic problem (Houser, 1992, p. xxxix) in a vein also taken up much later by Gadamer (1989) and Ricoeur (1981). One can only wonder how Peirce might have proceeded if he had been able to retain an academic appointment and productively focus on his work, instead of having been in the position of relying on income provided by his friends and by articles he wrote for popular periodicals. As Walker Percy puts it in his 1989 Jefferson Lecture in Washington, DC, Peirce laid the groundwork for coherent psychological and social sciences that have yet to be worked out (Percy, 1991; Houser, 1992, p. xxi).

Steps toward those sciences are suggested by Ricoeur, who, though not known as a mathematical philosopher, takes up themes more often associated with Peirce, aptly noting that the validation of interpretations involves

> a logic of uncertainty and of qualitative probability. ... The method of converging indices,[2] typical of the logic of subjective probability, gives a firm basis for a science of the individual deserving the name of science. A text is a quasi-individual, and the validation of an interpretation applied to it may be said, with complete legitimacy, to give a scientific knowledge of the text. (Ricoeur, 1981, p. 212)

The constellation of associated theory and evidence accumulated in this broad philosophical context complemented their experimental results in ways that inspired confidence in Jack and his colleagues. They saw they had good reasons for thinking they likely could succeed in advancing a new integration of science and commerce in education. Yet more intricate details were still to unfold, giving rise to an image Jack and I conjured of a kind of flower that not only perpetually bloomed but which changed the color, shape, and size of its petals over time. Once events started to take on lives of their own, we were often left in a state of wonder and amazement at how circumstances unfolded.

Seeing Language as a Measurement Model

Jack's intuitions, informed by trial-and-error experience, technical reading, history, and philosophy, led him to see that extending everyday language into a new psychological science would have to involve more than merely applying models to data in endless instrument calibration exercises. From his point of view, a huge business opportunity was opened up by the contrast between a fixation on inefficient data gathering and analysis in the mainstream, on the one hand, and the confidence obtained when repeated empirical experiments and theoretical predictions converged on a systematically repeating construct, on the other hand.

---

[2] Converging, not "conveyance of," as given in Ricoeur, 1981, p. 212. The original French (Ricoeur, 1986, p. 202) is plainly "convergence d'indices;" also see the translation given in Ricoeur, 1976, p. 79.

The magnitude of the opportunity Jack perceived is inversely related to the embedded depth of the data-centric focus on statistical analysis as necessary to measurement, which dominates the mainstream research culture in many fields. In a bizarre kind of cultural schizophrenia (Bateson, 1978; Fisher, 2021a; Wright, 1988), statistical methods are universally referred to as quantitative, even though those methods are hardly ever used to formulate and test the hypothesis that a unit quantity exists (Meehl, 1967; Michell, 1986). The deep entrenchment of the automatic association of numbers with quantities is counter-intuitive in that everyone knows counts of rocks are in no way indicative of the quantitative amount of rock possessed; your two rocks may have far more mass than my ten rocks. The same principle applies to test scores and ratings, such that everyone is well aware that these numbers vary in ways that depend on the difficulty or agreeability characteristics of the questions asked.

These obvious facts seem to have had very little impact on the popularity of statistical significance tests and techniques such as factor analysis. This remains so despite the failure of these methods to live up to their billing as the means by which scientific laws will be established, as has been pointed out for decades to no avail. Cohen (1994) notes that the spherical shape of the earth was not proven by comparing control and random samples producing a p-value less than 0.05, and Guttman (1985) takes up "the illogic of statistical inference for cumulative science." This theme is echoed by Duncan and Stenbeck (1988, p. 23), who say

> It was the determination of what observations to make, under what conditions, that required the genius of a Tycho Brahe, a Galileo, a Mendel, or a Pasteur, all of whom could have made good use of statistical methods, had they been available, but none of whose discoveries could have been made by a platoon of statisticians.

In an introductory chapter to a textbook that remained unpublished at the time of his death, Guttman (1994, p. 82) similarly pointed out

> Those who firmly believe that rigorous science must consist largely of mathematics and statistics have something to unlearn. Such a belief implies emasculating science of its basic substantive nature. Mathematics is contentless, and hence—by itself—not empirical science. As will be seen, rather rigorous treatment of content or subject matter is needed before some mathematics can be thought of as a possibly useful (but limited) partner for empirical science.

Thinking along the same lines, Thurstone (1959, p 10) observed that some students unable to formulate a mathematical expression may nonetheless be

> …well able to comprehend an essentially mathematical formulation of a psychological problem with its implications and experimental possibilities. Such a student may be more fertile with ideas than one who possesses considerable mathematical skill without the flexibility of mind that is essential in creative scientific work. More fortunate is the student who has all these aptitudes.

Thurstone emphasized mathematics as not being just a tool applied to problems, but the "very language" in which we think. The student who comprehends an everyday language expression of an essentially mathematical formulation of a psychological problem is then tapping into and extending the mathematics of identity (the ball is red) and comparative relations (she reads with more comprehension than they do) widespread in daily discourse. Remarking on this qualitative mathematics, Guttman (1994, p. 103) wryly noted that "it may be a surprise to most people to learn that they speak mathematically routinely without knowing it."

In fact, a close examination of the origins of mathematics and geometry in ancient Greece leads to a reconceptualization of what it means for fields of investigation to be genuinely scientific only to the extent they are mathematical, as Kant famously puts it (Heidegger, 1977, p. 249; Kisiel, 1973, p. 110; also see Danesi, 2017; Derrida, 1989; Fisher, 1992, 2003, 2004, 2010; Fisher & Stenner, 2011a; Harries, 2010). Thurstone (1959, p. 9) gets exactly to the point with his "distinction between statistics and mathematical psychology," where it becomes apparent that:

> A study can be quantitative without being mathematical. Merely to count noses or the answers in a test or seconds of reaction time or volume of secretion does not make a study either mathematical or scientific.

And so, it follows that, conversely, a study can be mathematical without being quantitative: "quantification is neither necessary nor sufficient for measurement" (Mari, et al., 2013, 2016; Fisher, 1992, 2003, 2004, 2010; Fisher & Stenner, 2011a).

In his consideration of the history of measurement, Kuhn (1977) arguably concurs, remarking on a recurring implication that corresponds with Guttman's (1994, p. 82) emphasis on the need for quantification to be preceded by "rather rigorous treatment of content:"

> ...much qualitative research, both empirical and theoretical, is normally prerequisite to fruitful quantification of a given research field. In the absence of such prior work, the methodological directive, 'Go ye forth and measure,' may well prove only an invitation to waste time. (Kuhn, 1977, p. 213)

Thurstone, Guttman, and Kuhn contend that measurement comes into its own as a productive activity only in the context of thorough qualitative and substantive understandings of phenomena. This perspective stands in marked contrast with the still-dominant opinion, stated explicitly by Campbell (1952, p. 134), that measurement is of vital importance in science because it is essential to the discovery of laws. But Kuhn (1977, p. 219; original emphasis), writing in 1961, asserts that

> The road from scientific law to scientific measurement can rarely be traveled in the reverse direction. To discover quantitative regularity one must normally know what regularity one is seeking and one's instruments must be designed accordingly; even then nature may not yield consistent or generalizable results without a struggle.

Empirical and theoretical study precedes quantification because knowledge of repeating and reproducible regularities is required before analogies with linear geometry can be substantiated (Black, 1962; Carreira, 2001; Hendriana & Rohaeti, 2017; Hesse, 1970; Kuhn, 1979; Nersessian, 2008). The formal structure of scientific laws provides an organizing principle applicable as a model only when useful approximations of substantive additivity can be obtained.

Controllable variation in phenomena is typically produced by means of a technology affording systematically repeatable experimental comparisons. Kuhn (1977, p. 90) then points out that the majority of those contributing to the emergence of the thermodynamic theory were engineers with hands-on experience working with engines. Price (1986) similarly remarks that

> Historically, we have almost no examples of an increase in understanding being applied to make new advances in technical competence, but we have many cases of advances in technology being puzzled out by theoreticians and resulting in the advancement of knowledge. It is not just a clever historical aphorism, but a general truth, that 'thermodynamics owes much more to the steam engine than ever the steam engine owed to thermodynamics.'
>
> …historically the arrow of causality is largely from the technology to the science. (p. 240)

Others (Bud & Cozzens, 1992; van Helden & Hankins, 1999; Ihde, 1991; Latour, 1987, 2005; Blok, et al., 2020) have also made much of "the historical and ontological priority of technology over science," as Ihde (1983) puts it. Others go so far as to say that:

> Instruments have a life of their own. They do not merely follow theory; often they determine theory, because instruments determine what is possible, and what is possible determines to a large extent what can be thought. (Hankins & Silverman 1999, p. 5)

In Ricoeur's (1967a) terms, "the symbol gives rise to thought." Kuhn's paradigmatic example of fruitful quantification is illuminating, then, for its unintended correspondence with the emergence of widespread networks of instruments traceable to metrological standards. First, he (Kuhn, 1977, p. 219) notes that:

> Sometime between 1800 and 1850 there was an important change in the character of research in many of the physical sciences, particularly in the cluster of research fields known as physics.

He then qualifies what he wants to say before continuing on to offer a conjecture:

> It would be absurd to pretend that mathematization was more than a facet. The first half of the nineteenth century also witnessed a vast increase in the scale of the scientific enterprise, major changes in patterns of scientific organization, and total reconstruction of scientific education. But these changes affected all the sciences in much the same way. They ought not to explain the characteristics that differentiate the newly mathematized sciences of the nineteenth century from other sciences of the same period. Though my sources are now impressionistic, I feel quite sure that there are such characteristics. Let me hazard the following prediction. Analytic, and in part statistical, research would show that physicists, as a group, have displayed since about 1840 a greater ability to concentrate their attention on a few key areas of research than have their colleagues in less completely quantified fields. In the same period, if I am right, physicists would prove to have been more successful than most other scientists in decreasing the length of controversies about scientific theories and increasing the strength of the consensus that emerged from such controversies. In short, I believe that the nineteenth-century mathematization of physical science produced vastly refined professional criteria for problem selection and that it simultaneously very much increased the effectiveness of professional verification procedures.

Now consider the importance of Kuhn's focus on the period of 1800 to 1850 and on the mention of 1840 in particular. Hacking (1983, p. 234) says that, in this passage, Kuhn "suggests 1840 as a date when measuring, as we now conceive of it, takes on its fundamental role." In his history of the meter, Alder (2002) says:

In 1837 the [French] government revived the metric system, both as a promise to modernize France and a public assertion that the new regime was a worthy successor to the first great Revolution. (p. 327)

The legislation, which was passed with overwhelming support in 1837, made the metric system obligatory throughout France and its colonies as of January 1, 1840. (p. 328)

By the time France restored the metric system in 1840 [after reversing it in 1820], it had already been obligatory for two decades in Holland, Belgium, and Luxembourg. (p. 330).

That both Kuhn and Alder mention the year 1840 is not a coincidence. The introduction of metrological standards in Western Europe in the early nineteenth century was a watershed moment in the evolution of science and commerce. As Alder points out, democratic revolutions were also implicated. That is a story for another time, but it points toward two issues at the heart of the matter.

First, technoscientific and social orders co-evolve, with quality-assured measurements, legal property rights and contracts, accounting standards, and communications networks each informing one another (Bernstein, 2004; Fisher, 2012b, 2020; Fisher & Stenner, 2011b; Jasanoff, 2004; Latour, 1987, 1993a, 2005; Shapin, 1994). As Kuhn points out, not only did the mathematization of the sciences take off in the early nineteenth century, but there were also vast changes in the scale of the scientific enterprise, patterns of scientific organization, and a total reconstruction of science education.

Echoing Kuhn and Price on the "arrow of causality" going from technology to science, Latour (1983) plays off Archimedes, saying "give me a laboratory and I will raise the world." Society is shaped by linguistic and scientific standards to the point that, as Alder (2002, p. 325) says, "Measures are more than a creation of society, they create society." Societies do not just make use of linguistic signs and symbols; instead, the ways in which things are symbolized shape shared social reality (Hutchins, 2014; Latour, 2005; Petracca & Gallagher, 2020; Ricoeur, 1981, p. 219; Thrift, 2008).

Second, as pointed out by Whitehead (1911, p. 61), civilization advances by creating and distributing new, useful, and highly simplified ways of performing complex operations, like flipping a switch to turn on a light. Individuals need not understand more than a small portion of the total knowledge incorporated in everyday language and in scientific instruments to make productive use of complex semantics and tools like clocks, thermometers, computers, smartphones, blenders, and automobiles. A good way to understand how it could be possible to perform important operations without thinking about them is to see them analogically, as they might appear if the social body had biological functions.

That is, in the same way that metabolic, neuromuscular, and digestive processes are performed with no conscious thought on the part of individuals, so, too, is the social mind embodied in the infrastructure of cognitive supports built into the external environment (Ihde & Malafouris, 2019; Meloni & Reynolds, 2021; Scarinzi, 2015). Human-crafted technologies like alphabets, phonemes, grammars, texts, and tools extend embodied subjectivity in ways that enable the performance of communications operations beyond individuals' available resources.

Intuitions informed by this kind of tacit knowledge (Polanyi, 1966) lift the intelligence of all without altering their individual cognitive abilities "one whit" (Dewey, 1954, p. 210). Whitehead (1925, p. 107) similarly remarked in 1925 that the then-recent quantum revolution in physics did not come about because individual scientists became more imaginative; instead, he says, the change in thinking was the result of advances in instrument design. Or, as Latour (2010b, p. 153) puts it in the context of Tarde's economics, "change the instruments, and you will change the entire social theory that goes with them."

Plainly, societies whose languages and technologies enhance their populations' capacities for sharing complex imaginative realms and levels of intelligence are more likely to be creative and productive than those lacking means of sharing higher order conceptualizations. New capacities to think together and communicate in common languages about how to obtain food and shelter would likely lift a society previously lacking those ideas to higher levels of reproductive success. Further advances in the quality of life would follow as additional systematically distributed products and processes were developed, such as mathematical modeling and manufacturing mechanical tools.

Shared social realities are structured by means of linguistic standards. And "The development from the spoken language ... through symbols and pictograms ... to what we now understand as written language is a perfect standardization process" (Weitzel, 2004, p. 11). Linguistic standards are the prototype for scientific standards. The power of being able to think and act together to accomplish complex goals ought to be a more deliberately conceived goal of research. The processes by which that power is obtained via science are metrological. As Latour (1987) puts it:

- "The point here is that the easiest means to enrol people in the construction of facts is to let oneself be enrolled by them! By pushing their explicit interests, you will also further yours. The advantage of this piggy-back strategy is that you need no other force to transform a claim into a fact; a weak contender can thus profit from a vastly stronger one" (p. 110).
- "Every time you hear about a successful application of a science, look for the progressive extension of a network" (p. 249).
- "The predictable character of technoscience is entirely dependent on its ability to spread networks further" (pp. 249–250).
- "Facts and machines are like trains, electricity, packages of computer bytes or frozen vegetables: they can go everywhere as long as the track along which they travel is not interrupted in the slightest. This dependence and fragility is not felt by the observer of science because 'universality' offers them the possibility of applying laws of physics, of biology, or of mathematics everywhere in principle. It is quite different in practice" (p. 250).
- "In all these mental experiments you will feel the vast difference between principle and practice, and that when everything works according to plan it means that you do not move an inch out of well-kept and carefully sealed networks" (p. 250).
- "Metrology is the name of this gigantic enterprise to make of the outside a world inside which facts and machines can survive" (p. 251).

- "Scientists build their enlightened networks by giving the outside the same paper form as that of their instruments inside. [They can thereby] travel very far without ever leaving home" (p. 251).
- "...our method would gain nothing in explaining 'natural' sciences by invoking 'social' sciences. There is not the slightest difference between the two, and they are both to be studied in the same way. Neither of them should be believed more nor endowed with the mysterious power of jumping out of the networks it builds" (p. 256).
- "That much more effort has to be invested in extending science than in doing it may surprise those who think it is naturally universal" (p. 250).
- "The paramount importance of metrology (like that of development and industrial research) gives us a measure, so to speak, of our ignorance" (p. 250).

In accord with Latour's metrological program, to generalize his insights, Stenner allied himself with others—measurement theoreticians, psychometric analysts, statisticians, researchers, educators, school district administrators, testing agencies, state departments of education, book publishers, curriculum theorists, management experts, marketing and sales specialists, accountants and lawyers—whose interests could be better realized in productive partnerships than they could be in isolation. Jack and his colleagues progressively expanded a network of relationships by piggy-backing the shared interests of all these groups on each other in ways that multiplied the value each obtained with the addition of each new domain of stakeholders.

A large literature has emerged in the last 40 years on this topic of actor-network alliances (in addition to the previously cited works, see, in particular, Akrich et al., 2002a, 2002b; Blok, Farias, & Roberts, 2020; Bowker, et al., 2015; Miller & O'Leary, 2007). Though Ackermann (1985) predates the emergence of actor-network theory, that book may have inspired a good deal of confidence in Jack. Past that, the one volume in science and technology studies (Wise, 1995) with which I am sure Jack was familiar did not emerge until he was already well underway in his work.

Jack seems to have had strong intuitions and a remarkable capacity for learning from his and others' mistakes. One of his basic operational precepts was to engage in dialogue with anyone who raised questions about his and his colleagues' work, since there was either something important to be learned or something important to be taught. Either way, reaching out was essential whenever a potential ally emerged displaying an interest—positive or negative—in the overall mission. A key rhetorical advantage Jack leveraged in expanding his network was the availability of a map of the reading comprehension terrain occupied by everyone involved in literacy education.

Maps and Mapping

A foreshadowing of the direction Jack took up is provided by Latour's (1987, p. 223; original emphases) question as to:

…how to act at a distance on unfamiliar events, places, and people? Answer: by somehow bringing home these events, places and people. How can this be achieved, since they are distant? By inventing means that (a) render them mobile so that they can be brought back; (b) keep them stable so that they can be moved back and forth without additional distortion, corruption or decay, and (c) are combinable so that whatever stuff they are made of, they can be cumulated, aggregated, or shuffled like a pack of cards. If those conditions are met, then a small provincial town, or an obscure laboratory, or a puny little company in a garage, that were at first as weak as any other place will become centres dominating at a distance many other places.

One of the primary tasks of explorers traveling to new regions is to map what they see so that others learn from their experience and can then find their way more easily than the first to arrive. So, too, does measurement research map the repeatable structural invariance of objectively existing constructs so that those arriving later can learn from those who came before. Science models new domains in ways that draw maps others can follow; metrological networks trace out paths on those maps that let everyone inhabiting the terrain see where they are in relation to everyone else, and in relation to where they have been and where they want to go. And, importantly, as anyone experienced in pilot-navigator communications can attest, maps are not always simple or easy to use. I'll have more to say about maps in relation to levels of complexity, below.

Though Jack and I had shared our readings of Ackermann (1985), I do not recall bringing Latour's work to Jack's attention until after he had progressed well into expanding a reading comprehension network. He did, however, closely follow Wright's (Wright & Stone, 1979, pp. 121, 205–206; Wright & Masters, 1982, pp. 124–126, 140–146, etc.) mapping of measurement variables. Rasch (1960, pp. 10, 124) had provided a provocative hint in his 1960 book, saying that scientific:

> …laws do not at all give an accurate picture of nature. They are very simplified descriptions of a very complicated reality. The laws describe an ideal universe, a model, on which reality may be mapped—leaving aside a lot of details. E.g. 'a heavy point swinging without friction on a weightless string' never existed in the real world, but at a certain stage of the process of knowledge it is a very useful model of a pendulum."

Rasch here makes a distinction between the formal, unrealistic idealizations incorporated in maps, where no one ever actually occupies a position in the sky at such an elevation with a comprehensive view of the ground, and the complications of concrete reality that never perfectly fits a model. Improved modeling of communication necessitates close attention to this distinction between the map and the territory it represents (Bateson, 1972, p. 186; Fisher, 2021a). Where communication theory is usually restricted to a focus on information content and signal transmission, comprehension processes must include semantic and syntactic elements (Bowman & Targowski, 1987). Beyond that, considerations of social interactions in which participants in dialogue reference maps and markers must also be taken into account.

Hutchins (1995; also see Latour, 1995a) describes ways in which socially distributed remembering and perception (Sutton, et al., 2010) are structured by means of maps seen as material artifacts embedded in the external world as cognitive infrastructure:

> The system for ship navigation…is based on formal manipulation of numbers and of the symbols and lines drawn on charts. It is a system that exploits the powerful idea of formal operations in many ways. But not all the representations that are processed to produce the computational properties of this system are inside the heads of the quartermasters. Many of them are in the culturally constituted material environment that the quartermasters share with and produce for each other. (Hutchins, 1995, p. 360)

Across a great many areas of life, the same kinds of culturally constituted material reference points connect technologically mediated maps with features in the world interpreted in linguistically standardized ways. Clocks correspond with the shifting shadows and light over the course of the day, thermometers with sensations of hot and cold. Pronunciations correspond with phonemic standards, which are connected to alphabets and grammar.

In all of these cases, the chicken and egg question emerges: which came first? While emergent usage surely precedes and is codified into technical standards, standards then also feedback onto usage, constraining and limiting it via overt, covert, and subconscious means until new categorizations and constructs emerge to set new standards. A major insight achieved by Jack was that opportunities for innovating and extending capacities for constructing shared social realities can be found in the structural invariances documented in applications of probabilistic measurement models.

Rasch and Wright laid out exactly the same consequences as those described by Latour. To map distant and previously unseen cognitive and social events and processes in ways that can be useful to those who come later, the representation has to be mobile and portable in a convenient kind of way, stable enough to be moved without introducing irrelevant sources of noise, and, most important, the local particulars of reading tests and the students taking them had to be combinable in endless variations. What are now known as construct maps, Wright maps, and kidmaps illustrate, in correspondence with Ackermann's distinctions, the respective formal theoretical, abstract instrumental, and concrete data levels of complexity in what is measured (Fisher et al., 2021). Generally speaking, concrete data contrasts with abstract information, and these both differ from formal knowledge. Data embedded in information systems entails operations at a lower level of conceptual complexity than is involved in knowledge infrastructures. Understanding is implied by the systematic implementation of knowledge.

The correspondence of these themes from the history of science with Jack Stenner's contributions to measurement in psychology suggests opportunities for advancing psychology and the social sciences that have hardly yet been clearly stated, much less taken up. The alliances formed by Jack and his colleagues are an exemplary instance of the commercial and scientific power made available by expanding metrological networks. But these alliances leveraging a common language of reading comprehension measurement have scarcely begun to create the kinds of effects produced by the introduction of metrological systems in the nineteenth century. Much more stands to be done to follow through into higher order integrations of the interests of stakeholders across sectors, following Bernstein's (2004) and Miller and O'Leary's (2007) analyses of the roles of property rights, scientific rationality, capital markets, and communications networks in creating prosperous ways of life (Fisher, 2012b, 2020; Fisher & Stenner, 2011b).

Generalized metrological networks are as possible in the social sciences as in the natural sciences because product, regulatory, scientific, legal, monetary, and accounting standards extend the prototype for all standards: language. The power of these standards is remarkable. That power is the operational core of the economy in general. Efficient markets depend to large extents on the creation and maintenance of standards, which evolve from language as higher order levels of socio-ecological complexity. Language thinks the world for us in advance, saving us from the trouble of having to work out new words and semantics for connecting newly conceived ideas in relation to useful things in the world.

Here again we encounter this labor-saving function of language and the economy of thought (Banks, 2004; Mach, 1919, pp. 481–495; Franck, 2002, 2019) created by shared linguistic standards and extended into the broader economy of manufactured capital and finance by scientific standards. Jack and his colleagues leveraged the efficiencies gained by setting up a new domain of scientifically supported semiotic connections. In so doing, they enacted a new model for advancing meaningful communications across a wide array of domains. Completing what they started will be challenging and complex, but do-able.

From Theory to Practice

In contrast to the usual perception of test data in education as idiosyncratically variable, as subjective, and as uncontrollably and inherently local, Jack saw that education has always and everywhere been structured, from its initial institution, around universally observed progressions in learning. Children acquire spoken language well enough to have their attention directed toward the distinctive shapes of letters in an alphabet; from there to associations between orally pronounced words, written words, and images; from there to texts composed of very common words and very short sentences; and then to longer and longer sentences and texts incorporating increasingly less common words.

This kind of progression in learning is as much an objective fact characterizing the real world as any physical entity or process. Rasch (1960, pp. 110–115) accordingly then asserted that children's reading abilities could be measured with the same kind of objectivity obtained in measuring their weight. Rasch spoke of how varying tests could be linked so they measure in a common metric (Jolander, 1957/2008), his son-in-law intentionally designed tests to measure at specific levels of difficulty (Prien, 1989), one of his students devised an explanatory model integrating predictive item component difficulties (Fischer, 1973), and Rasch (1972/2010) held that the social sciences could one day possess an "instrumentarium" akin to that available in the natural sciences.

Rasch did not, however, take the step to substantiating the suggestion that all instruments measuring reading comprehension might be equated to a common scale.

Neither did he advance the essential point as to the advantageous efficiencies obtained when item locations not only can be predicted from theory but their content can also be automatically generated as well. Jack understood, however, that the repeatable and reproducible structure of learning progressions spanning elementary and secondary education sets out broad limits within which reading comprehension had long shown itself as a persistent and objective phenomenon. Jack saw that the idea of a common scale might be controversial, but drew confidence from the fact that the entire edifice of education as a human cultural institution is founded on the universally shared consensus that consistent variation in reading comprehension across learners and across texts is real.

Jack also understood that much of what makes the idea of a common scale problematic is the historically demonstrated tendency in education to apply such scales indiscriminately. Given his work in Head Start and other early education endeavors, Jack was well aware that, too often, divergent response patterns in test data are seen only as inconvenient nuisances to be skipped over in favor of the larger actionable evidence available in test scores. Knowing that detailed, useful information on individuals' unique strengths and weaknesses could be obtained and leveraged in classroom applications gave Jack further confidence in the value of addressing each of education's separate stakeholder groups in their own languages and in the context of solutions to their primary problems.

And so, Jack thought through the implications of learning environments in detail, as might be guessed from his own personal koan invention, "Does the reader comprehend the text because the reader is able or because the text is easy?" (Stenner & Stone, 2004; Chap. 11). Furthermore, in taking a stylized yin-yang symbol as the logo for the Lexile Framework for Reading, Jack intentionally sought to suggest a vitally important principle fundamental to enhancing human institutions' capacities for nurturing authentic, healthy, satisfying relationships.

The Lexile logo omits one of the key features of many yin-yang symbols, a small dot in both of the complementary curves that represents the presence of the other in each. The abstract medium of the Lexile unit of measurement embodies a representation of the implicit existence of the unrealistic ideal in the concrete particulars, as well as of the concrete in the formal idealization. The noted Zen philosopher, Suzuki (1949, pp. 23–25), points out that "there is nothing infinite apart from finite things," and that "there was from the very beginning no need for a struggle between the finite and the infinite." One way of understanding the yin-yang symbol is, then, as representing the mutual implication of finite concrete physicality and infinite formal ideality. The measurement model sets up an axiomatic formulation of the infinite realms of all possible readers and all possible texts that accepts finite and imperfect concrete instances of particular students' responses to actual assessment items.

These opposites are embodied in the abstract medium of a standard in a way that also sets up the yin-yang symbolization of matched reader abilities and text complexities.

Most reflections on and implementations of the impacts of learning environments on outcomes, unfortunately, do not consider how to create a broader cultural context for generalizing complex understandings integrating the concrete, abstract, and formal in systems, and integrating all four of these levels in systems of systems, metasystematically. This perspective is, however, advanced in developmental psychology's model of hierarchical complexity (Commons et al., 1982; Commons & Richards, 2002) and skill theory (Fischer, 1980; Fischer & Farrar, 1987; Dawson-Tunik et al., 2005). Here, the cognitive role of standards embedded in the infrastructural scaffolding of the social environment is explicitly recognized as a key factor shaping educational outcomes. As Commons and Goodheart (2008) put it, "cultural progress is the result of developmental level of support."

Plainly stated, educators face the problem of matching children's reading abilities with the complexity of the available reading materials. Any given single elementary school classroom may have students varying across three to five grade levels of reading ability. A shared textbook may adequately target and engage with the middle third of the ability range, but that textbook will be too easy for, and so will bore, the upper third, and the bottom third will be lost. Jack and his colleagues wondered if and how it might be possible to set up instructional contexts in which everyone is studying the same topic but reading material targeting their individual comprehension level, and learning from any especially salient mistakes they might have made.

Operationalizing Wright's (1958, 1968a) concern with the need to learn more systematically from mistakes and his celebration of a possible "curriculum of failure" led to an emphasis on the formative processes by which assessment and instruction can be integrated. In formulating what he called the "two-sigma problem," Wright's Chicago colleague, Ben Bloom (1984) wondered how classroom achievement could be improved to a level two standard deviations higher, where the outcomes of one-on-one tutoring are typically found. Personalized learning technologies bringing together Bloom's mastery learning (Bloom, 1971; Guskey, 2009), formative feedback, and adaptive testing methods have been documented repeatedly as solving Bloom's two-sigma problem (DeMillo, 2015, p. 99).

Jack saw early on how learning environments might be improved by means of shared standards that—crucially—are adaptable to individuals and local circumstances, and that provide a medium through which unique voices can make themselves heard. As Jack said in a 2011 interview,

> As education moves toward more personalized learning, today's one-size-fits-all textbook will become a thing of the past. Our metrics support machine-generated item types that adapt to test takers on the fly, and provide more accurate measures of growth (Rivero, 2011).

Dewey (1954) articulated the nature of the situation, saying:

> The function which science has to perform in the curriculum is that which it has performed for the race: emancipation from local and temporary incidents of experience, and the opening of intellectual vistas unobscured by the accidents of personal habit and predilection. The logical traits of abstraction, generalization, and definite formulation are all associated with this function. In emancipating an idea from the particular context in which it originated and giving it a wider reference the results of the experience of any individual are put at the disposal of all.... Thus ultimately and philosophically science is the organ of general social progress. (p. 230)

> Ideals and standards formed without regard to the means by which they are to be achieved and incarnated in flesh are bound to be thin and wavering.... Our Babel is not one of tongues

but of the signs and symbols without which shared experience is impossible.... A more intelligent state of social affairs, one more informed with knowledge, more directed by intelligence, would not improve original endowments one whit, but it would raise the level upon which the intelligence of all operates.... Uniformity and standardization may provide an underlying basis for differentiation and liberation of individual potentialities. They may sink to the plane of unconscious habituations, taken for granted in the mechanical phases of life, and deposit a soil from which personal susceptibilities and endowments may richly and stably flower. (pp. 141, 210, 215–216)

So, when Commons and Goodheart's (2008) title echoes Dewey, saying "Cultural progress is the result of developmental level of support," the implication is that developmental levels of support advancing cultural progress must give voice to local creative improvisations, to individual strengths and weaknesses, as well as to generalized comparisons. In efforts outside of developmental psychology and philosophy, historians of science (Bowker, et al., 2015; Galison, 1997; Haraway, 1996; Latour, 1987, 2005; O'Connell, 1993; Star & Griesemer, 1987; Scott, 1998) have similarly articulated arguments prioritizing the need for embedded relational media and infrastructures capable of satisfying "the competing requirements of openness and malleability, coupled with structure and navigability," as Star and Ruhleder (1996, p. 132) put it.

To succeed, ideals and standards must pragmatically accept and integrate local situations that never conform to modeled expectations. The conceptual, mathematical formalism of probabilistic models and construct specification equations is less the "immutable mobile" described by Latour (1986, 1987) than it is a "boundary object" (Bowker, et al., 2015; Star & Griesemer, 1989) situated in a "trading zone" (Galison, 1997, 1999; Galison & Stump, 1996) where it is valued and exchanged in different terms by different groups (Fisher & Wilson, 2015; Fisher, 2020). Data never perfectly fit models; models are not meant to be true, but to be useful (Rasch, 1960/1980, pp. 34–35; Box, 1979).

Usefulness is determined in practice by the ways formal models structure abstract standards and accommodate the infinitely varying concrete situations from which the models and standards are created. When consensus can be reached on shared terms for meaningful communication, assurances and certifications are supported by routing checks on the system through obligatory passage points, such as metrological, accounting, regulatory, or legal requirements. The due diligence of cross-sector agreements on how and when these obligations are met is essential to creating shared social realities.

For instance, what comes next in any given instructional program will, of course, be very different for different students enrolled in different schools using different curricula. Even when students have the same measured reading comprehension, their patterns of strengths and weaknesses surely will differ, as also will the texts they read and the test items they respond to. And what comes next will also vary across teachers, even when their classes have the same average measurement, and so on, for principals, parents, researchers, legal advocates, administrators, policymakers, etc. Just as is the case for looking at reading comprehension tests in isolation, narrow slices of experience across diverse perspectives may seem hopelessly confusing.

Even when the larger view is achieved, ideals and generalities remain inherently and always caught up in local instances of playful back and forth.

The complexity and difficulty of the issues involved in constituting measurable and manageable boundary objects should not be underestimated. Jack understood that an essential part of the task that keeps it manageable involves keeping the big picture in mind. The uncertainties arising in narrowly defined contexts must inevitably overwhelm variation. Individual differences obtained within disconnected situations will present the appearance of unmanageable inconsistency. Focusing on single assessments of reading comprehension administered to students of similar reading ability is akin to wearing blinders or accepting an unnecessary myopia as an inviolable reality.

Hierarchical Complexity as a Basis for Pragmatic Action

Language prethinks the world for us in advance, saving us the trouble of having to work out new words and semantics for connecting newly conceived ideas in relation to useful things in the world. This labor-saving economical function of language's semiotic triangle—things, words, and ideas—sets up the structural capacities for science's analogous assemblages of data, instruments, and theories. These three levels of complexity (concrete, abstract, and formal) are then augmented in the model of hierarchical complexity (Commons & Richards, 2002) by four post-formal levels (systematic, metasystematic, paradigmatic, and cross-paradigmatic). Innovators in the history of science typically attain metasystematic integrations generalizing capacities for mediating locally situated applications and axiomatic ideals in transdisciplinary systems of systems (Commons & Bresette, 2006; Commons, Ross, & Bresette, 2011). Each level of complexity constitutes another layer of integrations incorporated in a boundary object's heterogeneous and distributed functions (Fisher, 2020, 2021a). The mediations effected make it possible to negotiate locally situated meanings on the fly, doing so in ways that do not require homogenizing diversity into monotonous sameness.

Organizing alliances coordinating theory and data by means of standardized instrumentation provides human activity with the media for remotely and effectively communicating and managing infinite arrays of concrete, material things, and processes. Coordinating such networks across social sectors requires effective ways of communicating and aligning shared interests by means of agreed-upon standards and obligatory passage points. Such coordinations are not undertaken in the data-focused world of Rasch analysis, which concentrates strictly on calibrating abstract instruments from concrete observations, successfully integrating two levels of hierarchical complexity, data set by data set. Jack saw how to continue the process into higher order levels of complexity by

- employing formal theory to predict item calibrations, which allowed him to integrate axiomatic models, unit standards, and observed data (three levels of complexity) within multiple data- and instrument-specific studies;

- demonstrating the applicability of the theory across any given testing organization's or state department of education's samples and examinations, which integrate concrete data, abstract instrument standards, and formal theory within systems, at the fourth level of hierarchical complexity; and
- promoting the adoption of the resulting frame of reference by multiple educational, testing, and publishing systems, which integrated the systems in a system of systems (i.e., metasystematically).

And here we begin to glimpse some of the consequences of bringing into common metrics the self-organized, repeating, and objective assertion of empirically and theoretically reproducible constructs. These constructs assert their structure or its absence independent of the wishes or desires of the researcher. This kind of objective, independent existence provides a more elegant, philosophically sophisticated, and methodologically satisfying basis for human, social sciences than has previously been available in modern and postmodern contexts (Latour, 2009, 2010a, 2011, 2015).

Operationalizing these kinds of involuntary, unwilled modes of being sets the stage for a paradigm shift in the methods of psychology and the social sciences. Hierarchically complex knowledge infrastructures of the kind Jack played a key role in creating may comprise the "new science" projected by Star and Ruhleder (1996, p. 132; also see Eveland & Bikson, 1987; Feldman, 1987). They described this new science as "highly challenging technically," requiring "new forms of computability that are both socially situated and abstract enough to travel across time and space." Though Star and Ruhleder said (in 1996) "we cannot yet imagine" the forms and conventions that would attend the organic and natural emergence of an infrastructure of this kind, they cited others (Goguen, 1994) who had already anticipated that such "abstract situated data types" could potentially lead to "the reconciliation of technical and social issues." Later work addressed this theme in an explicitly semiotic modeling process (Xu & Feng, 2004), distinguishing concrete data, abstract information, and formal knowledge, though not in relation to the model of hierarchical complexity or to Rasch's probabilistic models. Wright (1997b) provides a start in this direction, however.

Nearly 100 years have passed with little change away from the centrally planned data analytics dominating psychology and the social sciences. The enduring persistence of methods that are not only widely recognized as insufficient but for which superior replacements are readily available is remarkable. The ongoing situation is testimony to the futility of trying to transform cultures without transforming the systems of incentives and rewards structuring the values in play. But those transformations will be enacted as socio-ecological functionality of a new kind takes root and grows. It is only reasonable to expect that fundamental enhancements in communications will have consequences extending far beyond education, psychology, and the social sciences.

As pragmatically ideal and flexibly rigorous boundary objects become more widely appreciated as viable, feasible, and desirable, new chapters in human history will be written. Actor-network theory's efficacy in explaining the growth of science, economies, and technologies (Blok, et al., 2020; Bowker, et al., 2015) will be augmented by practical methods for creating and managing hierarchical and cross-sector ecological complexity (Fisher, 2020, 2021a; Fisher & Cano, 2022). Heterogeneously distributed abstract situated data types are already deployed in metrological networks in education and health care (Fisher & Stenner, 2016 (Chap. 22); He, Li, & Kingsbury, 2016; Massof & Bradley, 2022; Morrison & Fisher, 2018, 2019, 2021). As these ecosystems are nourished and grow, they will create populations of end users habituated to and benefiting from improved cross-sector communications and new economies of scale. That said, of course, as the history of electrical standards shows (Hunt, 1994), the emergence of these markets will not be without struggles, setbacks, and failures.

But, as time passes, the advantages of coordinating visualizations, plans, skills, resources, and incentives (Knoster et al., 2000) across horizontally distributed and hierarchically complex domains will organically and naturally emerge and become a different way of life. Where, today, highly technical skills across a wide range of areas must be brought together and managed to effect even narrowly defined complex results, as with sociocognitive instrument design, calibration, and standard setting, metasystematic and paradigmatic integrations of infrastructural systems will remove the need to invest large resources in mounting disconnected and redundant local metrological networks. Quality-assured quantitative estimation will, in due course, recursively feedback collectively projected, qualitatively meaningful knowledge structures in psychology and the social sciences. These fields will then begin to succeed, in the terms used by Whitehead (1911, p. 61), in advancing civilization by extending the number of important operations that can be performed without thinking about them.

Population-level critical thinking capable of recognizing and acting on exceptions and anomalies must find some way of systematically facilitating flexible ways of living up to the maxim, "Think global, act local." As Linssen (1958, p. 304) puts it in the context of Zen philosophy,

> In the game of life, let us use the cards we have in hand.
>
> These cards are our faculties of thought, love and action. No one is expected to do the impossible.

The cards we have in our hands include the immensely useful symbols of language, which can be organized to aid and advance our faculties of thought, love, and action. In keeping with a return to a shared reality seen with a new vision, a different perspective on standards and their roles comes into view. Instead of seeing standards as rigidly imposed, a wide range of philosophers taking markedly different approaches recognize that everyday language provides a model in which standards serve as the media through which meanings at the moment are negotiated, and creative improvisations are expressed.

A primary function of standards and invariances is, then, the display of anomalies (Cook, 1914/1979; Kuhn, 1977, pp. 205–206; Rasch, 1960, pp. 10, 124). Cook (1914/1979) is particularly insightful, noting that

- "we can but use the instruments we have" (p. 431),
- "without the mathematical expression as a guide we should be unable to take note of the aberration" (pp. 434–435),
- "we shall use the theory or law chiefly to discover the exception" (p. 431),
- "the hidden meaning of such deviations will be more significant than any mere conformity to the law itself" (p. 431), and
- "the really important thing is the exception" (p. 434).

Kuhn (1977, pp. 205–206) similarly says,

To the extent that measurement and quantitative technique play an especially significant role in scientific discovery, they do so precisely because, by displaying significant anomaly, they tell scientists when and where to look for a new qualitative phenomenon. To the nature of that phenomenon, they usually provide no clues. When measurement departs from theory, it is likely to yield mere numbers, and their very neutrality make them particularly sterile as a source of remedial suggestions. But numbers register the departure from theory with an authority and finesse that no qualitative technique can duplicate, and that departure is often enough to start a search. Neptune might, like Uranus, have been discovered through an accidental observation; it had, in fact, been noticed by a few earlier observers who had taken it for a previously unobserved star. What was needed to draw attention to it and to make its discovery as nearly inevitable as historical events can be was its involvement, as a source of trouble, in existing quantitative observation and existing theory. It is hard to see how either electron spin or the neutrino could have been discovered in any other way.

Rasch (1960, pp. 10, 124) remarks on the same theme, saying

Once a law has been established within a certain field then the law itself may serve as a tool for deciding whether or not added stimuli and/or objects belong to the original group.

In many fields this type of predictions can be made with an astonishing degree of precision, so close indeed, that aberrations between observations and calculated values have given rise to discovery of previously unknown factors, e.g. the presence of hitherto unknown planets.

Accepting the unrealistic nature of scientific laws and models while simultaneously making use of them to structure meaningful communications and the display of individual level exceptions to the rule requires a highly complex ability to coordinate formal, abstract, and concrete perspectives simultaneously. And if it is difficult for individuals to coordinate these somewhat divergent and somewhat convergent perspectives in a personal pragmatic idealism, it is all the more challenging to figure out how to embed that complexity systematically, metasystematically, paradigmatically, and cross-paradigmatically in the social world.

A way forward is suggested by Latour (2004):

Social sciences may become as scientific…as the natural sciences, on the condition that they run the same risks, which means rethinking their methods and reshaping their settings from top to bottom on the occasion of what those they articulate say. [The] …general principle becomes: devise your inquiries so that they maximize the recalcitrance of those you interrogate. (p. 217)

In Jack's work, as in theory-informed Rasch measurement modeling in general, the recalcitrance of the constructs interrogated is maximized by providing a medium that may be capable of giving voice to emerging harmonies and dissonances. Identified measurement models (Koopmans & Reiersøl, 1950; San Martin, Gonzalez, & Tuerlinckx, 2009; San Martin & Rolin, 2013; Fisher, 2021b) focus on autonomous structural invariances to create contexts in which people and organizations may push back via the consistencies of their responses to affirm or contradict the coherence and comparability of the questions asked. By focusing attention on the lived experiences of readers as they develop from the bottom to the top of the continuum of the learning progressions incorporated into the educational enterprise, Jack and his colleagues rethought the methods of the social sciences, reshaping them on the basis of what students themselves say.

The fit of data to measurement models is evaluated by a variety of statistical and principal components analyses (Karabatsos, 2003; Smith, 2002; Smith, 1996; Smith & Plackner, 2009) designed to anticipate structural invariances in individualized learning or developmental trajectories, and unique special strengths and weaknesses impacting outcomes. The difficult order of test items emerges from examinees' experiences and is not imposed from the outside by the preconceptions or theories of the test designer or researcher. The construct measured finds its voice in the medium of the instrument. Though inevitably colored to some extent by the local features of the sample, the phrasings of the items, the administration process, the researcher's theoretical orientation, etc., repeated investigations varying these unique particulars can and do converge on structural invariances (contextualized within uncertainty ranges and confidence intervals) estimated to differing levels of precision.

This is a way, then, in which psychology and the social sciences have come to join the natural sciences in "Rendering talkative what was until then mute," and so, also, in honoring the use of the word "logos" in the names of the various disciplines (Latour, 2004, pp. 217–218). The use of Rasch's models in psychology and sociology brings new ways of enabling disciplines to live up to the meaning of their various "-ologies." This is especially true when empirical experiments are complemented by explanatory models, predictive theories, and construct specification equations (Stenner & Smith, 1982 (Chap. 3); Stenner, Smith, & Burdick, 1983 (Chap. 4); Stenner, et al., 2013 (Chap. 18); Stenner, et al., 2016; Carpenter, Just, & Shell, 1990; de Boeck & Wilson, 2004; Embretson, 1999, 2010; Fischer, 1973; Green & Kluever, 1992; Green & Smith, 1987; Latimer, 1982; Whitely, 1981), which combine to create new efficiencies, as in the natural sciences:

> ...laboratory settings where propositions can be articulated in a non-redundant fashion. [where]...talking and writing is not a property of scientists uttering statements about mute entities of the world, but a property of the well-articulated propositions themselves, of whole disciplines. (Latour, 2004, p. 218)

Whether in conversation or in measurement, dialogues that constantly refer back to the subject matter, allowing the course of the exchange to itself feedback indications of what comes next, reveal "the immanent logic of the subject matter that is unfolded in the dialogue" (Gadamer, 1989, p. 367). Here,

> What emerges in its truth is the logos, which is neither mine nor yours and hence so far transcends the interlocutors' subjective opinions that even the person leading the conversation [Socrates, in Plato's dialogues] knows that he does not know. As the art of conducting a conversation, dialectic is also the art of seeing things in the unity of an aspect—i.e. , it is the art of forming concepts through working out the common meaning. (Gadamer, 1989, p. 367)

In accord with the anticipatory sense of a qualitative mathematical "thinking-cap" described by Butterfield (1957, pp. 16–17), Heidegger (1977), Cook (1914/1979), Rasch (1960), and Kant (1929/1965, p. 20), finding a common meaning is not a process in which a tool serving this aim is produced and applied. Instead, language finds its true value in the moment of use as a medium agreed upon in advance to play the role of the vehicle of shared thinking.

Thus, the more genuine a conversation is, the less does anyone conduct it (Gadamer, 1989, p. 338). Someone skilled in the art of questioning may well appear to have the weaker opinion, as the full strength of the object is allowed to show itself, and questions about it are not suppressed. The common meaning identified and agreed upon in any given moment exhibits the unity of being a particular aspect of a conceptual ideal. Different circumstances will necessarily involve potentially incommensurable local particulars. This is true even in the meaning of something as mundane as saying, "Pull up a chair," since no chair embodies every possible feature of the concept, and the word might be used when no chairs are actually available.

Although Gadamer mistakenly conceives of mathematical understanding as perfect and certain, he is correct in making the point that

> the true being of things is to be investigated 'without names' means that there is no access to truth in the proper being of words as such—even though, of course no questioning, answering, instructing, and differentiating can take place without the help of language. (Gadamer, 1989, p. 414)

Gadamer (1989, p.415) recognizes mathematical symbolism's formal axioms as a kind of nonphonetic writing that represents relationships generally, without naming the specific entities or processes involved. And he recognizes things have a truth about them apart from any given language while simultaneously accepting the necessity of relying on language despite its shortcomings. Language simultaneously and systematically performs conceptual ideals, instrumental abstractions, and concrete particulars as a semiotic assemblage.

And here we come to the relevance of Galison's (1997, pp. 843–844) surprising conclusion, reported at the end of his study of the material culture of microphysics. To wit: conceptual and communicative disorder plays an essential role in the coherence and success of science. He traced that complex blend of unity and disunity to the inner structures of everyday language:

> It seems to be a part of our general linguistic ability to set broader meanings aside while regularizing different lexical, syntactic, and phonological elements to serve a local communicative function. So too does it seem in the assembly of meanings, practices, and theories within physics. (Galison, 1997, p. 49)

The same sequence occurs as we move from Rasch's projection of uniform idealized relationships between reading comprehension, reading ability, and text complexity to measurement. Jack's pragmatic idealism embodies the follow through from the experience of formless pure concepts back into words and things. Jack formulated and tested the mathematical hypothesis that reading comprehension could be conceived as a pure contentless and wordless form, and that this unrealistic formal ideal could be brought to bear on abstract words and concrete observations in useful and meaningful ways.

In no way does that usefulness require any one level of complexity to serve as the basis for a homogenizing reduction, as Galison (1999, p. 143) points out. Each respective community of research or practice focuses attention on the objects of its interests: concrete in the experimental, executive, and managerial worlds of laboratories, classrooms, and clinics focused on data; abstract in the metrological, legislative, and standards worlds of instrument makers and consensus processes; and formal in the explanatory, judicial, and axiomatic worlds of theoreticians and modelers. Further distinctions can be made at the systematic and other higher levels of complexity and will need to be taken up to methodically cultivate participatory social ecologies analogous to, and expanding on, those created by Jack and his colleagues for reading instruction.

Lingering a Moment Longer in Mindful Attentiveness

Letting reading comprehension be what it is, independent of any particular way of representing or observing it, is an example of a kind of Zen doing by not-doing that accords with the yin-yang symbolization of reader and text in dialogue. The not-doing happens in standing back for a moment before proceeding directly into data analysis. In this pause, the researcher looks closely for possible evidence of an objective, independently asserted structure that cannot be willfully or methodically imposed from the outside.

A key lesson taught by the works of Jack Stenner collected in this volume concerns how we learn the limits of a measured phenomenon as a form of life or mode of being. Those limits become evident by providing things themselves with media through which they may find their voices. Seeing all things, from rocks to chemical processes, as alive is a philosophical theme explored by Peirce and Hegel that asks how things might come to be included more fully as participants in humanity's shared social reality (Bennett, 2010, p. 104; Fielding, 2003; Latour, 1991; Latour & Wiebel, 2005; Maran, 2007; Oele, 2017). Measurement modeling and metrological infrastructures provide an array of means for populating what Latour calls a "parliament of things."

Complementary masculinist and feminist perspectives on this event of meaningful representative governance come to bear. In Latour's (1987, pp. 87–93) and Gadamer's (1989, p. 367) masculinist language, we test the strength of the measures we propose to pass in our laboratory legislatures, so to speak. These collectively formed patterns in data document the constraints of the systems by which a rule of law may be exported successfully from the laboratory into a shared social reality. When the sources and characteristics of the measured construct's strengths are understood, the laboratory environment in which it stands forth as a thing, as a fabricated artifact (Ihde, 1991), can be inferred as both derived from and applicable to the larger world.

In Haraway's (1996, pp. 439–440) feminist language, in contrast, the focus is on creating livable worlds from questions of pattern that also speak to embedded relational infrastructures capable of serving as "the prophylaxis for both relativism and transcendence." Instead of focusing on exercises of strength, Haraway, in effect, looks to the capacities of the mathematical matrix of relationships modeled in the laboratory as a womb in which a new form of life is conceived and gestated before being midwifed into the world and nurtured to maturity.

Here resides a kind of maternal patience attending to the growth and development of something new in the world. Lingering in watchful anticipation enacts a nurturing respect for life that contrasts dramatically with the way commonly used methods move directly toward a kind of mechanical dissection that cannot but terminate the new form of life before it can be born. Leaping to the conclusion that numbers are inherently and automatically always measured quantities mindlessly disconnects symbolic representations from any possible way of meaningfully connecting with human social reality. In the emotional psychology of contemporary culture, should not these systematically disempowering methods—routinely applied in so many areas of research and practice involving human, social, and environmental issues—be considered as analogous to a kind of conceptual infanticide or genocide? In enacting so methodically and routinely what Ricoeur (1974) called the "violence of the premature conclusion," do not we cut ourselves off from essential ways of respecting and valuing the ethical importance of pregnancy (Kristeva, 1980, 2014)? What kinds of consequences for healing might follow if humanity would heed Latour's (2011) recommendation that we "care for our technologies as we do our children"?

The contrasting masculinist and feminist points of view on mindfully lingering in attentive care for the objects of discourse both require a poetic openness to noticing and listening. This openness is notably absent in the hurried push in most psychological research toward presentations of merely numeric results masquerading as measured quantities (Bateson, 1978; Fisher, 2021a; Wright, 1994). Tallying correct answers or summing ratings and treating these scores as measurements commits a kind of schizophrenic violence. Everyone is well aware that the counts are not quantities, since the scores are universally understood to change meaning across tests and surveys. In addition, many critics deplore the ways unique individual differences are homogenized and reduced to false uniformity within any given scoring system (Maraun, 1996; Michell, 1986). The same number is assumed to mean the same thing even though structural evidence as to that meaning is usually not provided, and even though disparate uncontrolled variations can often be discerned.

Measurement models and methods may, however, be attuned to letting things show themselves as they are. In contrast to the immediate scoring and analysis of counts, calibrating instruments read for signs of measured quantities must allow things themselves the agency of showing the way. Measuring then becomes an active doing that fundamentally and simultaneously does not do what normally is taken for granted as basic procedures in measurement research and practice. Letting things themselves speak in their own voices across samples and instruments requires capacities for listening to what measured constructs have to say as to what goes with what, and how different media for culturing social forms of life vary.

Jack was oriented toward this critical perspective before he encountered Rasch's models for measurement. He was a co-author of three early papers not included in this collection that explicitly addressed issues of cross-level meaning (Hayman, Rayder, Stenner, & Madey, 1979) and cross-sample factor structure invariance (Katzen-meyer & Stenner, 1975, 1977). Jack's intellectual roots in these fundamentally important and mutually implicating areas help inform one's reading of his thinking about reading comprehension as it evolved over time, culminating in widely reproduced variations on the reading comprehension construct map (McCabe, 2006; Peabody, O'Neill, & Peterson, 2015; Stenner, 1998; Wildomar, 2014).

Cross-Level Meanings and Cross-Sample Invariances in "Ecologies of Mind"

The hypothesis considered in one of Jack's early publications (Hayman, et al., 1979) was "The utility of a set of evaluation data varies inversely with the number of organizational levels between the action the data describe and the decision process they are intended to influence" (Hayman, et al., 1979, p. 31). Without naming it as such, this hypothesis concerns the ecological fallacy (Pollet, Stulp, Henzi, & Barrett, 2015; Rousseau, 1985) and the atomistic fallacy (Diez-Roux, 1998). The former involves making false and illogical inferences about individuals from population-level data, and the latter, making false inferences about populations from individuals (Forer & Zumbo, 2011; Zumbo & Forer, 2011). Rousseau (1985) notes that

- "To reduce the risk of misspecification, we need to develop theories and research designs that allow us to take issues of level into account" (p. 10).
- "Multi-level models specify patterns of relationships replicated across levels of analysis. Hence they are concerned with generalizations that may be made from phenomena observed at one level to those occurring at another" (p. 22).
- "Theories must be built with explicit description of the levels to which generalization is appropriate" (p. 29).
- "...level-specific construct validity must be determined for any construct taken from one level and applied to another. Researchers are often cavalier in their use of constructs from one level as surrogates for those at other levels" (p. 29).

Bateson (1972, p. 493) saw cross-level fallacies of these kinds as committing a fundamental epistemological error, one in which thinking "is separated from the structure in which it is immanent, such as human relationship, the human society, or the ecosystem." Creating an "ecology of mind," to adopt Bateson's (1972) phrase, requires capacities for systematically distinguishing levels of complexity. The point of Katzenmeyer and Stenner's (1975, 1977) testing for the invariance of item hierarchies across samples of examinees is, in fact, to investigate whether level-specific forms of construct validity and generalization might be distinguished.

Levels of complexity vary in ways usefully characterized by the differences between maps and territories. "Differences that make a difference," another of Bateson's memorable phrases, are important to include in a map, while those that do not are excluded (Bateson, 1972, pp. 183–186; 1991, pp. 200–201). Determining which differences matter, and which do not, is a major goal of mapping, just as it is also part of the general modeling goal of simplifying matters without oversimplifying them. Different kinds of maps (topographic, road, tourist, walking, etc.) emphasize different features, and, importantly, maps' simplifications never remove the often-difficult challenges encountered in locating one's position relative to the desired goal.

These challenges follow from trying to reconcile what is represented abstractly from a position that no one in fact ever occupies with what is seen in the concrete world (Hutchins, 1995, 2014; Latour, 1995a). Jack understood that mapping psychological and social variables (Stone, Wright, & Stenner, 1999; Chap. 8) requires distinguishing levels of complexity capable of serving purposes analogous to those served by geographic maps. What Jack learned from these efforts has since been taken up by others in similar mapping and theory development projects, many of which also draw from extensive interactions with the works of Benjamin Wright (Best, 2008; Fisher & Wright, 1994; Melin, Cano, & Pendrill, 2021; Peabody, O'Neill, & Peterson, 2015; Wilson, 2005).

The distinction between maps and territories hinges on the way the latter exist at a concrete level of real-world things and data, while maps are abstractions. As Bateson (1972, pp. 182–186) puts it, when we speak directly about concrete things, we use denotative forms of expression, such as "the cat is on the mat." When we refer to what has been learned, we make metalinguistic statements referring to abstractions, like "the word 'cat' cannot scratch." Then, when we move on to talking about theories of learning, we make formal metacommunicative statements: "my telling you where to find your cat was friendly."

Most research in psychology and education does not distinguish these levels of complexity. Counts of correct test answers, for instance, support valid and legitimate denotative statements concerning how many times a student understood what was asked and could respond appropriately. But when these counts are treated as measured quantities, as they often are, Bateson's (1978) and Wright's (1994) distinction between denotative counts and metalinguistic quantities is ignored (Fisher, 2021a). The meaning of the same numeric count of correct answers must necessarily vary across groups of students and across different collections of test questions. The map tracks the territory so closely they become identical, making the map useless apart from the specifics of the features of one local learning environment's students and test questions.

Log interval formulations of additive conjoint measurement models are used to test the hypothesis that a unit quantity exists, doing so in ways that achieve fundamental measurement (Andrich, 1978, 1988; Fisher & Wright, 1994; Narens & Luce,

1986; Newby, Connor, Grant, et al., 2009; Rasch, 1960, 1977; Wilson, 2005, 2013; Wright, 1968b, 1977, 1984, 1997a). Measurement modeling of this kind successfully maintains the distinction between denotative and metalinguistic levels of complexity (Fisher, 2021a; Fisher, Oon, & Benson, 2021). Here, differences that make differences, that sustain distinctions between consistently identifiable levels of performance, are identified. The quantitatively scaled map that is drawn now is an abstraction quite different from the concrete territory occupied by the numeric data. Now, instead of a one-to-one correspondence between features on the map and features in the territory, local particulars that do not accumulate into differences that make a difference are skipped over. Different students can answer different questions without compromising the estimation of their reading ability locations on the map, to within the range of uncertainty, just as different questions can be posed to different students without compromising the comparability of the item locations on the map.

In psychometrics in general, and in Rasch model applications in particular, however, these experimental evaluations are almost always repeated anew for each separate sample of students or collection of test items. Justifications for the validity of the quantities relative to the counts are left tied to the specifics of the sample and the items involved. Just as the denotative and metalinguistic levels were left undifferentiated when counts are mistaken for quantities, we see here a failure to distinguish between metalinguistic and metacommunicative statements. The consistency of sample and item characteristics across data sets is left unexamined and unleveraged as a means of explaining what has been learned about the learning documented at the metalinguistic scale level of complexity.

An explanatory model predicting student measurements and item calibrations on quantitative scales (Stenner, 1996 (Chap. 6); Stenner & Stone, 2010 (Chap. 13); Carpenter, Just, & Shell, 1990; De Boeck & Wilson, 2004; Embretson, 2010; Fischer, 1973; Green & Kluever, 1992; Green, Kluever, & Wright, 1994; Green & Smith, 1987; Melin, Cano, & Pendrill, 2021; Prien, 1989; Wright & Stone, 1979) is needed to achieve the formal, metacommunicative level of complexity (Fisher, 2020, 2021a). Now, a theory of learning contextualizes what was learned about learning, which in turn contextualizes the facts learned. An axiomatic model of the measured construct now informs the quantitative scaling process and the interpretation of the concrete numeric data. Understanding the measured construct in this way is a kind of synthetic learning, where a capacity to reproduce the construct at will from theory is realized. Jack implicitly agreed with Feynman's (1988) recognition that, "what I cannot create, I do not understand," and so strove to demonstrate how an understanding of reading comprehension could be expressed creatively.

Wider recognition of Jack's accomplishments must inevitably remain limited as long as mainstream research methods continue to conflate levels of complexity, remain ensnared in the crisis of reproducibility caused by the fixation on statistical significance testing (Fiedler, 2017; Hardwicke, Thibault, Kosie, et al., 2021; Pashler & Wagenmakers, 2012), and fail to implement metrological solutions (Cano, et al., 2018, 2019; Fisher, 2022; Fisher & Stenner, 2016; Mari, et al., 2021; Pendrill, 2019; Wilson & Fisher, 2019) to problems of validity generalization (Schmidt & Hunter, 1977). The situation will likely be resolved only in the wake of decisive demonstrations supplanting existing systems of incentives and rewards with new, more satisfying ways of embedding relational structures in educational environments. That kind of a shift in power relations requires the emergence of more organically cohesive social ecologies. Hope for positive outcomes is not unreasonable, given the practical power of thinking collectively in shared languages. An additional factor to consider is the phenomenologically superior meaningfulness obtained when common languages are themselves structured as recursive expressions of collectively projected data patterns. We may yet be able to activate new languages poetically and scientifically, effectively and efficiently functioning as vehicles of thought in the pragmatically ideal terms so many philosophers have sought to articulate for so long.

Such combinations of coordinated behaviors and decisions rooted in spontaneously occurring human cognitive and social processes ought to easily out-perform today's mainstream paradigm's disconnected and incommensurable pseudo-systems. Even so, proving that and cultivating ecosystems in which human institutions partner with social and cognitive forms of life in new economic relationships are hugely formidable challenges. Those challenges will become better understood with the unfolding of multiple ongoing projects (Massof & Bradley, 2022; Morrison & Fisher, 2018, 2019, 2021; Quaglia et al., 2016–2019, 2019–2022) that are emerging alongside the established and still-growing reading comprehension ecosystems initiated by Jack and his colleagues.

Getting Down to Business

The repeated demonstration of how student experiences of the difficulty of reading comprehension test items can be predicted from sentence length and the commonality of the words used contributes to the formation of a new economy of thought in education. The principle is as old as education itself, which begins from the alphabet and moves from there to word-picture associations, and from there to increasingly complex text and ideas. The predictable association of semantic and syntactic components in text complements the explanatory power of data by offering independent validation that the object of measurement is understood well enough to synthesize it. But instead of restricting themselves to by-project equations of specific tests, Jack and his team envisioned a larger market, the one implied by the measurement model.

That is, in parameterizing the infinite universes of all possible English reading students' abilities and all possible English reading passage difficulties, the potential for a virtual connectivity of every member of both sets with all of the others comes into view. Student ability and item difficulty parameters in probabilistic measurement models are, then, heuristic fictions in principle capable of telling the stories of any possible readers and collections of reading comprehension items. The potential viability of this scenario was tested in the 1970s by the Anchor Test Study, which equated seven reading tests in administrations involving 350,000 students in all 50

U.S. states (Jaeger, 1973). But the resulting "National Reference Scale for Reading" (Rentz & Bashaw, 1977) was rendered obsolete as soon as the test was finished because the test publishers changed the items. Unless another large and expensive study was undertaken, there would be no way to connect the existing results with the new ones.

With the introduction of explanatory models predicting variation in item difficulty, the commercial opportunity for creating a new market in literacy capital emerged. Though difficulties were encountered along the way, it was not long before reading test and curricula publishers, state departments of education, school districts, and book and magazine publishers were avidly interested in this new way of connecting readers with text.

In a 2012 presentation to the Pearson Global Research Conference held in Fremantle, Australia, Jack spoke of personalized learning platforms (Stenner, Swartz, Hanlon, et al., 2012). Here, instruction and assessment are integrated, with unique items generated from the lesson being studied, and then scored, in real time, in a single use application (see Appendix B on EdSphere in Chap. 24; Stenner, Stone, & Fisher, 2018). In this pioneering application of automatic item generation (Attali, 2018; Barney & Fisher, 2016; Bejar, Lawless, Morley, et al., 2003; Embretson, 1999; Gierl & Haladyna, 2012; Gierl & Lai, 2012; Kosh, Simpson, Bickel, et al., 2019; Poinstingl, 2009; Sonnleitner, 2008), the student may receive immediate feedback designed with the intention of being able to support developmental processes of cognitive growth. Opportunities for revealing misconceptions commonly made in any given area of study are built into assessments and are highlighted in Socratically suggestive feedback encouraging the student to think again (Black, Wilson, & Yao, 2011; Chien, Linacre, & Wang, 2011; Fisher, 2013; Hattie & Timperley, 2007; Masters, Adams, & Lokan, 1994; Wilson & Scalise, 2003; Wilson & Toyama, 2018; Wright, Mead, & Ludlow, 1980). Testing, in this context, ceases to be a one-time hoop to jump through and cram for, and instead becomes an integral part of the assessment of, for, and as learning (Black & Wiliam, 1998, 2018; Black, Wilson, & Yao, 2011; Fisher, 2013). In this work, what Ben Wright (1958, 1968a) called "a curriculum of failure" is at work providing safe contexts for learning from mistakes. Interestingly, the adaptive feedback mechanism was foreshadowed in ruminations about computerized learning by Wright's physics lab supervisor, John Platt (1970, pp. 139–141).

Studies in multiple fields (Locoro, Fisher, & Mari, 2021; Massof, Ahmadian, Grover, et al., 2007; Morrison & Fisher, 2018, 2019, 2021; Melin, et al., 2021; Powers, Fisher, & Massof, 2016) are exploring the power of systematically integrating (a) formal theories explaining variation in item calibrations and person measurements, (b) empirically calibrated and equated instruments measuring in quality-assured standardized units with known uncertainties, and (c) individual response data reported as instructionally or clinically actionable feedback. These projects are all akin to Stenner's ecologized environment for reading comprehension in that multiple stakeholders are included in the development and deployment of common languages.

Via investigations of these kinds, the instruments of psychology and the social sciences will become a medium for giving voice not only to things themselves, but also to individuals and communities that have been excluded from status quo methods. Having extensions of everyday language built into comparisons of outcomes across groups and over time, as well as in the monitoring and facilitating of growth and healing, may be able to simultaneously support critical inquiry, quality improvement, and equitable budgeting of resources.

When theory succeeds in retrospectively predicting item calibrations, following the already-established principles of item banking and anchoring, Jack saw that it becomes possible to prospectively estimate new items' calibrations from theory, administer them, and validate them by checking to see if they produced the expected proportions of correct responses. Successful pilots of this kind set up the personalized learning technology Jack initially called Oasis, in which unique new cloze reading comprehension items were created from the text read by a student as they read (Stenner, et al., 2012). Every second or third sentence had a word blanked out. When the student clicked on the blank, a list of four words (the correct word with three uniquely included distractors) would appear. The student would choose the correct word from the list and continue.

Upon finishing the reading assignment, the application provides feedback on the student's responses, identifying special strengths and weaknesses by means of unexpected correct or incorrect responses to especially difficult or easy items. Transitional misconceptions scored as partially correct are set in context for further reflection as a means of capitalizing on the cognitive priming of the moment. Other students reading the same passage would have other words blanked out and would see different distractors when the same words were blanked out, but would nonetheless have measurements expressed in a comparable metric. A record of the transaction would be registered by the software for incorporation in growth charts and reports for use by teachers, parents, researchers, administrators, and others (Stenner et al., 2012; Fisher, 2013; Williamson, 2016, 2018; Williamson, Fitzgerald, Stenner, 2013, 2014).

Linguistically materializing meaningful events and processes in shared signs and symbols that make visible connections between things and thought frees the cognitive process from identification with any particular thinker. This is a consequence of having, as Scott (1998, p. 357) puts it, taken language itself as a model for the kinds of institutional systems needed for improving the human condition. Grasping the stochastic patterns in the observations cohering into something namable requires arriving at a conceptual determination of what the thing itself is. Semiotically speaking, of course, no concrete thing ever embodies every possible feature of a formal conceptual ideal, and so the best that can be done by abstract words is to serve as a standardized medium that is flexible enough to adapt to immediate circumstances by disappearing into the referent.

This semiotic perspective fits squarely in the oft-repeated maxim that models are not meant to be true, but to be useful. Box (1979, p. 202) and Rasch (1960, pp. 37–38; 1973/2011) are repeatedly cited in this vein, but rarely with an appreciation for the pragmatic idealism that operationalizes boundary objects (Bowker, et al., 2015; Star & Griesemer, 1989; Star & Ruhleder, 1996). It is imperative that practical methods accept the unrealistic geometrical form of scientific models as eminently applicable idealizations differentially structuring shared social reality. No amount of data or observations are ever sufficient to lead to the empirical derivation of the laws of science. Instead, the form of human thinking in the language is set up to begin from qualitative, implicitly mathematical models to see if and when observational data, calibrated instrumentation, and explanatory theory can be envisioned and ultimately organized in meaningful and useful relational ensembles (Overton, 1998). It is hard to imagine more conclusive proof of the practical commercial value of this kind of modeling than the international ecosystem of relationships Jack and his colleagues created around reading instruction.

Prospects for the Future

Heidegger (1991, p. 24) remarks that "What is great and constant in the thinking of a thinker simply consists in its expressly giving word to what always already resounds." In education, what always already resounds is the enduring value of learning through what we already know. And there, in the process of learning, we also see that the meaning "which can already be found in the older philosophers is seen only when one has newly thought it out for himself" (Heidegger, 1967, p. 79).

In collaborations with extensive networks of allies, Jack was a key figure giving word to what already resounds, not just in education but across a wide range of fields. He thought through for himself essential insights already put on the record by previous thinkers and made them available to others. He followed through in imaginative and original ways to collaboratively produce results recognized immediately as valuable and useful by tens of millions of students, teachers, parents, publishers, psychometricians, and others, though they themselves would have been incapable of envisioning, planning, staffing, resourcing, or incentivizing those results.

And Jack would be the first to recognize those results were possible only by means of collective efforts. The theory of action implied by text interpretation (Ricoeur, 1981, 1991a) and by the role of metrology in the history of science and commerce (Latour, 1987, pp. 247–257; Latour, 1991, 2005) suggests that we understand, inhabit, and dwell in the world projected temporally in front of us (Fisher & Stenner, 2005):

> …we understand ourselves only by the detour of the signs of humanity deposited in the cultural work (Ricoeur, 1991a, p. 87).
>
> To understand is to understand oneself in front of the text… It is exposing ourselves to the text and receiving from it an enlarged self (Ricoeur, 1991a, p. 88).

We learn about who we are as individuals and communities by testing ourselves against what the world and others in it say to us. We express our identities via the stories we tell about ourselves, prefiguring the future, configuring the present, and refiguring the past as we go. We are shaped moment by moment by the way our world horizons are altered, as what is said and done changes our expectations and colors our self-images. Each encounter offers new opportunities for knowing ourselves, broadening the range of possible ways we might signify who we are.

Jack Stenner and his colleagues have provided a model art and science that both demonstrates and enacts humanity's universalizations and individualizations of its stories. Science, just as much as classic myths, popular literature, art, and films, recounts events in which everyone can see themselves, even when the story told never actually happened to anyone in particular. Jack's qualitative quantities similarly give voice to the "variations on an invariant" (Ricoeur, 1991b, p. 196) that are always already resounding as we configure, refigure, and prefigure who we are in relation to who we have been and who we want to be.

Following Stenner's lead in co-evolving artful science and commerce, we might yet find our way to what Dewey (1954, p. 210) called "A more intelligent state of social affairs, one more informed with knowledge, more directed by intelligence...[that] would raise the level upon which the intelligence of all operates." Given the fullness of the way he opened himself to the world of the text, Jack Stenner not only received a remarkably enlarged self, but he also shared it on previously unimagined scales, raising the level upon which the intelligence of millions now operates.

Overview of Jack's Articles

The following chapters are organized chronologically from earliest to latest. In the course of reformatting the articles, a number of minor editorial revisions were made; these involved correcting typos, adding citations for uncited references (or dropping them), adding references for citations lacking them, clarifying direct quotes from sources, adding missing subscripts in equations, etc. Chapters 1–4 present Jack's early efforts at understanding how to consolidate a formal level of complexity by integrating concrete data and calibrated instruments via a predictive theory and an explanatory model. The first two chapters apply traditional psychometric methods in efforts aimed at potentially standardizing and theoretically explaining growth in learning. A probabilistic Rasch model of measurement positing unidimensional invariance is not introduced until Chap. 3 (Stenner & Smith, 1982). As can be anticipated from the first two chapters, the scaling model is immediately connected with the need for an explanatory account of what is quantified.

Stenner and Smith's focus is, moreover, shifted away from the usual concentration on between-person variation toward the relation of item characteristics and item scores. Applications of explanatory models may, of course, focus on the person or item characteristics, or both (Carpenter, et al., 1990; de Boeck & Wilson, 2004; Embretson, 1999, 2010; Fischer, 1973; Green & Smith, 1987; Latimer, 1982; Whitely, 1981). Stenner and Smith's analytic method for validating theoretical predictions is based on linear regression, which was shown by Green and Smith (1987) to produce results equivalent to those obtained from Fischer's (1973) linear logistic test model.

No matter which method is used, empirical and theoretical approaches converge on mutually validating results (pro or con). Either way, different data sets involving different samples responding to different items designed to measure the same thing are shown to provide evidence supporting or contradicting the linear comparability of empirically estimated and theoretically predicted item locations. Repeated analyses of data sets reproducing linearly equivalent instrument calibrations formally integrate the concrete and abstract levels of hierarchical complexity via the application of the model. The explanatory construct model then integrates the concrete-abstract-formal assemblage at the systems level, when different data sets, test instruments, and instances of construct specifications producing comparable results are put to work in any given sphere of application.

Chapters 5–13 mark the transition from these kinds of systems integrating concrete, abstract, and formal levels of complexity into systems of systems. Now, alliances of publishers, school districts, state departments of education, curriculum developers, and testing agencies are being formed. At this metasystematic level, measurement expertise, instrument calibration capacities, and a wide array of complementary skill sets are not evenly distributed across stakeholder sectors. Each group of partners in this alliance lacks the technical skills possessed by the others, but all of them must be brought to bear to coordinate a system of systems aligning instruction, test design, item writing, examination administration, data analysis, instrument calibration, curriculum development, book publishing, specification equations, etc. These alliances require the coordination of stakeholders' varying, somewhat divergent, and somewhat convergent interests in an overarching boundary object. The business side of the work performed by Jack and his colleagues involved, first, negotiation of the form of the passage points through which quality-assured representations would flow, and then the creation and management of those representations. Chapter 5's (Stenner et al., 1988) response to criticisms conveys some of the feelings of the kinds of questions that arose in the course of those negotiations (also see Burdick et al., 2014; Stenner & Burdick, 1997; Stenner et al., 2001).

Chapters 14–24 take on yet broader, paradigmatic and cross-paradigmatic concerns. Many earlier chapters, and Chaps. 6 (Stenner, 1996) and 13 (Stenner & Stone, 2010) in particular, connect the quality of measurement information and knowledge being constructed with that obtained in the natural sciences (as was also addressed in articles not included here, such as Cano et al, 2016). In this final group of chapters, the general implications of that connection are explored in relation to methodology, entrepreneurial business models, commerce, the economy, aesthetics, and governance.

Several years have already passed since the most recent of the articles collected here was published. New developments are proceeding at a rapid pace. To pick but two examples, first, Loubert and colleagues (Loubert, Regnault, Sebille, et al., 2020) contemplate clinical trials as measurement systems affected by multiple sources of uncertainty, expanding on the topic of uncertainty budgets raised in Chap. 23 (Fisher & Stenner, 2017).

Second, Melin and colleagues (Melin, Cano, & Pendrill, 2021) tap some of the deepest roots in Jack's oeuvre. They expand on the subject of information entropy implicated in reading comprehension (Fisher & Stenner, 2016, p. 493; Chap. 22) to augment the theory of the construct specification equation. They, moreover, take on this task in the context of the measurement of short-term memory and attention span, extending Stenner's and Stone's work in that area (Stenner & Smith, 1982;

Stenner & Stone, 2003; Stone, 2002; Wright & Stone, 1979). Finally, beyond theory and research, Jack played a role as a thought partner in advancing that cognitive construct as a viable candidate for a new reference standard unit system in Europe (Cano, Melin, Fisher, Stenner, Pendrill, et al., 2018). Research in this area continues (Cano, et al., 2019; Melin, Cano, Göschel, et al., 2021; Quaglia, Pendrill, Melin, et al., 2016–2019, 2019–2022).

If that standard is approved and adopted, Jack will have made substantial contributions to creating the first quality-assured, metrologically traceable unit of psychological measurement ever introduced. This achievement will mark a major milestone on the road traveled by language as the vehicle of thought. In the same way, technical standards for the measurement of time, temperature, length, weight, etc. have all been brought into everyday language in convenient yet rigorously defined terms, so, too, might Jack's work in reading and mathematics education also lead to similar results in the not-too-distant future (Fitzgerald, Elmore, Koons, et al., 2015; Fitzgerald, Elmore, Hiebert, et al., 2016; Simpson, Kosh, Elmore, et al., 2015; Williamson, 2016, 2018; Williamson, Fitzgerald, & Stenner, 2013, 2014).

An overarching theme unifying the various developmental phases in Jack's thinking follows Dewey's (1954, p. 210) insights into how:

> Meanings run in channels formed by instrumentalities of which, in the end, language, the vehicle of thought as well as of communication, is the most important. A mechanic can discourse of ohms and amperes as Sir Isaac Newton could not in his day. Many a man who has tinkered with radios can judge of things which Faraday did not dream of.

Measurement research and practice are inexorably following Jack Stenner's lead in bringing new levels of shared knowledge into everyday language's channels of meaning. Explanatory models in multiple fields are being designed, tested, validated, and used to structure locally situated applications communicated in common metrics. As these efforts proceed, new meanings will run in channels formed by the shared languages in which we think and communicate. When that happens, teachers, clinicians, managers, and analysts at the front lines of practice will come to discourse on technical matters and judge things that today's experts do not dream of.

Paraphrasing Rasch (1960/1980, p. xx), the challenges are indeed huge, but with the clear formulations of the problems we have in hand, and with the model of successful implementations provided by Jack and his allies, it does seem possible to meet those challenges. In fact, in this context, given humanity's history of resilient creativity, it is not just *reasonable* to hope, dream, and work for a better future;

instead, *passionately engaging* in those hopes, dreams, and tasks may be key to fulfilling humanity's evolutionary potential.

William P. Fisher Jr.
Living Capital Metrics LLC
Sausalito, CA, USA

Research Institute of Sweden
Gothenburg, Sweden

BEAR Center, Graduate School of Education
University of California
Berkeley, USA

References

Ackermann, R. J. (1985). *Data, instruments, and theory.* Princeton University Press.

Akrich, M., Callon, M., & Latour, B. (2002a). The key to success in innovation Part I: The art of interessement. *International Journal of Innovation Management,6*(2), 187–206.

Akrich, M., Callon, M., & Latour, B. (2002b). The key to success in innovation Part II: The art of choosing a good spokesperson. *International Journal of Innovation Management,6*(2), 207–225.

Alder, K. (2002). *The measure of all things.* The Free Press.

Andersen, E. B. (1977). Sufficient statistics and latent trait models. *Psychometrika,42*(1), 69–81.

Anderson, M., Deely, J., Krampen, M., Ransdell, J. M., Sebeok, T. A., & Von Uexküll, T. (1984). A semiotic perspective on the sciences: Steps toward a new paradigm. *Semiotica,50*(1/2), 7–47.

Andrich, D. (1978). A rating formulation for ordered response categories. *Psychometrika,43*(4), 561–573.

Andrich, D. (1988). *Sage University paper series on Quantitative applications in the social sciences. Vol. series no. 07-068: Rasch models for measurement.* Sage Publications.

Andrich, D., & Douglas, G. A. (Eds.). (1982). Rasch models for measurement in educational and psychological research [Special issue]. *Education Research and Perspectives, 9*(1), 5–118.

Attali, Y. (2018). Automatic item generation unleashed: An evaluation of a large-scale deployment of item models. In *International Conference on Artificial Intelligence in Education* (pp. 17–29). Springer.

Ballard, E. G. (1978). *Man and technology.* Duquesne University Press.

Banks, E. (2004). The philosophical roots of Ernst Mach's economy of thought. *Synthese,139*(1), 23–53.

Barney, M., & Fisher, W. P., Jr. (2016). Adaptive measurement and assessment. *Annual Review of Organizational Psychology and Organizational Behavior,3*, 469–490.

Bateson, G. (1972). *Steps to an ecology of mind.* University of Chicago Press.

Bateson, G. (1978). Number is different from quantity. *CoEvolution Quarterly,17*, 44–46. [Reprinted from pp. 53–58 in Bateson, G. (1979). *Mind and nature: A necessary unity.* E. P. Dutton.]

Bateson, G. (1991). *A sacred unity: Further steps to an ecology of mind.* HarperOne.

Bejar, I., Lawless, R. R., Morley, M. E., Wagner, M. E., Bennett, R. E., & Revuelta, J. (2003). A feasibility study of on-the-fly item generation in adaptive testing. *The Journal of Technology, Learning, and Assessment,2*(3), 1–29.

Bennett, J. (2010). *Vibrant matter: A political ecology of things.* Duke University Press.

Bernasconi, R. (1989). Seeing double: *Destruktion* and deconstruction. In D. P. Michelfelder & R. E. Palmer (Eds.), *Dialogue & deconstruction: The Gadamer-Derrida encounter* (pp. 233–250). State University of New York Press.

Bernstein, W. J. (2004). *The birth of plenty*. McGraw-Hill.

Best, W. R. (2008). A construct map that Ben Wright would relish. *Rasch Measurement Transactions,22*(3), 1169–1170.

Black, M. (1962). *Models and metaphors*. Cornell University Press.

Black, P., & Wiliam, D. (1998). Assessment and classroom learning. *Assessment in Education,5*(1), 7–74.

Black, P., & Wiliam, D. (2018). Classroom assessment and pedagogy. *Assessment in Education: Principles, Policy & Practice,25*(6), 1–25.

Black, P., Wilson, M., & Yao, S. (2011). Road maps for learning: A guide to the navigation of learning progressions. *Measurement: Interdisciplinary Research and Perspectives, 9,* 1–52.

Blok, A., Farias, I., & Roberts, C. (Eds.). (2020). *The Routledge companion to actor-network theory*. Routledge.

Bloom, A. (1987). *The closing of the American mind*. Simon & Schuster.

Bloom, B. S. (1971). Mastery learning. In J. H. Block (Ed.), *Mastery learning: Theory and practice* (pp. 47–63). Holt.

Bloom, B. S. (1984). The two sigma problem: The search for methods of group instruction as effective as one-to-one tutoring. *Educational Researcher,13*(6), 4–16.

Bowker, G., Timmermans, S., Clarke, A. E., & Balka, E. (Eds). (2015). *Boundary objects and beyond: Working with Leigh Star*. MIT Press.

Bowman, J. P., & Targowski, A. S. (1987). Modeling the communication process: The map is not the territory. *The Journal of Business Communication,24*(4), 21–34.

Box, G. E. P. (1979). Robustness in the strategy of scientific model building. In R. L. Launer & G. N. Wilkinson (Eds.), *Robustness in statistics* (pp. 201–235). Academic Press Inc.

Brier, S. (2013). Cybersemiotics: A new foundation for transdisciplinary theory of information, cognition, meaningful communication and the interaction between nature and culture. *Integral Review,9*(2), 220–263.

Bud, R., & Cozzens, S. E. (Eds.). (1992). *SPIE Institutes: Vol. 9. Invisible connections: Instruments, institutions, and science* (R. F. Potter, Ed.). SPIE Optical Engineering Press.

Burdick, D. S., Stone, M. H., & Stenner, A. J. (2006). The combined gas law and a Rasch reading law. *Rasch Measurement Transactions,20*(2), 1059–1060.

Burdick, H., Swartz, C. W., Stenner, A. J., Fitzgerald, J., Burdick, D., & Hanlon, S. T. (2014). Technological assessment of composing: Response to reviewers. *Literacy Research and Instruction,53*(3), 184–187.

Butterfield, H. (1957). *The origins of modern science* (revised ed.). The Free Press.

Cajori, F. (1985). *A history of mathematics*. Chelsea Publishing Co.

Campbell, N. R. (1952). *What is science?* Dover.

Campbell, J. G. (1992). *A critical survey of some recent philosophical theories of metaphor*. [Diss]. Washington University in St. Louis.

Cano, S., Melin, J., Fisher, W. P., Jr., Stenner, A. J., Pendrill, L., & EMPIR NeuroMet 15HLT04 Consortium. (2018). Patient-centred cognition metrology. *Journal of Physics: Conference Series,1065*, 072033.

Cano, S., Pendrill, L., Melin, J., & Fisher, W. P., Jr. (2019). Towards consensus measurement standards for patient-centered outcomes. *Measurement,141*, 62–69.

Cano, S., Vosk, T., Pendrill, L., & Stenner, A. J. (2016). On trial: The compatibility of measurement in the physical and social sciences. *Journal of Physics: Conference Series,772*, 012025.

Caputo, J. D. (1997). A commentary. In J. D. Caputo (Ed.), *Deconstruction in a nutshell: A conversation with Jacques Derrida* (pp. 31–202). Fordham University Press.

Carreira, S. (2001). Where there's a model, there's a metaphor. *Mathematical Thinking and Learning,3*(4), 261–287.

Carpenter, P. A., Just, M. A., & Shell, P. (1990). What one intelligence test measures. *Psychological Review,97*, 404–431.

Cartwright, N. (1983). *How the laws of physics lie*. Oxford University Press.
Chien, T.-W., Linacre, J. M., & Wang, W.-C. (2011). Examining student ability using KIDMAP fit statistics of Rasch analysis in Excel. In H. Tan & M. Zhou (Eds.), *Communications in computer and information science: Vol. 201. Advances in information technology and education, CSE 2011 Qingdao, China Proceedings, Part I* (pp. 578–585). Springer.
Choppin, B. (1968). An item bank using sample-free calibration. *Nature,219*, 870–872.
Choppin, B. (1976). Developments in item banking. In R. Sumner (Ed.), *Monitoring national standards of attainment in schools* (pp. 216–234). NFER.
Cohen, J. (1994). The earth is round (p < 0.05). *American Psychologist,49*, 997–1003.
Commons, M. L., & Bresette, L. M. (2006). Illuminating major creative scientific innovators with postformal stages. In C. Hoare (Ed.), *Handbook of adult development and learning* (pp. 255–280). Oxford University Press.
Commons, M. L., & Goodheart, E. A. (2008). Cultural progress is the result of developmental level of support. *World Futures: The Journal of New Paradigm Research,64*(5–7), 406–415.
Commons, M. L., & Richards, F. A. (2002). Four postformal stages. In J. Demick & C. Andreoletti (Eds.), *Handbook of adult development*. Plenum Press.
Commons, M. L., Richards, F. A., & Kuhn, D. (1982). Systematic and metasystematic reasoning: A case for levels of reasoning beyond Piaget's stage of formal operations. *Child Development,53*(4), 1058–1069.
Commons, M. L., Ross, S. N., & Bresette, L. M. (2011). The connection between postformal thought, stage transition, persistence, and ambition and major scientific innovations. In D. Artistico, J. Berry, J. Black, D. Cervone, C. Lee, & H. Orom (Eds.), *The Oxford handbook of reciprocal adult development and learning* (pp. 287–301). Oxford University Press.
Cook, T. A. (1914/1979). *The curves of life*. Dover.
Cupples, L. M. (2019). Measure development and the hermeneutic task. *Synthese*, 1–16.
Danesi, M. (2017). Semiotics as a metalanguage for the sciences. In K. Bankov & P. Cobley (Eds.), *Semiotics and its masters* (pp. 61–81). DeGruyter.
Dawson-Tunik, T. L., Commons, M., Wilson, M., & Fischer, K. (2005). The shape of development. *The European Journal of Developmental Psychology,2*, 163–196.
De Boeck, P., & Wilson, M. (Eds.). (2004). *Explanatory item response models*. Springer.
Delandshere, G., & Petrosky, A. R. (1994). Capturing teachers' knowledge. *Educational Researcher,23*(5), 11–18.
de Marchi, N. (Ed.). (1993). *Non-natural social science*. Duke University Press.
DeMillo, R. A. (2015). *Revolution in higher education*. MIT Press.
Derrida, J. (1976). *Of grammatology* (G. C. Spivak, Trans.). Johns Hopkins University Press.
Derrida, J. (1981). *Positions* (A. Bass, Trans.). University of Chicago Press.
Derrida, J. (1989). *Edmund Husserl's origin of geometry*. University of Nebraska Press.
Derrida, J. (2003). Interview on writing. In G. A. Olson & L. Worsham (Eds.), *Critical intellectuals on writing* (pp. 61–69). State University of New York Press.
De Soto, H. (2000). *The mystery of capital*. Basic Books.
Dewey, J. (1954). *The public and its problems*. Swallow Press.
Dewey, J. (2012). *Unmodern philosophy and modern philosophy* (P. Deen, Ed.). Southern Illinois University Press.
Diez-Roux, A. V. (1998). Bringing context back into epidemiology: Variables and fallacies in multilevel analysis. *American Journal of Public Health,88*, 216–222.
Duncan, O. D. (1984). *Notes on social measurement: Historical and critical*. Russell Sage Foundation.
Duncan, O. D., & Stenbeck, M. (1988). Panels and cohorts: Design and model in the study of voting turnout. In C. C. Clogg (Ed.), *Sociological methodology 1988* (pp. 1–35). American Sociological Association.
Einstein, A. (1954). Physics and reality. In C. Seelig & others (Eds.), *Ideas and opinions* (pp. 290–323). Bonanza Books.
Embretson, S. E. (1999). Generating items during testing: Psychometric issues and models. *Psychometrika,64*(4), 407–433.

Embretson, S. E. (2010). *Measuring psychological constructs: Advances in model-based approaches*. American Psychological Association.

Eveland, J. D., & Bikson, T. K. (1987). Evolving electronic communication networks: An empirical assessment. *Office Technology and People,3*(2), 103–128.

Feldman, M. (1987). Electronic mail and weak ties in organizations. *Office Technology and People,3*(2), 83–101.

Ferguson, M. (1980). *The Aquarian conspiracy: Personal and social transformation in the 1980s*. Martin's Press.

Feynman, R. (1988, February 15). Richard Feynman's blackboard at the time of his death. In *CalTech Image Archive*. Retrieved December 31, 2019, from California Institute of Technology: http://archives-dc.library.caltech.edu/islandora/object/ct1%3A551

Fiedler, K. (2017). What constitutes strong psychological science? The (neglected) role of diagnosticity and a priori theorizing. *Perspectives on Psychological Science,12*(1), 46–61.

Fielding, H. (2003). Questioning nature: Irigaray, Heidegger and the potentiality of matter. *Continental Philosophy Review,36*(1), 1–26.

Fischer, G. H. (1973). The linear logistic test model as an instrument in educational research. *Acta Psychologica,37*, 359–374.

Fischer, K. W. (1980). A theory of cognitive development: The control and construction of hierarchies of skills. *Psychological Review,87*, 477 531.

Fischer, K. W., & Farrar, M. J. (1987). Generalizations about generalization: How a theory of skill development explains both generality and specificity. *International Journal of Psychology,22*(5–6), 643–677.

Fisher, W. P., Jr. (1988). Truth, method, and measurement: The hermeneutic of instrumentation and the Rasch model [Diss]. *Dissertation Abstracts International (Dept. of Education, Division of the Social Sciences, University of Chicago), 49,* 0778A. (376 pages, 23 figures, 31 tables).

Fisher, W. P., Jr. (1990). *Mangoes, metaphors, and a measure of meaning*. Presented at the African Studies Association, Baltimore, MD, November.

Fisher, W. P., Jr. (1992). Objectivity in measurement. In M. Wilson (Ed.), *Objective measurement: Theory into practice* (Vol. I, pp. 29–58). Ablex Publishing Corporation.

Fisher, W. P., Jr. (1994). Counting as metaphor. *Rasch Measurement Transactions,8*(2), 358.

Fisher, W. P., Jr. (1995a). Generating truth from fiction. *Rasch Measurement Transactions,8*(4), 401.

Fisher, W. P., Jr. (1995b, Aprilb). *Metaphor as virtual measurement*. University of California.

Fisher, W. P., Jr. (2002). "The Mystery of Capital" and the human sciences. *Rasch Measurement Transactions,15*(4), 854.

Fisher, W. P., Jr. (2003). Mathematics, measurement, metaphor, metaphysics: Parts I & II. *Theory & Psychology,13*(6), 753–828.

Fisher, W. P., Jr. (2004). Meaning and method in the social sciences. *Human Studies: A Journal for Philosophy and the Social Sciences,27*(4), 429–454.

Fisher, W. P., Jr. (2010). Reducible or irreducible? Mathematical reasoning and the ontological method. In M. Garner, G. Engelhard, W. P. Fisher, & M. Wilson (Eds.), *Advances in Rasch measurement* (Vol. 1, pp. 12–44). JAM Press.

Fisher, W. P., Jr. (2012a). Metaphor as measurement and vice versa: Love is a rose. In A. Maul (Chair), *Metaphors and measurement: An invited symposium on validity*. International Meeting of the Psychometric Society, Lincoln, Nebraska, July 12. http://www.slideshare.net/wpfisherjr/fisher-imps2012c-metaphor

Fisher, W. P., Jr. (2012b, June 1). What the world needs now: A bold plan for new standards [Third place, 2011 NIST/SES World Standards Day paper competition]. *Standards Engineering, 64*(3), 1 & 3–5. http://ssrn.com/abstract=2083975

Fisher, W. P., Jr. (2013). Imagining education tailored to assessment as, for, and of learning: Theory, standards, and quality improvement. *Assessment and Learning,2*, 6–22.

Fisher, W. P., Jr. (2015). A probabilistic model of the law of supply and demand. *Rasch Measurement Transactions,29*(1), 1508–1511.

Fisher, W. P., Jr. (2018, January 14). Modern, postmodern, amodern. *Educational Philosophy and Theory, 50,* 1399–1400. (Rpt. in Peters, M., Tesar, M., Jackson, L., & Besley, T. (Eds.). (2020). *What comes after postmodernism in educational theory?* (pp. 104–105). Routledge.)

Fisher, W. P., Jr. (2020). Contextualizing sustainable development metric standards: Imagining new entrepreneurial possibilities. *Sustainability,12*(9661), 1–22.

Fisher, W. P., Jr. (2021a). Bateson and Wright on number and quantity. *Symmetry, 13*(1415).

Fisher, W. P., Jr. (2021b). Separation theorems in econometrics and psychometrics. *Journal of Interdisciplinary Economics, OnlineFirst,* 1–32.

Fisher, W. P., Jr. (2022). Measurement systems, brilliant processes, and exceptional results in healthcare. In W. P. Fisher, Jr. & S. J. Cano (Eds.), *Person-centered outcome metrology* (in press). Springer.

Fisher, W. P., Jr., & Cavanagh, R. (2016). Measurement as a medium for communication and social action, II. In Q. Zhang & H. H. Yang (Eds.), *Pacific Rim Objective Measurement Symposium (PROMS) 2015 Conference Proceedings* (pp. 167–182). Springer.

Fisher, W. P., Jr., Oon, E.P.-T., & Benson, S. (2021). Rethinking the role of educational assessment in classroom communities. *Educational Design Research,5*(1), 1–33.

Fisher, W. P., Jr., & Stenner, A. J. (2005, April 15). The model of the text: A basis for objectivity in the measurement of reading ability. In *Philosophical issues in education (2).* American Educational Research Association, Philosophical Studies in Education SIG.

Fisher, W. P., Jr., & Stenner, A. J. (2011a). Integrating qualitative and quantitative research approaches via the phenomenological method. *International Journal of Multiple Research Approaches,5*(1), 89–103.

Fisher, W. P., Jr., & Stenner, A. J. (2011b, September 2). A technology roadmap for intangible assets metrology. In *Fundamentals of measurement science.* International Measurement Confederation (IMEKO) TC1-TC7-TC13 Joint Symposium, Jena, Germany. http://www.db-thueringen.de/ser vlets/DerivateServlet/Derivate-24493/ilm1-2011imeko-018.pdf

Fisher, W. P., Jr., & Stenner, A. J. (2013). On the potential for improved measurement in the human and social sciences. In Q. Zhang & H. Yang (Eds.), *Pacific Rim Objective Measurement Symposium 2012 Conference Proceedings* (pp. 1–11). Springer.

Fisher, W. P., Jr., & Stenner, A. J. (2016). Theory-based metrological traceability in education: A reading measurement network. *Measurement,92,* 489–496.

Fisher, W. P., Jr., & Stenner, A. J. (2017, September 18). Towards an alignment of engineering and psychometric approaches to uncertainty in measurement. In *18th International Congress of Metrology, 12004* (pp. 1–9). https://doi.org/10.1051/metrology/201712004

Fisher, W. P., Jr., & Stenner, A. J. (2018a). On the complex geometry of individuality and growth: Cook's 1914 'Curves of Life' and reading measurement. *Journal of Physics Conference Series,1065,* 072040.

Fisher, W. P., Jr., & Stenner, A. J. (2018b). Ecologizing vs modernizing in measurement and metrology. *Journal of Physics Conference Series, 1044*(012025).

Fisher, W. P., Jr., & Wilson, M. (2015). Building a productive trading zone in educational assessment research and practice. *Pensamiento Educativo: Revista De Investigacion Educacional Latinoamericana,52*(2), 55–78.

Fisher, W. P., Jr., & Wright, B. D. (Eds.). (1994). Applications of probabilistic conjoint measurement. *International Journal of Educational Research, 21*(6), 557–664.

Fitzgerald, J., Elmore, J., Hiebert, E. H., Koons, H. H., Bowen, K., Sanford-Moore, E. E., & Stenner, A. J. (2016). Examining text complexity in the early grades. *Phi Delta Kappan,97*(8), 60–65.

Fitzgerald, J., Elmore, J., Koons, H., Hiebert, E., Bowen, K., Sanford-Moore, E., & Stenner, A. (2015). Important text characteristics for early-grades text complexity. *Journal of Educational Psychology,107*(1), 4–29.

Forer, B., & Zumbo, B. D. (2011). Validation of multilevel constructs. *Social Indicators Research,103*(2), 231–265.

Franck, G. (2002). The scientific economy of attention. *Scientometrics,55*(1), 3–26.

Franck, G. (2019). The economy of attention. *Journal of Sociology,55*(1), 8–19.

French, A. P., & Kennedy, P. J. (Eds.). (1985). *Niels Bohr: A centenary volume.* Harvard University Press.

Gadamer, H.-G. (1980). *Dialogue and dialectic.* Yale University Press.

Gadamer, H.-G. (1989). *Truth and method.* Crossroad.

Gadamer, H.-G. (1991). Gadamer on Gadamer. In H. J. Silverman (Ed.), *Continental philosophy: Vol. IV. Gadamer and hermeneutics* (pp. 13–19). Routledge.

Gadamer, H.-G. (1994). *Heidegger's ways.* SUNY Press.

Galison, P. (1997). *Image and logic: A material culture of microphysics.* University of Chicago Press.

Galison, P. (1999). Trading zone: Coordinating action and belief. In M. Biagioli (Ed.), *The science studies reader* (pp. 137–160). Routledge.

Galison, P., & Stump, D. J. (1996). *The disunity of science: Boundaries, contexts, and power.* Stanford University Press.

Gasché, R. (1987). Infrastructures and systemacity. In J. Sallis (Ed.), *Deconstruction and philosophy: The texts of Jacques Derrida* (pp. 3–20). University of Chicago Press.

Gasché, R. (2014). "A certain walk to follow": Derrida and the question of method. *Epoché: A Journal for the History of Philosophy, 18*(2), 525–550.

Gierl, M. J., & Haladyna, T. M. (2012). *Automatic item generation: Theory and practice.* Routledge.

Gierl, M. J., & Lai, H. (2012). The role of item models in automatic item generation. *International Journal of Testing, 12*(3), 273–298.

Goguen, J. (1994). Requirements engineering as the reconciliation of technical and social issues. In M. Jirotka & G. J. J. (Eds.), *Requirements engineering: Social and technical issues* (pp. 27–56). Academic Press.

Gould, S. J. (1996). *The mismeasure of man.* Norton.

Goyal, S., Heidari, H., & Kearns, M. (2019). Competitive contagion in networks. *Games and Economic Behavior, 113*, 58–79.

Green, K. E., & Kluever, R. C. (1992). Components of item difficulty of Raven's Matrices. *Journal of General Psychology, 119*, 189–199.

Green, K. E., Kluever, R. C., & Wright, B. D. (1994). Predicting item difficulties from item characteristics. *Rasch Measurement Transactions, 8*(2), 354.

Green, K. E., & Smith, R. M. (1987). A comparison of two methods of decomposing item difficulties. *Journal of Educational Statistics, 12*(4), 369–381.

Guskey, T. R. (2009). Mastery learning. In T. L. Good (Ed.), *21st century education: A reference handbook* (Vol. 1, pp. 194–202).

Guttman, L. (1985). The illogic of statistical inference for cumulative science. *Applied Stochastic Models and Data Analysis, 1*, 3–10.

Guttman, L. (1994). The mathematics in ordinary speech. In S. Levy (Ed.), *Louis Guttman on theory and methodology: Selected writings* (pp. 103–119). Dartmouth Publishing Company.

Hacking, I. (1983). *Representing and intervening.* Cambridge University Press.

Hankins, T. L., & Silverman, R. J. (1999). *Instruments and the imagination.* Princeton University Press.

Haraway, D. J. (1996). Modest witness: Feminist diffractions in science studies. In P. Galison & D. J. Stump (Eds.), *The disunity of science* (pp. 428–441). Stanford University Press.

Hardwicke, T. E., Thibault, R. T., Kosie, J. E., Wallach, J. D., Kidwell, M. C., & Ioannidis, J. P. (2021). Estimating the prevalence of transparency and reproducibility-related research practices in psychology (2014–2017). *Perspectives on Psychological Science, 17*(1), 239–251.

Harries, K. (2010). 'Let no one ignorant of geometry enter here': Ontology and mathematics in the thought of Martin Heidegger. *International Journal of Philosophical Studies, 18*(2), 269–279.

Hattie, J., & Timperley, H. (2007). The power of feedback. *Review of Educational Research, 77*(1), 81–112.

Hayman, J., Rayder, N., Stenner, A. J., & Madey, D. L. (1979). On aggregation, generalization, and utility in educational evaluation. *Educational Evaluation and Policy Analysis, 1*(4), 31–39.

He, W., Li, S., & Kingsbury, G. G. (2016). A large-scale, long-term study of scale drift: The micro view and the macro view. *Journal of Physics Conference Series, 772*, 012022.

Heelan, P. A. (1983a). Natural science as a hermeneutic of instrumentation. *Philosophy of Science,50*, 181–204.

Heelan, P. A. (1983b). *Space perception and the philosophy of science.* University of California Press.

Heelan, P. A. (1984, May 13). *Issues in philosophy of natural science.* [Stephen Toulmin, respondent]. Presented at the Conference on Continental and Anglo-American Philosophy: A New Relationship. University of Chicago Divinity School and Department of Philosophy.

Heelan, P. A. (1994). Galileo, Luther, and the hermeneutics of natural science. In T. J. Stapleton (Ed.), *The question of hermeneutics* (pp. 363–374). Kluwer Academic Publishers.

Heidegger, M. (1962). *Being and time.* Harper & Row.

Heidegger, M. (1967). *What is a thing?* Regnery/Gateway.

Heidegger, M. (1977). Modern science, metaphysics, and mathematics. In D. F. Krell (Ed.), *Basic writings [reprinted from What is a thing? pp. 66–108]* (pp. 243–282). Harper & Row.

Heidegger, M. (1982). *The basic problems of phenomenology.* Indiana University Press.

Heidegger, M. (1991). *The principle of reason.* Indiana University Press.

Hendriana, H., & Rohaeti, E. E. (2017). The importance of metaphorical thinking in the teaching of mathematics. *Current Science,113*(11), 2160–2164.

Hesse, M. (1970). *Models and analogies in science.* Notre Dame University Press.

Hodas, N. O., & Lerman, K. (2014). The simple rules of social contagion. *Scientific Reports,4*, 4343.

Houser, N. (1992). Introduction. In N. Houser & C. Kloesel (Eds.), *The essential Peirce: Selected philosophical writings, volume 1 (1867–1893)* (pp. xix–xli). Indiana University Press.

Hunt, B. J. (1994). The ohm is where the art is: British telegraph engineers and the development of electrical standards. *Osiris: A Research Journal Devoted to the History of Science and Its Cultural Influences, 9,* 48–63.

Husserl, E. (1965). Philosophy as rigorous science (Q. Lauer, Trans.). In *Phenomenology and the crisis of philosophy* (pp. 69–147). Harper & Row.

Hutchins, E. (1995). *Cognition in the wild.* MIT Press.

Hutchins, E. (2014). The cultural ecosystem of human cognition. *Philosophical Psychology,27*(1), 34–49.

Ihde, D. (1983). The historical and ontological priority of technology over science. In *Existential technics* (pp. 25–46). State University of New York Press.

Ihde, D. (1991). *Instrumental realism.* Indiana University Press.

Ihde, D., & Malafouris, L. (2019). Homo faber revisited: Postphenomenology and material engagement theory. *Philosophy & Technology,32*(2), 195–214.

Jaeger, R. M. (1973). The national test equating study in reading (The Anchor Test Study). *Measurement in Education,4*, 1–8.

Jasanoff, S. (2004). *States of knowledge.* Routledge.

Jolander, F. (1957/2008). Something about bridge-building [test-equating] techniques—a sensational new creation by Dr. Rasch (C. Kreiner, Trans.). [Reprinted from *Folkeskolen [The Danish Elementary School Journal].* *Rasch Measurement Transactions, 21*(4), 1129–1130.]

Kant, I. (1929/1965). *Critique of pure reason.* St. Martin's Press.

Karabatsos, G. (2003). A comparison of 36 person-fit statistics of item response theory. *Applied Measurement in Education,16*, 277–298.

Katzenmeyer, W. G., & Stenner, A. J. (1975). Strategic use of random subsample replication and a coefficient of factor replication. *Educational and Psychological Measurement,35*, 19–29.

Katzenmeyer, W. G., & Stenner, A. J. (1977). Estimation of the invariance of factor structures across sex and race with implications for hypothesis testing. *Educational and Psychological Measurement,37*, 111–119.

Kearney, R. (1984). *Dialogues with contemporary Continental thinkers.* Manchester University Press.

Kisiel, T. (1973). The mathematical and the hermeneutical. In E. G. Ballard & C. E. Scott (Eds.), *Martin Heidegger: In Europe and America* (pp. 109–120). Martinus Nijhoff.

Knoster, T. P., Villa, R. A., & Thousand, J. S. (2000). A framework for thinking about systems change. In R. A. Villa & J. S. Thousand (Eds.), *Restructuring for caring and effective education* (pp. 93–128). Paul H. Brookes.

Koopmans, T. C., & Reiersøl, O. (1950). The identification of structural characteristics. *The Annals of Mathematical Statistics, XXI,* 165–181.

Kosh, A., Simpson, M. A., Bickel, L., Kellog, M., & Sanford-Moore, E. (2019). A cost-benefit analysis of automatic item generation. *Educational Measurement: Issues and Practice,38*(1), 48–53.

Kristeva, J. (1980). *Desire in language: A semiotic approach to literature and art.* Columbia University Press.

Kristeva, J. (2014). Reliance, or maternal eroticism. *Journal of the American Psychoanalytic Association,62*(1), 69–85.

Kuhn, T. S. (1977). *The essential tension.* University of Chicago Press.

Kuhn, T. S. (1979). Metaphor in science. In A. Ortony (Ed.), *Metaphor and thought* (pp. 409–419). Cambridge University Press.

Lakoff, G., & Johnson, M. (1980). *Metaphors we live by.* University of Chicago Press.

Latimer, S. L. (1982). Using the linear logistic test model to investigate a discourse-based model of reading comprehension. *Rasch Models for Measurement in Educational and Psychological Research. A Special Issue of Education Research and Perspectives, 9*(1), 73–94.

Latour, B. (1983). Give me a laboratory and I will raise the world. In K. D. Knorr-Cetina & M. Mulkay (Eds.), *Science observed* (pp. 141–170). Sage Publications.

Latour, B. (1986). Visualization and cognition: Thinking with eyes and hands. *Knowledge and Society: Studies in the Sociology of Culture past and Present,6,* 1–40.

Latour, B. (1987). *Science in action.* Harvard University Press.

Latour, B. (1990). Postmodern? No, simply amodern: Steps towards an anthropology of science. *Studies in History and Philosophy of Science,21*(1), 145–171.

Latour, B. (1991). The impact of science studies on political philosophy. *Science, Technology, & Human Values,16*(1), 3–19.

Latour, B. (1993a). *We have never been modern.* Harvard University Press.

Latour, B. (1993b). Pasteur on lactic acid yeast: A partial semiotic analysis. *Configurations,1*(1), 129–146.

Latour, B. (1995a). *Cogito ergo sumus!* Or psychology swept inside out by the fresh air of the upper deck: Review of Hutchins' *Cognition in the Wild,* MIT Press, 1995. *Mind, Culture, and Activity: An International Journal,3*(192), 54–63.

Latour, B. (1995b). Moderniser ou écologiser? A la recherche de la septième Cité. *Ecologie Politique, 13,* 5–27. (Rpt. in Castree, N., & Willems-Braun, B. (Eds.). (1998). To modernize or to ecologize? That's the question. In *Remaking reality: Nature at the millennium* (pp. 221–242). Routledge.)

Latour, B. (2004). How to talk about the body? The normative dimension of science studies. *Body & Society,10*(2–3), 205–229.

Latour, B. (2005). *Reassembling the social: An introduction to Actor-Network-Theory.* Oxford University Press.

Latour, B. (2009). *On the modern cult of the factish gods* (H. MacLean & C. Porter, Trans.). Duke University Press.

Latour, B. (2010a). A compositionist manifesto. *New Literary History,41,* 471–490.

Latour, B. (2010b). Tarde's idea of quantification. In M. Candea (Ed.), *The social after Gabriel Tarde: Debates and assessments* (pp. 145–162). Routledge.

Latour, B. (2011). Love your monsters: Why we must care for our technologies as we do our children. *Breakthrough Journal,2,* 21–28.

Latour, B. (2014, February 26). *On some of the affects of capitalism.* Lecture given at the Royal Academy. http://www.bruno-latour.fr/sites/default/files/136-AFFECTS-OF-K-COPENH AGUE.pdf

Latour, B. (2015). Fifty shades of green. *Environmental Humanities,7,* 219–225.

Latour, B., & Lépinay, V. A. (2010). *The science of passionate interests.* Prickly Paradigm Press.

Latour, B., & Weibel, P. (2005). *Making things public: Atmospheres of democracy*. MIT Press.
Latour, B., & Woolgar, S. (1979). *Laboratory life: The social construction of scientific facts*. Sage.
Lenoir, T. (1998). *Inscribing science*. Stanford University Press.
Linacre, J. M. (2000). Was the Rasch model almost the Peirce model? *Rasch Measurement Transactions, 14*(3), 756–757.
Linssen, R. (1958). *Living Zen*. George Allen & Unwin.
Locoro, A., Fisher, W. P., Jr., & Mari, L. (2021). Visual information literacy. *IEEE Access, 9*, 71053–71071.
Loubert, A., Regnault, A., Sebille, V., Hardouin, J.-B., Melin, J., Cano, S., & Fisher, W. P., Jr. (2020, October 2). *Contemplating clinical trials as measuring systems: A new perspective for measurement uncertainty associated with treatment effect demonstration*. Presented at the 27th ISOQOL Conference https://www.isoqol.org/events/27th-annual-conference/. https://www.loom.com/share/c464a9fb3b8049219c8b4c6c4ddb0a35
Mach, E. (1919). *The science of mechanics: A critical and historical account of its development* (T. J. McCormack, Trans.; 4th ed.). The Open Court Publishing Co.
Maran, T. (2007). Towards an integrated methodology of ecosemiotics: The concept of nature-text. *Sign Systems Studies, 35*(1/2), 269–294.
Maraun, M. D. (1996). Meaning and mythology in the factor analysis model. *Multivariate Behavioral Research, 31*(4), 603–616.
Mari, L., Maul, A., Torres Irribarra, D., & Wilson, M. (2013). Quantification is neither necessary nor sufficient for measurement. *Journal of Physics Conference Series, 459*(1).
Mari, L., Maul, A., Torres Irribara, D., & Wilson, M. (2016). Quantities, quantification, and the necessary and sufficient conditions for measurement. *Measurement, 100*, 115–121.
Mari, L., & Wilson, M. (2014). An introduction to the Rasch measurement approach for metrologists. *Measurement, 51*, 315–327.
Mari, L., Wilson, M., & Maul, A. (2021). *Measurement across the sciences*. Springer.
Massof, R. W., Ahmadian, L., Grover, L. L., Deremeik, J. T., Goldstein, J. E., Rainey, C., Epstein, C., & Barnett, G. D. (2007). The activity inventory: An adaptive visual function questionnaire. *Optometry & Vision Science, 84*, 763–774.
Massof, R. W., & Bradley, C. (2022). An adaptive strategy for measuring patient-reported outcomes. In W. P. Fisher, Jr. & S. J. Cano (Eds.), *Person-centered outcome metrology* (in press). Springer.
Masters, G. N. (1984). Constructing an item bank using partial credit scoring. *Journal of Educational Measurement, 21*(1), 19–32.
Masters, G. N., Adams, R. J., & Lokan, J. (1994). Mapping student achievement. *International Journal of Educational Research, 21*(6), 595–610.
McCabe, B. (2006). *Letter to Illinois educators: Reporting Lexile measures in Illinois Standards Achievement Test results* [Division Administrator, Student Assessment Division, Illinois State Board of Education, Springfield]. https://www.isbe.net/Documents/lexile.pdf
Meehl, P. E. (1967). Theory-testing in psychology and physics: A methodological paradox. *Philosophy of Science, 34*(2), 103–115.
Melin, J., Cano, S. J., Göschel, L., Fillmer, A., Lehmann, S., Hirtz, C., Flöel, A., & Pendrill, L. R. (2021a, September 3). *Metrological references for person ability in memory tests*. Presented at the XXIII IMEKO World Congress "Measurement: sparking tomorrow's smart revolution".
Melin, J., Cano, S., & Pendrill, L. (2021b). The role of entropy in construct specification equations (CSE) to improve the validity of memory tests. *Entropy, 23*(2), 212.
Meloni, M., & Reynolds, J. (2021). Thinking embodiment with genetics: Epigenetics and postgenomic biology in embodied cognition and enactivism. *Synthese, 198*(11), 10685–10708.
Merrell, F. (2011). Tom's often neglected other theoretical source. In P. Cobley, J. Deely, K. Kull & S. Petrilli (Eds.), *Semiotics continues to astonish: Thomas A. Sebeok and the doctrine of signs* (pp. 251–279). De Gruyter.
Michell, J. (1986). Measurement scales and statistics. *Psychological Bulletin, 100*, 398–407.
Miller, P., & O'Leary, T. (2007). Mediating instruments and making markets. *Accounting, Organizations, and Society, 32*(7–8), 701–734.

Morrison, J., & Fisher, W. P., Jr. (2018). Connecting learning opportunities in STEM education. *Journal of Physics: Conference Series, 1065*(022009).

Morrison, J., & Fisher, W. P., Jr. (2019). Measuring for management in Science, Technology, Engineering, and Mathematics learning ecosystems. *Journal of Physics: Conference Series, 1379*(012042).

Morrison, J., & Fisher, W. P., Jr. (2021). Caliper: Measuring success in STEM learning ecosystems. *Measurement: Sensors, 18,* 100327.

Narens, L., & Luce, R. D. (1986). Measurement: The theory of numerical assignments. *Psychological Bulletin,99*(2), 166–180.

Nersessian, N. J. (2002). Maxwell and "the method of physical analogy." In D. Malament (Ed.), *Reading natural philosophy* (pp. 129–166). Open Court.

Nersessian, N. J. (2008). *Creating scientific concepts.* MIT Press.

Newby, V. A., Conner, G. R., Grant, C. P., & Bunderson, C. V. (2009). The Rasch model and additive conjoint measurement. *Journal of Applied Measurement,10*(4), 348–354.

Nöth, W. (Ed.). (1990). *Handbook of semiotics.* Indiana University Press.

Nöth, W. (2018). The semiotics of models. *Sign Systems Studies,46*(1), 7–43.

Nuzzo, A. (1999). The idea of 'method' in Hegel's science of logic—A method for finite thinking and absolute reason. *Hegel Bulletin,20*(1–2), 1–17.

Nuzzo, A. (2011). Thinking being: Method in Hegel's logic of being. In S. Houlgate & M. Bauer (Eds.), *A companion to Hegel* (pp. 109–138). Blackwell.

Nuzzo, A. (2018). *Approaching Hegel's logic, obliquely: Melville, Moliere.* SUNY Press.

O'Connell, J. (1993). Metrology: The creation of universality by the circulation of particulars. *Social Studies of Science,23,* 129–173.

Oele, M. (2017). Folding nature back upon itself. In G. Kuperus & M. Oele (Eds.), *Ontologies of nature* (pp. 47–66). Springer.

Overton, W. F. (1998). Developmental psychology: Philosophy, concepts, and methodology. *Handbook of Child Psychology,1,* 107–188.

Pashler, H., & Wagenmakers, E. J. (2012). Editors' introduction to the special section on replicability in psychological science: A crisis of confidence? *Perspectives on Psychological Science,7*(6), 528–530.

Pastor-Satorras, R., Castellano, C., Van Mieghem, P., & Vespignani, A. (2015). Epidemic processes in complex networks. *Reviews of Modern Physics,87*(3), 925.

Peabody, M., O'Neill, T. R., & Peterson, L. E. (2015). Illustrating the psychometric construct of family medicine on the American Board of Family Medicine's Examinations. *Rasch Measurement Transactions,29*(2), 1516–1519.

Peirce, C. S. (1868/1992). Some consequences of four incapacities. *Journal of Speculative Philosophy, 2,* 140–157. (Rpt. in Houser, N., & Kloesel, C. (Eds.). (1992). *The essential Peirce: Selected philosophical writings, volume I (1867–1893)* (pp. 28–55). Indiana University Press.)

Peirce, C. S. (1878/1992). Illustration of the logic of science. Fourth paper: The probability of induction. *Popular Science Monthly, 12,* 705–718. (Rpt. in Houser, N., & Kloesel, C. (Eds.). (1992). *The essential Peirce: Selected philosophical writings, volume I (1867–1893)* (pp. 155–69). Indiana University Press.)

Peirce, C. S. (1879/2002). Note on the theory of the economy of research. In P. Mirowski & E.-M. Sent (Eds.), *Science bought and sold: Essays in the economics of science* (pp. 183–190). University of Chicago Press.

Pendrill, L. R. (2019). *Quality assured measurement: Unification across social and physical sciences.* Springer.

Pendrill, L., & Fisher, W. P., Jr. (2015). Counting and quantification: Comparing psychometric and metrological perspectives on visual perceptions of number. *Measurement,71,* 46–55.

Percy, W. (1991). The fateful rift: The San Andreas Fault in the modern mind. In P. Samway (Ed.), *Signposts in a strange land* (pp. 271–291). Farrar.

Petersen, A. (1968). *Quantum physics and the philosophical tradition.* MIT Press.

Petracca, E., & Gallagher, S. (2020). Economic cognitive institutions. *Journal of Institutional Economics,16*(6), 747–765.

Platt, J. R. (1961, September 1). Social chain reactions. *Bulletin of the Atomic Scientists: Man and His Habitat, Part II, 17,* 365–386. (Rpt. in Platt, J. R. (1966). *The step to man* (pp. 39–52). Wiley.)

Platt, J. R. (1970). *Perception and change: Projections for survival.* University of Michigan Press.

Poinstingl, H. (2009). The Linear Logistic Test Model (LLTM) as the methodological foundation of item generating rules for a new verbal reasoning test. *Psychology Science Quarterly,51,* 123–134.

Polanyi, M. (1966). *The tacit dimension.* Doubleday & Co.

Pollet, T. V., Stulp, G., Henzi, S. P., & Barrett, L. (2015). Taking the aggravation out of data aggregation. *American Journal of Primatology,77*(7), 727–740.

Powers, M., Fisher, W. P., Jr., & Massof, R. W. (2016). Modeling visual symptoms and visual skills to measure functional binocular vision. *Journal of Physics Conference Series,772*(1), 012045.

Price, D. J. de Solla (1986). Of sealing wax and string. In *Little science, big science—And beyond* (pp. 237–253). Columbia University Press.

Prien, B. (1989). How to predetermine the difficulty of items of examinations and standardized tests. *Studies in Educational Evaluation,15,* 309–317.

Quaglia, M., Pendrill, L., Melin, J., Cano, S., & 15HLT04 NeuroMET Consortium. (2016–2019). *Innovative measurements for improved diagnosis and management of neurodegenerative diseases* (EMPIR NeuroMET) (36 pp.). EURAMET. https://www.lgcgroup.com/our-programmes/empir-neuromet/neuromet-landing-page/

Quaglia, M., Pendrill, L., Melin, J., Cano, S., & 18HLT09 NeuroMET2 Consortium. (2019–2022). *Publishable Summary for 18HLT09 NeuroMET2: Metrology and innovation for early diagnosis and accurate stratification of patients with neurodegenerative diseases* (EMPIR NeuroMET) (5 pp.). EURAMET. https://www.lgcgroup.com/our-programmes/empir-neuromet/neuromet-landing-page/

Rasch, G. (1960). *Probabilistic models for some intelligence and attainment tests* (Reprint, with Foreword and Afterword by B. D. Wright, University of Chicago Press, 1980). Danmarks Paedogogiske Institut.

Rasch, G. (1972/2010). Retirement lecture of 9 March 1972: Objectivity in social sciences: A method problem (Cecilie Kreiner, Trans.). *Rasch Measurement Transactions, 24*(1), 1252–1272.

Rasch, G. (1973/2011). All statistical models are wrong! Comments on a paper presented by Per Martin-Löf, at the Conference on Foundational Questions in Statistical Inference, Aarhus, Denmark, May 7–12, 1973. *Rasch Measurement Transactions, 24*(4), 1309.

Rasch, G. (1977). On specific objectivity: An attempt at formalizing the request for generality and validity of scientific statements. *Danish Yearbook of Philosophy,14,* 58–94.

Rentz, R. R., & Bashaw, W. L. (1977). The National Reference Scale for Reading: An application of the Rasch model. *Journal of Educational Measurement,14*(2), 161–179.

Ricoeur, P. (1965). *History and truth.* Northwestern University Press.

Ricoeur, P. (1967a). Conclusion: The symbol gives rise to thought. In P. Ricoeur (Ed.), *The symbolism of evil* (pp. 347–357). Beacon Press.

Ricoeur, P. (1967b). *Husserl: An analysis of his phenomenology.* Northwestern University Press.

Ricoeur, P. (1974). Violence and language. In D. Stewart & J. Bien (Eds.), *Political and social essays by Paul Ricoeur* (pp. 88–101). Ohio University Press.

Ricoeur, P. (1976). *Interpretation theory: Discourse and the surplus of meaning.* Texas Christian University Press.

Ricoeur, P. (1977). *The rule of metaphor.* University of Toronto Press.

Ricoeur, P. (1979). The metaphorical process as cognition, imagination, and feeling. In S. Sacks (Ed.), *On metaphor* (pp. 141–157). University of Chicago Press.

Ricoeur, P. (1981). *Hermeneutics and the human sciences.* Cambridge University Press.

Ricoeur, P. (1986). *Du texte à l'action: Essais d'hermeneutique, II.* Seuil.

Ricoeur, P. (1991a). Life in quest of narrative. In D. Wood (Ed.), *On Paul Ricoeur: Narrative and interpretation* (pp. 20–33). Routledge.

Ricoeur, P. (1991b). Narrative identity. In D. Wood (Ed.), *On Paul Ricoeur: Narrative and interpretation* (pp. 188–199). Routledge.

Risser, J. (1989). The two faces of Socrates: Gadamer/Derrida. In D. P. Michelfelder & R. E. Palmer (Eds.), *Dialogue & deconstruction: The Gadamer-Derrida encounter* (pp. 176–185). State University of New York Press.

Rivero, V. (2011, June 10). Interview: Jack Stenner takes education beyond the metrics. *Edtech Digest.* Retrieved April 6, 2022, from http://edtechdigest.wordpress.com/2011/06/10/interview-jack-stenner-takes-education-beyond-the-metrics/

Rousseau, D. M. (1985). Issues of level in organizational research: Multi-level and cross-level perspectives. *Research in Organizational Behavior, 7*(1), 1–37.

San Martin, E., Gonzalez, J., & Tuerlinckx, F. (2009). Identified parameters, parameters of interest, and their relationships. *Measurement: Interdisciplinary Research and Perspectives, 7*(2), 97–105.

San Martin, E., & Rolin, J. M. (2013). Identification of parametric Rasch-type models. *Journal of Statistical Planning and Inference, 143*(1), 116–130.

Scarinzi, A. (2015). *Aesthetics and the embodied mind.* Springer.

Schmidt, F. L., & Hunter, J. E. (1977). Development of a general solution to the problem of validity generalization. *Journal of Applied Psychology, 62*(5), 529–540.

Scott, J. C. (1998). *Seeing like a state.* Yale University Press.

Sebeok, T. A. (1991). *Sign is just a sign (Advances in semiotics).* Indiana University Press.

Sebeok, T. A. (2001). *Signs: An introduction to semiotics.* University of Toronto Press.

Sethy, S. S. (2021). Logical paradoxes. In *Introduction to logic and logical discourse* (pp. 71–79). Springer.

Shapin, S. (1994). *A social history of truth: Civility and science in seventeenth-century England.* University of Chicago Press.

Simpson, M. A., Kosh, A., Elmore, J., Bickel, L., Stenner, A. J., Fisher, W. P., Jr., Sanford-Moore, E., Koons, H., Enoch-Marx, M., Kellog, M., Rauch, S., Leathers, R., Lines, D., & Burdick, D. S. (2015, April 19). *Family group item generation theory: Large scale implementation* (M. A. Simpson, Chair). National Council on Measurement in Education.

Smith, E. V., Jr. (2002). Detecting and evaluating the impact of multidimensionality using item fit statistics and principal component analysis of residuals. *Journal of Applied Measurement, 3*(2), 205–231.

Smith, R. M. (1996). A comparison of methods for determining dimensionality in Rasch measurement. *Structural Equation Modeling, 3*(1), 25–40.

Smith, R. M., & Plackner, C. (2009). The family approach to assessing fit in Rasch measurement. *Journal of Applied Measurement, 10*(4), 424–437.

Sonnleitner, P. (2008). Using the LLTM to evaluate an item-generating system for reading comprehension. *Psychology Science Quarterly, 50*(3), 345–362.

Spivak, G. C. (1990). *The post-colonial critic: Interviews, strategies, dialogue.* Routledge.

Star, S. L., & Griesemer, J. R. (1989). Institutional ecology, 'translations', and boundary objects. *Social Studies of Science, 19*(3), 387–420.

Star, S. L., & Ruhleder, K. (1996). Steps toward an ecology of infrastructure. *Information Systems Research, 7*(1), 111–134.

Staten, H. (1984). *Wittgenstein and Derrida.* University of Nebraska Press.

Stenner, A. J. (1996). *Measuring reading comprehension with the Lexile Framework.* MetaMetrics, Inc.

Stenner, A. J. (1998). The Lexile Framework for Reading: A map to higher levels of achievement. *Popular Measurement, 1*(1), 9–11.

Stenner, A. J., & Burdick, D. S. (1997, January 3). *The objective measurement of reading comprehension.* Retrieved April 6, 2022, from http://files.eric.ed.gov/fulltext/ED435978. [Response to technical questions raised by the California Department of Education Technical Study Group]. MetaMetrics, Inc.

Stenner, A. J., Burdick, D., Sanford, E., & Burdick, H. (2001). A response to "Assessing the Lexile Framework: Results of a panel discussion." In S. White & J. Clement (Eds.), *Assessing the Lexile Framework: Results of a panel meeting* (pp. 46–55). U.S. Department of Education, National Center for Educational Statistics Working Paper No. 2001-08. Retrieved April 6, 2022, from http://nces.ed.gov/pubs2001/200108.pdf

Stenner, A. J., Fisher, W. P., Jr., Stone, M. H., & Burdick, D. S. (2013). Causal Rasch models. *Frontiers in Psychology: Quantitative Psychology and Measurement,4*(536), 1–14.

Stenner, A. J., Fisher, W. P., Jr., Stone, M. H., & Burdick, D. S. (2016). Causal Rasch models in language testing: An application rich primer. In Q. Zhang (Ed.), *Pacific Rim Objective Measurement Symposium (PROMS) 2015 Conference Proceedings* (pp. 1–14). Springer.

Stenner, A. J., & Smith, M., III. (1982). Testing construct theories. *Perceptual and Motor Skills,55*, 415–426.

Stenner, A. J., Smith, M., III., & Burdick, D. S. (1983). Toward a theory of construct definition. *Journal of Educational Measurement,20*(4), 305–316.

Stenner, A. J., & Stone, M. (2003). Item specification vs. item banking. *Rasch Measurement Transactions, 17*(3), 929–930.

Stenner, A. J., & Stone, M. (2004, May 4). *Does the reader comprehend the text because the reader is able or because the text is easy?* Presented at the International Reading Association, Reno, Nevada, USA.

Stenner, A. J., & Stone, M. (2010). Generally objective measurement of human temperature and reading ability: Some corollaries. *Journal of Applied Measurement,11*(3), 244–252.

Stenner, A. J., Stone, M. H., & Fisher, W. P., Jr. (2018). The unreasonable effectiveness of theory based instrument calibration in the natural sciences: What can the social sciences learn? *Journal of Physics Conference Series, 1044*(012070).

Stenner, A. J., Swartz, C., Hanlon, S., & Emerson, C. (2012, February). *Personalized learning platforms.* Presented at the Pearson Global Research Conference, Fremantle, Western Australia.

Stone, M. H. (2002). *Knox's cube test—Revised.* Stoelting.

Stone, M. H., Wright, B., & Stenner, A. J. (1999). Mapping variables. *Journal of Outcome Measurement,3*(4), 308–322.

Surber, J. O. (2011). Hegel's philosophy of language: The unwritten volume. In S. Houlgate & M. Bauer (Eds.), *A companion to Hegel* (pp. 243–261). Blackwell.

Sutton, J., Harris, C. B., Keil, P. G., & Barnier, A. J. (2010). The psychology of memory, extended cognition, and socially distributed remembering. *Phenomenology and the Cognitive Sciences,9*(4), 521–560.

Suzuki, D. T. (1949). *Essays in Zen Buddhism.* Rider & Co., for The Buddhist Society.

Szymanski, T. (1995). Zen koans and the art of mathematics. *Research and Teaching in Developmental Education,12*(1), 75–77.

Tasić, V. (2001). *Mathematics and the roots of postmodern thought.* Oxford University Press.

Thrift, N. (2008). Pass it on: Toward a political economy of propensity. *Emotion, Space, and Society,1*(2), 83–96.

Thurstone, L. L. (1959). Psychology as a quantitative rational science. *Midway Reprint Series* (pp. 3–11). University of Chicago Press. (Rpt. from Thurstone, L. L. (1937). *Science (Chicago, Illinois), LXXXV*(March 5), 228–232.)

Toulmin, S. E. (1982). The construal of reality: Criticism in modern and postmodern science. *Critical Inquiry,9*, 93–111.

van Helden, A., & Hankins, T. L. (Eds.). (1994). *Instruments* (Vol. 9) (Osiris: a research journal devoted to the history of science and its cultural influences). University of Chicago Press.

Webster, L. J. (2000). Jack Stenner: The Lexile king. *Popular Measurement,3*, 12–15.

Weitzel, T. (2004). *Economics of standards in information networks.* Physica.

Wheeler, J. A. (1980). Law without law. In P. Medawar & J. Shelley (Eds.), *Structure in science and art* (pp. 132–154). Excerpta Medica.

Wheeler, J. A. (1982). Particles and geometry. In Unified theories of elementary particles: Critical assessments and prospects [Special issue]. *Lecture Notes in Physics,160*, 189–217.

Wheeler, J. A. (1984). Bits, quanta, meaning. In A. Giovanni, F. Mancini, & M. Marinaro (Eds.), *Problems in theoretical physics* (pp. 121–141). University of Salerno Press.

Whitehead, A. N. (1911). *An introduction to mathematics.* Henry Holt and Co.

Whitehead, A. N. (1925). *Science and the modern world.* Macmillan.

Whitely, S. E. (1981). Measuring aptitude processes with multicomponent latent trait models. *Journal of Educational Measurement,18*, 67–84.

Wildomar Elementary School. (2014). *System 44 and Read 180 programs: All about a Lexile*. Retrieved March 28, 2022, from Wildomar Elementary School, Wildomar, California, USA: https://wes.leusd.k12.ca.us/apps/pages/index.jsp?uREC_ID=375353&type=d&pREC_ID=868048

Williamson, G. L. (2016). Novel interpretations of academic growth. *Journal of Applied Educational and Policy Research, 2*(2), 15–35.

Williamson, G. L. (2018). Exploring reading and mathematics growth through psychometric innovations applied to longitudinal data. *Cogent Education, 5*(1464424), 1–29.

Williamson, G. L., Fitzgerald, J., & Stenner, A. J. (2013). The Common Core State Standards' quantitative text complexity trajectory: Figuring out how much complexity is enough. *Educational Researcher, 42*(2), 59–69.

Williamson, G. L., Fitzgerald, J., & Stenner, A. J. (2014). Student reading growth illuminates the common core text-complexity standard. *The Elementary School Journal, 115*(2), 230–254.

Wilson, M. R. (2005). *Constructing measures: An item response modeling approach*. Lawrence Erlbaum Associates.

Wilson, M. R. (2013). Seeking a balance between the statistical and scientific elements in psychometrics. *Psychometrika, 78*(2), 211–236.

Wilson, M. R., & Fisher, W. P., Jr. (2016). Preface: 2016 IMEKO TC1-TC7-TC13 joint symposium: Metrology across the sciences: Wishful thinking? *Journal of Physics Conference Series, 772*(1), 011001.

Wilson, M. R., & Fisher, W. P., Jr. (2019). Preface of special issue, Psychometric Metrology. *Measurement, 145*, 190.

Wilson, M. R., & Scalise, K. (2003). Reporting progress to parents and others: Beyond grades. In J. M. Atkin & J. E. Coffey (Eds.), *Everyday assessment in the science classroom*. NSTA Press.

Wilson, M. R., & Toyama, Y. (2018). Formative and summative assessments in science and literacy integrated curricula: A suggested alternative approach. In A. L. Bailey, C. A. Maher, & L. C. Wilkinson (Eds.), *Language, literacy, and learning in the STEM disciplines* (pp. 231–260). Routledge.

Wise, M. N. (1995). Precision: Agent of unity and product of agreement. Part III—"Today precision must be commonplace." In M. N. Wise (Ed.), *The values of precision* (pp. 352–361). Princeton University Press.

Wittgenstein, L. (1958). *Philosophical investigations*. Macmillan.

Wright, B. D. (1958). On behalf of a personal approach to learning. *The Elementary School Journal, 58*, 365–375.

Wright, B. D. (1968a). The Sabbath lecture: Love and order. In A. R. Nielsen (Ed.), *Lust for learning* (pp. 65–68). New Experimental College Press.

Wright, B. D. (1968b). Sample-free test calibration and person measurement. In *Proceedings of the 1967 Invitational Conference on Testing Problems* (pp. 85–101). ETS.

Wright, B. D. (1977). Solving measurement problems with the Rasch model. *Journal of Educational Measurement, 14*(2), 97–116.

Wright, B. D. (1984). Despair and hope for educational measurement. *Contemporary Education Review, 3*(1), 281–288.

Wright, B. D. (1988). Georg Rasch and measurement. *Rasch Measurement Transactions, 2*(3), 25–32.

Wright, B. D. (1994). Measuring and counting. *Rasch Measurement Transactions, 8*(3), 371.

Wright, B. D. (1997a). A history of social science measurement. *Educational Measurement: Issues and Practice, 16*(4), 33–45, 52.

Wright, B. D. (1997b). Semiotics and scientific method. *Rasch Measurement Transactions, 11*(1), 539–540.

Wright, B. D., & Bell, S. R. (1984). Item banks: What, why, how. *Journal of Educational Measurement, 21*(4), 331–345.

Wright, B. D., & Masters, G. N. (1982). *Rating scale analysis*. MESA Press.

Wright, B. D., Mead, R. J., & Ludlow, L. H. (1980). *KIDMAP: Person-by-item interaction mapping* (MESA Memorandum #29) (6 pp.). MESA Press. http://www.rasch.org/memo29.pdf

Wright, B. D., & Stone, M. H. (1979). *Best test design*. MESA Press.
Wright, B. D., & Stone, M. H. (1999). *Measurement essentials*. Wide Range, Inc.
Xu, H., & Feng, J. (2004). A semiotic insight into model mapping. In K. Liu (Eds.), *Virtual, distributed and flexible organisations*. Springer.
Zumbo, B. D., & Forer, B. (2011). Testing and measurement from a multilevel view. In J. Bovaird, K. K. Geisinger, & C. Buckendahl (Eds.), *High stakes testing in education*. APA Press.

Contents

The Standardized Growth Expectation: Implications for Education Evaluation

A. Jackson Stenner, June D. Bland, Earl L. Hunter, and Mildred L. Cooper

Abstract The purpose of this paper is to review some assumptions underlying the use of norm-referenced tests in educational evaluations and to provide a prospectus for research on these assumptions as well as other questions related to norm-referenced tests. Specifically, the assumptions which will be examined are (1) expressing treatment effects in a standard score metric permits aggregation of effects across grades, (2) commonly used standardized tests are sufficiently comparable to permit aggregation of results across tests, and (3) the summer loss observed in Title I projects is due to an actual loss in achievement skills and knowledge. We wish to emphasize at the outset that our intent in this paper is to raise questions and not to present a coherent set of answers.

Throughout this paper we make use of an index termed the "standardized growth expectation" (SGE). The SGE is defined to be the amount of growth (expressed in standard deviation form) that a student must demonstrate over a given treatment interval to maintain his/her relative standing in the norm group (Stenner et al., 1977). The SGE rests on the assumption that a student will attain the same *raw score* on the pretest and posttest if no learning has taken place between testings. If the pretest raw score is equivalent to a national percentile of 50 and the same raw score is entered into the corresponding posttest percentile table, the resulting percentile score will be less than 50. The difference between the pretest percentile and the posttest percentile expressed in standard score form is termed the SGE. Stated another way, the SGE is the amount that a student at a particular pretest percentile is assumed to learn over a

Paper Presented at the Annual Meeting of The American Educational Research Association. Toronto, Canada. 1978.

A. J. Stenner (✉)
NTS Research Corporation, Durham, NC, USA

J. D. Bland · E. L. Hunter · M. L. Cooper
Washington DC Public School, Durham, NC, USA

1

period of time or, conversely, the loss in relative standing that such a student would suffer if he/she learned nothing during the time period.

An example may help to clarify the procedures used to calculate the SGE. Table 1 presents a raw score percentile conversion table for beginning of first grade and end of first grade on the Total Reading scale of the Comprehensive Test of Basic Skills, Form S. The average (50th percentile) beginning first grade student attains a raw score of 31 on Total Reading. Under the assumption that this illustrative average student learns nothing in the first grade, he/she would be expected to again obtain a raw score of 31 on the posttest. Whereas a raw score of 31 is equivalent to a beginning first grade percentile of 50, it represents an end of first grade percentile of 9. If both percentiles are converted to Z scores and subtracted, the result is an SGE of 1.39 (i.e., the 50th percentile equals a Z score of zero, whereas the 9th percentile equals a Z score of -1.39). In other words, if an average student learns nothing about reading during the first grade, he/she would be expected to lose 1.39 standard deviation units in relative standing because that is the amount of standardized growth exhibited by the national norm group during the first grade.[1]

1 Grade-To-Grade Variation in SGEs

Some educational evaluations which employ norm-referenced achievement tests share a common assumption, namely that observed treatment effects (e.g., differences between standardized means of observed treatment group posttest scores and expected treatment group posttest scores) are comparable across grades. Stated another way, it has been assumed that a one-third standard deviation difference between experimental and control students' reading comprehension has the same meaning whether observed at the second, fifth, or seventh grade levels. It has also been assumed, with apparent logic, that if a special program consistently demonstrates larger treatment effects in the primary as opposed to intermediate grades, then compensatory efforts should be concentrated at the lower level. In fact, the twelve-year history of ESEA Title I documents a nationwide trend toward focusing increasing amount of compensatory education efforts on primary students. Numerous evaluation studies have supported this movement through findings that larger treatment effects are possible with younger students. One question raised in this paper is whether or not there is a built-in-bias in our evaluation methodology and/or instrumentation that insures finding more "exemplary" programs at the primary grade levels.

---

[1] The SGE differs slightly depending upon where in the pretest distribution the raw score is selected to be entered into the posttest percentile distribution. The difference is of interest in its own right, but introducing the issue in the present paper would unnecessarily confuse the presentation.

Table 1 Raw Score to Percentile Table for Beginning and End of First Grade on CTBS, Level B Total Reading

Beginning of first grade		End of first grade	
Raw score	Percentile	Raw score	Percentile
73–84	99	84	99
86–72	98	84	98
65–67	97	84	97
61–64	96	84	96
59–60	95	84	95
57–58	94	83	94
55–56	93	83	93
53–54	92	82	92
52	91	82	91
31	50	59	50
31	49	58	49
31	48	58	48
31	47	57	47
31	46	56	46
30	45	55	45
29	44	54	44
29	43	53	43
29	42	53	42
29	41	52	41
20	10	32	10
19	9	31	9
18	8	30	8
18	7	29	7
18	6	28	6
18	5	27	5
17	4	25–26	4
16	3	24	3
15	2	21–23	2
0–14	1	0–23	1

We raise this issue of cross grade comparisons because educational policy may rest upon the legitimacy of just such comparisons. For example, a review of seventy-three school desegregation studies concluded that the critical period for desegregation (to maximize black students' achievement) is prior to third grade (Crain and Mahard, 1977). Black students desegregated beyond the third grade tend to show lower achievement test gains. Of the ten studies involving first and second graders, eight showed that desegregation produced higher achievement levels and two showed no effect. Only nine of twenty-one studies showed higher achievement among third and fourth grade students and for students in grades five through twelve, only sixteen of thirty-one studies showed any achievement gain attributable to desegregation. Eleven of the seventy-three studies reviewed by Crain and Mahard were not analyzed by grade. Taken at face value, these findings suggest that younger students benefit more from desegregation (in terms of achievement) than do older students. Crain and Mahard (1977) conclude: "The review of these studies is inconclusive or debatable on nearly every point except that desegregation in the early grades is superior to desegregation in the later grades" (p. 19). It is precisely this kind of conclusion based upon cross grade comparisons of norm-referenced achievement tests that may be invalid.

Table 2 presents standardized growth expectations for five commonly used norm-referenced achievement tests. The full year SGEs show a consistent decrement with each grade. The negative relationship between grade and SEG hold for both reading and mathematics across all five tests. The differences between second grade SGEs and eighth grade SGEs averaged across tests exceeds one-half standard deviation. Interestingly, the largest losses in SGEs for both Total Reading and Total Math occur during the third grade period. Several recent studies on test score decline (cf Wirtz, 1977) have concluded that test scores begin to drop at the fourth grade level. It might be rewarding to investigate the possibility that SGE decrements are causally implicated in reported test score declines. The Stanford Achievement Test, for example, exhibit almost a fifty percent decrement in Total Reading SGE from second to third grade. Similarly, the ITBS, CAT-77, and CTBS-S all show decrements approaching twenty-five percent.

Insight into the implications that these grade-to-grade differences may have for educational evaluation is gained by realizing that a treatment effect of one-third standard deviation (often employed as a threshold value for an educationally meaningful or practically significant effect size) represents a 33% increase above expectation for second graders on the MAT Total Reading and a 200% increase above expectation for eighth graders. If the ongoing instructional process is incapable of producing more than 0.15 SD's of growth on MAT Total Reading among eighth graders, then it seems somewhat unrealistic to expect an eighth grade compensatory program or desegregation effort to demonstrate a treatment effect of 0.33 SD's.

The conclusion implied above is that any cross grade comparisons of treatment effects expressed as national percentiles, standard scores, NCEs or grade equivalents are usually inappropriate. Although treatment effects expressed in standard score form may be smaller at the eighth grade level than at the second grade level, the statistical significance (e.g., F ratio) of these effects may be the same for both grade levels (assuming equal sample sizes) because of the increased pre-post correlation at the eighth grade level. Thus to apply an arbitrary treatment effect criterion of 0.33 SD's (or any other uniformly applied criterion based upon standardized scores) when screening for exemplary projects or reviewing research studies, unfairly discriminates against upper grade projects. A metric for practical or education significance which

Table 2 Standardized Growth Expectations for Selected Tests and Grades (50%tile) (Standard Deviation Units)

Total reading

Spring to Spring Grade	Stanford	ITBS	CAT-77	CTBS	Metropolitan
1.7–2.7	1.72	0.95	1.08	1.04	?
2.7–3.7	0.64	0.74	0.74	0.74	0.71
3.7–4.7	0.56	0.74	0.61	0.52	0.71
4.7–5.7	0.44	0.74	0.47	0.41	?
5.7–6.7	0.33	0.57	0.36	0.30	0.44
6.7–7.7	0.30	0.47	0.30	0.30	0.30
7.7–8.7	0.30	0.41	0.38	0.28	0.41
8.7–9.7	0.20	0.33	0.25	0.23	?
Total math					
1.7–2.7	1.17	1.17	1.23	1.17	?
2.7–3.7	0.99	0.84	0.99	0.99	0.84
3.7–4.7	0.58	0.91	0.71	0.56	0.88
4.7–5.7	0.64	0.71	0.64	0.61	?
5.7–6.7	0.44	0.67	0.49	0.36	0.50
6.7–7.7	0.33	0.49	0.36	0.30	0.47
7.7–8.7	0.41	0.47	0.47	0.30	0.38
8.7–9.7	0.28	0.33	0.28	0.33	?

would be comparable across grades cannot be formulated without consideration of the fact that pre-post correlations increase with grade.[2]

Following are five hypotheses regarding decrement in SGE as grade increases. It is highly likely that several of these alternative explanations combine to account for cross grade differences. Although the first two hypotheses are intuitively more appealing than the others, much more study of the merits of each explanation is recommended.

Domain Expansion Hypothesis: *With each increase in grade the relevant domain (e.g., reading or math) expands in terms of the number of concepts encompassed by the domain.* The result of an expanding domain is that a fixed number of items will be less and less representative; proportionately fewer items can be allocated to any given span of concepts and objectives. As the range of concepts and objectives covered by a test increases, the SGE decreases and edumetric validity is reduced (cf Carver, 1974). The poorer the match between what is taught at a given grade level and what is tested, the less sensitive the test is to growth, and the lower the SGE.

---

[2] Perhaps a more methodologically defensible metric would be the standard deviation of the pre-post residuals. The metric should be comparable across grades since it is adjusted for pre-post correlations.

Shifting Constructs Hypothesis: *The levels of some tests are not well articulated and with each succeeding grade stable organizing influences other than reading or math achievement increasingly determine students' scores on norm-referenced tests.* For example, if reasoning ability becomes progressively more confounded with reading and math achievement scores as grade increases, and reasoning ability grows at an increasingly lower rate than reading and math achievement, then the confounded reading and math SGEs would be expected to decline as confounding increases. As what is measured by norm-referenced reading and math tests changes, the edumetric validity of these tests may be reduced.

Learning Curve Hypothesis: *The deteriorating SGE is due to an actual slowing in the rate of learning similar to the way height slows down from birth to eighteen years of age.* According to this hypothesis younger students have a greater capacity for learning and this capacity deteriorates with age.

Unequal Interval Hypothesis: *Standard deviation units are not equal interval across grade.* Imagine a rubber band marked into ten equal intervals representing the one standard deviation SGE at second grade on the MAT Total Reading. Now imagine the rubber band stretched to the point that the distance between any two marks is equal to the entire length of the unstretched rubber band. In this way we can see how growth at the seventh grade on the MAT Total Reading (SGE = 0.10) might equal growth at the second grade (SGE = 1.0). If this hypothesis were accepted, the validity of cross grade comparisons would be questionable, but so would just about all other comparisons of interest in educational evaluation.

Instructional Emphasis Hypothesis: *Upper grade teachers do not emphasize reading and math as much as lower grade teachers and, as a consequence, students learn less and subsequently show less growth on norm-referenced reading and mathematics tests.* As upper grade teachers concentrate less on reading and math instruction than primary grade teachers, the SGE decreases.

The five hypotheses are rank ordered from most likely to least likely (in our opinion) as explanations for the observed decrement in SGE as grade increases. The first two hypotheses state that as grade increases, the edumetric validity of NRTs decreases. The second three hypotheses offer explanations which, although not related to the edumetric properties of NRTs, cannot be discounted without further research. At present all the hypotheses and the rank ordering are exercises in speculation. However, we are confident that variation in the SGE across grades represents an important phenomenon which may have implications for both policy makers and evaluation specialists. Until the grade-to-grade fluctuations in SGE are better understood, researchers might refrain from sweeping policy recommendations based upon cross grade comparisons of norm-referenced achievement test scores.

2   Test-To-Test Variation

Almost as striking as the grade-to-grade variations in full year SGEs within a test are the test-to-test differences within a grade. Examination of Table 2 reveals numerous

Table 3 Standardized Growth Expectations for Selected Tests and Grades (50%tile) (Standard Deviation Units) School Year

Total reading

Fall to Spring Grade	Stanford	ITBS	CAT-77	CTBS	Metropolitan
2.1–2.7	0.95	0.74	0.56	0.51	1.00
3.1–3.7	0.58	0.47	0.49	0.41	0.36
4.1–4.7	0.49	0.41	0.30	0.30	0.30
5.1–5.7	0.38	0.38	0.25	0.28	0.30
6.1–6.7	0.30	0.33	0.20	0.23	0.25
7.1–7.7	0.28	0.28	0.18	0.18	0.10
8.1–8.7	0.18	0.28	0.18	0.18	0.15
9.1–9.7	0.18	0.12	0.10	?	?

Total math

2.1–2.7	0.92	0.92	0.77	0.61	1.17
3.1–3.7	0.84	0.58	0.84	0.55	0.84
4.1–4.7	0.50	?	0.52	0.30	0.58
5.1–5.7	0.61	0.44	0.44	0.33	0.30
6.1–6.7	0.33	0.38	0.33	0.31	0.15
7.1–7.7	0.30	0.28	0.25	0.15	0.05
8.1–8.7	0.25	0.30	0.10	0.15	0.10
9.1–9.7	0.20	0.20	0.12	?	?

instances of SGEs being thirty to forty percent higher for some tests than for others. When we shift focus to school-year SGEs (see Table 3), the differences across tests are even more dramatic. School-year SGEs frequently vary among tests by as much as fifty to sixty percent with isolated instances of SGEs for some tests being three to six times as large as those of other tests.

All other things being equal, the higher the match between what is learned and what is tested (i.e., the higher the edumetric validity) the higher the SGE. An SGE near zero means that either nothing was taught, or something was taught but nothing was learned, or the test did not reflect what was taught and/or learned. A large SGE suggests that something was learned and the test reflects well whatever was learned. Presumably, criterion-referenced tests are superior to norm-referenced tests precisely because they provide a better match between what is taught/learned and what is tested. The SGE may provide a simple index for evaluating the claims made on behalf of criterion-referenced tests that they are superior evaluation tools. If CRTs demonstrate higher SGEs than NRTs, then these claims are likely valid. (The last section on the edumetric ratio addresses this issue more thoroughly.) A properly developed CRT should have greater fidelity to the curriculum and, consequently, larger SGEs. The SGE may be an effective means of assessing, a priori, tests' probable sensitivity to instruction.

We offer four hypotheses for the variation in SGEs across tests. Again we order the hypotheses in terms of our present thinking regarding the probability that each hypothesis will be sustained in future studies.

Edumetric Hypothesis: *Norm-referenced tests differ in the extent to which they reflect stable between-individual differences (Carver's psychometric dimension) and the extent to which they reflect within-individual growth (Carver's edumetric dimension).* A test may possess exemplary psychometric properties (e.g., high internal consistency and a good p value distribution) but be insensitive to what students learn over a given treatment interval. Such a test will have a low SGE but otherwise indistinguishable from other norm-referenced tests. The reader is encouraged to re-examine Table 2 in light of this hypothesis.

Procedures Hypothesis: *Test publishers use vastly different approaches to interpolation/extrapolation and make different assumptions regarding summer growth, thus artificially creating SGE differences.* The fact that full year SGEs are much more comparable between tests than are school year or summer SGEs suggests that publishers differ considerably in the assumptions they make about summer growth.

Norm Group Hypothesis: *The composition of the norm groups for the various tests differ to such an extent that the SGEs are affected.* Suppose, for example, that the Stanford Achievement Test (SAT) norm group was substantially brighter than the Metropolitan Achievement Test (MAT) norm group. The result would be that the Stanford Achievement Test norms would reflect more growth and, consequently, the SGEs for the SAT would be larger than those for the MAT. We should note that findings from the Anchor Test Study, for at least four of the tests considered in this paper, do not account for the large SGE differences across tests.

Cohort Hypothesis: *Although the norm groups for the various tests were selected in essentially similar ways because the tests were normed in different years, the samples may have differed in rate of achievement.* Teachers are fond of claiming that, like fine wines, there are "vintage years" in which a particular group of students just seems brighter, however, the pattern of SGE differences across tests (taking into consideration the year each test was normed) is not consistent with this hypothesis.

Of the four hypotheses just presented, the procedures and edumetric hypotheses seem most compelling. The fact that full-year SGEs are substantially more comparable across tests than either school-year or summer SGEs suggests that publishers make different assumptions about what students learn during the summer period. Apparently publishers of the Stanford Achievement Test assume that very little reading or math achievement growth should be expected of a fiftieth percentile student, whereas publishers of the MAT seem to assume a large amount of summer growth.[3] It seems probable that evaluation findings will vary depending upon which test is used, how closely different publishers' assumptions regarding summer growth coincide with empirical findings, and whether fall to spring or spring to spring testing dates are employed.

---

[3] The fact that both the Stanford and Metropolitan claim to have empirically determined fall and spring norms makes the substantial differences in summer SGEs for those two tests all the more puzzling.

According to the edumetric hypothesis, some norm-referenced tests are more sensitive to student growth in reading and math than other tests. Those tests with low SGEs measure well the between-individual differences which become more and more stable as students get older, but do a relatively poor job of measuring what students learn during a particular treatment interval. Most users of NRTs, particularly evaluation specialists, are primarily interested in measuring achievement growth. The SGE differences across tests seem to indicate that commonly available NRTs differ considerably in their edumetric validity, i.e., sensitivity to instruction.

The implications for educational evaluation of sustaining the edumetric hypothesis are substantial indeed. First of all, assuming the validity of this hypothesis, it is little wonder that most of our school effects studies have accounted for such minuscule proportions of variance with instructional process measures (cf Cooley & Lohnes, 1976). The problem may not rest with so-called "weak treatments" but rather with measurement instruments that are systematically biased against showing either significant treatment–control differences or substantial process-outcome relationships. When the SGE is as small as 0.15 standard deviations, as is the case with several tests at the eighth grade level, is it any wonder we find very few "exemplary" eighth grade reading and math programs or that Coleman et al. (1966) could find so few school variables that correlated with STEP Reading Test scores. Similarly, it is perhaps no coincidence that at those grade levels where the SGEs are largest and, presumably the edumetric validity of the tests is highest, we find a higher frequency of "exemplary" projects. An evaluation study that employs an NRT with a low SGE may be a priori doomed to add yet another conclusion of "no significant difference" to the literature on school effects.

3 Summer Loss Phenomenon

Several recent studies have highlighted the fact that Title I students achieve above expectation during the regular school year and lose in relative standing during the summer months (Pelavin and David, 1977, Stenner et al., 1977). Title I projects that use fall to spring testing dates often report substantial treatment effects, whereas projects that use fall to fall or spring to spring testing dates often report no treatment effects (Pelavin and David, 1977). In general, there has been limited appreciation for the different conclusions regarding treatment effects that result from simply varying testing dates. For example, tentative procedures in the OE Title I Evaluation System call for aggregating treatment effects without regard for testing dates. Similarly, the Joint Dissemination Review Panel typically evaluates reported treatment effects without considering testing dates.

Table 4 presents standardized growth expectations for the summer period (spring to fall). Except for the SAT, all tests exhibit substantial growth expectations over the summer period. We suggest that an edumetrically valid achievement test should have a large SGE over the school year and a small summer SGE. However, since the size of both school-year and summer SGEs can be manipulated by making different assumptions about summer growth, the data presented in this paper cannot speak directly to this point. If empirical data could be collected at three points in time (fall, spring, fall) for all commonly used NRTs, then the ratio of summer growth to school-year growth might address the question of comparative edumetric validity. Under such an analysis, when SGEs for the summer period approach or exceed SGEs for the school year, a test's sensitivity to instruction must be questioned. Large summer SGEs would suggest that the construct being measured by the test evidences growth whether or not the student is in school. Such an instrument would not only be relatively less sensitive to instruction-related achievement growth but would also presumably be insensitive to special project treatment effects. Again we emphasize that given the lack of multiple empirical norming points, the summer – school-year ratios given in Table 5 may simply reflect variation in publishers' assumptions about summer growth.

The large summer SGEs exhibited by four of the five tests examined in this paper raises questions about how much of the report summer loss among Title I students is due to absolute loss in achievement and how much is due to assumptions made by publishers. If Title I students actually lose raw score points over the summer period then we must conclude that there is an absolute loss in acquired skills and knowledge. If, however, there is no raw score change from spring to fall, that the Title I summer loss is relative rather then absolute and is a function of publisher assumptions. Discussions with other researchers studying this phenomenon suggest that there is some doubt as to whether the absolute achievement loss among Title I

Table 4 Standardized Growth Expectations (50%tile) for Selected Tests and Grades (Standard Deviations)

Total reading					
Spring to Fall Grade Period	Stanford	ITBS	CAT-77	CTBS	Metropolitan
2.7–3.1	0.10	0.38	0.21	0.36	0.41
3.7–4.1	0.02	0.25	0.30	0.29	0.36
4.7–5.1	0.02	0.23	0.20	0.15	?
5.7–6.1	0.05	0.25	0.15	0.12	0.18
6.7–7.1	0.00	0.20	0.15	0.12	0.25
7.7–8.1	0.10	0.15	0.20	0.10	0.36
8.7–9.1	0.02	0.10	0.15	?	0.15
Total math					
2.7–3.1	0.15	0.15	0.23	0.52	0.36
3.7–4.1	0.05	0.38	0.25	0.23	0.25
4.7–5.1	0.00	0.28	0.25	0.28	?
5.7–6.1	0.10	0.25	0.18	0.20	0.30
6.7–7.1	0.02	0.23	0.12	0.20	0.25
7.7–8.1	0.15	0.20	0.28	0.15	0.36
8.7–9.1	0.20	0.15	0.20	?	0.10

Table 5 Ratio of Summer Growth Expectation to School Year Expectation (50%tile)

Total reading

Grade	Stanford	ITBS	CAT-77	CTBS	Metropolitan
2	0.11	0.51	0.41	0.70	0.41
3	0.03	0.53	0.67	0.56	1.00
4	0.04	0.56	0.65	0.50	?
5	0.13	0.66	0.60	0.42	0.60
6	?	0.61	0.76	0.52	1.00
7	0.36	0.54	1.13	0.56	3.60
8	0.11	0.36	0.86	?	1.00

Total math

2	0.16	0.16	0.29	0.85	0.30
3	0.06	0.66	0.30	0.42	0.30
4	?	0.64	0.48	0.93	?
5	0.16	0.57	0.40	0.67	1.00
6	0.60	0.61	0.37	0.65	1.67
7	0.50	0.71	1.11	1.00	7.20
8	0.18	0.50	1.14	?	1.0

students is as large as is commonly believed. According to the arguments presented in this paper, the amounts of both absolute and relative loss may depend upon the NRT employed in the evaluation.

4   The Edumetric Ration

The fact that students do not attend school year around suggests a means for computing edumetric validities for commonly used norm-referenced and criterion-referenced tests. An edumetrically valid test, i.e., a test which is sensitive to instruction, should evidence proportionately higher SGEs during the school year than during the summer. If we assume a nine month school year and a three month summer, then any test purporting to measure what is taught in school (e.g., reading comprehension and math computation) should evidence a ratio of "school-year SGE" to "summer period SGE" larger than 3:1 (for convenience, we term this value the edumetric ratio). On the other hand, a test of nonverbal reasoning might evidence an edumetric ratio near 3:1, indicating that nonverbal reasoning (or what Cattell (1971) calls Fluid Ability) grows at a constant rate largely unaffected by school experiences.[4] Tests purporting to measure skills and objectives taught in school which show edumetric

---

[4] Edumetric ratios computed separately for different socio-economic groups might provide insight into how differences in out-of-school and in-school experiences impact on achievement.

ratios near "three" would probably prove highly insensitive to treatment effects and might be expected to evidence near zero correlations with variables similar to those employed by Coleman et al., (1966).

The edumetric ratio may also provide a means of externally validating criterion-referenced test items. Typically CRT validation efforts rely heavily on content analysis and judgements of curriculum experts regarding the match between curriculum and what a test item presumably measures. Edumetric validity of such items is assumed when judges agree on what concept or objective an item is measuring. We suggest that rating consensus is insufficient evidence to conclude that a test item is edumetrically valid. One more methodologically defensible approach might be to compute edumetric ratios on a set of items judged to be measuring a particular concept or objective and include on the final instrument only those items with high ratios.

Lastly, a comparison of SGEs for a widely used achievement and ability test offer some additional insights into the aptitude-achievement distinction (Green, 1974). Judging from theory and publisher test descriptions, one would expect achievement tests to have higher SGEs than ability tests. For example, the technical manual for the Cognitive Abilities Test (Thorndike & Hagen, 1971) states "...The test can be characterized by the following statements and these characteristics describe behavior that is important to measure for understanding an individual's education and work potential: (1) The tasks deal with abstract and general concepts, (2) In most cases, the tasks require the interpretation and use of symbols, (3) In large part, it is relationships among concepts and symbols with which the examinee must deal, (4) The tasks require the examinee to be flexible in his basis for organizing concepts and symbols, (5) Experience must be used in new patterns, and (6) Power in working with abstract materials is emphasized, rather than speed" (p. 25). Contrast the above description with that given in the technical manual for the Iowa Test of Basic Skills, "...The ITBS provides for comprehensive and continuous measurement of growth in the fundamental skills: vocabulary, reading, the mechanics of writing, methods of study, and mathematics. These skills are crucial to current day-to-day learning activities as well as to future educational development" (p. 3). In the ability test manual, phrases such as "educational potential," "general concepts," and "interpretation and use of symbols" are used whereas the achievement test manual uses such terms as "growth," "fundamental skills," "diagnosis," and "skill improvement." Clearly the impression one gets from these two manuals is that the Cognitive Abilities Test measures something more stable and less sensitive to school experiences than the ITBS; an impression which is not sustained by the SGE data.

Table 6 contrasts SGEs for the Cognitive Abilities Test and the ITBS. A first observation is that ITBS-Reading SGEs are comparable to CAT-Verbal SGEs. Thus, the ITBS-Reading appears to be almost as sensitive to instruction as the CAT-Verbal. Whether comparability between the two is due to the fact that the achievement test is actually more an ability test or the ability test is just a relabeled achievement test, or the distinction between verbal ability and reading is a sham, merits further

Table 6 Standardized Growth Expectations for the Iowa Test of Basic Skills and the Cognitive Abilities Test

	ITBS reading comprehension	Cognitive abilities test verbal	ITBS total math	Cognitive abilities test quantitative	Cognitive abilities test nonverbal
3.7–4.7	0.74	0.74	0.91	0.71	0.38
4.7–5.7	0.74	0.61	0.71	0.52	0.38
5.7–6.7	0.57	0.52	0.67	0.38	0.23
6.7–7.7	0.47	0.41	0.49	0.36	0.20
7.7–8.7	0.41	0.30	0.47	0.36	0.23
8.7–9.7	0.33	0.30	0.33	0.28	0.23

study.[5] One conclusion appears disconcertingly clear, the ITBS-Reading appears to be only slightly more edumetrically valid than the CAT-Verbal. How serious this predicament is depends on whether one elects to fault the CAT for being too much like an achievement test or the ITBS for being too much like an ability test.

The CAT-Quantitative appears to be less edumetrically valid than the ITBS-Total Math, but more sensitive to instruction than the CAT-Nonverbal. Since the CAT-Quantitative items loaded highly on the nonverbal factor and failed to define a quantitative factor (Thorndike & Hagen, 1971) one is left with the possibility that the Quantitative items are simply a mixture of items similar to ITBS-Total Math items and nonverbal reasoning items. Had the Quantitative Scale held more true to its label, we suspect that the SGEs would more closely approximate those for ITBS-Total Math. Finally, the CAT-Nonverbal evidences the lowest SGE. Whether the nonverbal growth expectations for the summer period are proportional to the school year, indicating little school effect, is a question requiring further study.

5   A Prospectus for Research

A major thesis of this paper is that policy decisions based upon grade-to-grade and test-to-test comparisons rest on a potentially shaky foundation. If the SGE index is meaningful and the analyses based upon it are valid, then a potentially large number of research findings merit re-examination. Granting the far-reaching policy implications inherent in our assertions and the need to establish quickly whether or not these assertions are valid, we offer the following research agenda.

---

[5] We are not suggesting that just because two tests have similar SGEs they are necessarily measuring the same thing. We are suggesting that evidence of comparable SGEs when added to information that disattenuated inter-test correlations approach 1.00 provides a pretty strong case for the fact that the two tests measure the same psychological construct.

- Submit the SGE concept to comprehensive analysis by measurement specialists focusing upon the conceptual basis for the index and assumptions underlying its computation.
- Compute SGEs for all subtests of commonly used achievement and ability test marketed during the past twenty years, and compare SGEs across grades and subtests. Some form of multi-method, multi-trait analysis might prove useful in such a substudy.
- Conduct a comprehensive content analysis of commonly used NRTs to determine the extent to which item content, type, or format contribute to variability in SGEs across tests.
- Conduct a meta analysis of reported treatment effects across a wide range of studies to determine whether treatment effects are correlated with SGEs. Preliminary investigations suggest that this may be a particularly fruitful area for further investigation.
- Conduct a logical and empirical analysis of the summer loss phenomenon found among Title I students. Estimate, if possible, what proportions of the loss are relative and absolute, and examine ways these proportions differ depending upon which NRT is used. Also conduct an item analysis to determine which skills evidence the largest losses over the summer.
- Conduct a preliminary investigation of the relationship between shifting edumetric validity and the Scholastic Aptitude Test score decline.
- Compute SGEs for a sample of criterion-referenced tests and investigate the claim that the SGE provides a useful index for comparing edumetric validities of CRTs and NRTs.
- Conduct a logical and empirical examination of the effects of out-of-level testing on the edumetric validity of NRTs.

The above research agenda will first address the utility and validity of the SGE concept and then proceed to examine selected implications of sustaining the edumetric hypothesis. The current nationwide interest in basic skills testing makes the topic of the proposed research particularly policy relevant at this time.

References

Carver, Ronald P. (1974). Two Dimensions of Tests—Psychometric and Edumetric. *American Psychologist*, 512–518.

Cattell, R. B. (1971). *Abilities: Their Structure, Growth and Action*. Houghton-Mifflin.

Coleman, J. S., et al. (1966). *Equality of Educational Opportunity*. U.S. Office of Education.

Cooley, W. W., & Lohnes, P. R. (1976). *Evaluation Research in Education*. Irvington Publishers Inc.

Crain, Robert L., and Mahard, Rita E. (1977). Desegregation and Black Achievement. National Review Panel on School Desegregation

Green, Donald R. (1974). *The Aptitude-Achievement Distinction*. CTB/McGraw Hill.

Pelavin, Sol H., and David, Jane. (1977). An Analysis of Longitudinal Data from Compensatory Education Programs. Stanford Research Institute.

Stenner, A. Jackson, Riegel, N. Blyth, Feifs, Helmuts A., and Davis, B. Steven. (1977). *Evaluation of the ESEA Title I Program of the Public Schools of the District of Columbia.*

Stenner, A. Jackson, Strang, Ernest W., and Baker, Robert F. (1978). Technical Assistance in Evaluating Career Education Projects—Volume II. DHEW/Office of Education, Contract No. 300760312.

Thorndike, R. L., & Hagen, E. (1971). *Cognitive Abilities Test.* Houghton Mifflin.

Construct Definition Methodology and Generalizability Theory Applied to Career Education Measurement

A. Jackson Stenner and Richard J. Rohlf

Abstract The field of career education measurement is in disarray. Evidence mounts that today's career education instruments are verbal ability measures in disguise. A plethora of trait names such as career maturity, career development, career planning, career awareness, and career decision making have, in the last decade, appeared as labels to scales comprised of multiple choice items. Many of these scales appear to be measuring similar underlying traits and certainly the labels have a similar sound or "jingle" to them. Other scale names are attached to clusters of items that appear to measure different traits and at first glance appear deserving of their unique trait names, e.g., occupational information, resources for exploration, work conditions, personal economics. The items of these scales look different and the labels correspondingly are dissimilar or have a different "jangle" to them.

As instrument developers and users we commit the "jingle" fallacy (Green, 1974) when we give the same or nearly the same name to clearly distinct underlying traits. Similarly, we commit the "jangle" fallacy when different labels are assigned to essentially the same underlying trait. When a trait label such as Career Maturity is assigned to a set of items which in fact measures verbal ability, we have committed the jangle fallacy. Whenever we find evidence that two similarly name scales are only moderately correlated, there exists the possibility of the jingle fallacy.

Whether or not a given scale is a measure of verbal ability as opposed to career maturity is, of course, a question of validity, i.e., is the scale actually a measure of "what it is intended to measure"—or is it? This chapter asserts that the current state of affairs in career education measurement exists because of the lack of carefully defined and operationalized career education constructs and will suggest a theory and methodology that researchers and practitioners will, hopefully, find useful in their continuing efforts to develop and refine measurement in the field of career education.

From U.S. Department of Education. ED204374. 1979.

A. J. Stenner (✉) · R. J. Rohlf
NTS Research Corporation, Durham, NC, USA

© The Author(s) 2023
W. P. Fisher and P. J. Massengill (eds.), *Explanatory Models, Unit Standards, and Personalized Learning in Educational Measurement*,
https://doi.org/10.1007/978-981-19-3747-7_2

1   Construct Definition

Constructs are the means by which science orders observations. We take it on faith that the universe of our observations can be ordered and subsequently understood with a comparatively small number of constructs or inferred organizing influences. Observations are aggregated and constructs created through the mental process of abstraction and induction. When we observe a group of children and describe some of the children as more aggressive than others, we employ a construct. We create the construct "aggression" by observing that certain behaviors tend to vary together and this pattern of covariation among observations we come to designate as aggression. In describing the differences in behavior among children, we might conclude that one child is much more aggressive than other children. We arrive at this conclusion informally by summing up the frequency of observed aggressive acts and we use the total score as an index of each child's level of aggression. These total scores are then compared and we arrive at decisions about each child.

This process of weighting individual observations, aggregating the observations into a total score and then checking the quality of the construct score by determining how well the total score can predict the original observations happens so fast and so frequently and works so well in our everyday lives, that there is seldom need to reflect critically on the process itself. The search for pattern or regularity among observations is, it seems, just as central to our daily lives as it is to scientific activity. Perhaps because the process of observation, abstraction and construct formation is so fundamental to daily functioning, it is taken for granted in behavioral science research. Often observations in the form of questionnaire items and test questions are aggregated without adequately examining the assumptions and implications inherent in the summation and averaging procedures. The simple fact that observations are combined and a total score computed means that we entertain a hypothesis that the observations are in some way related to one another. If the observations are uncorrelated, then combining them into a total score is a meaningless undertaking, since the total or construct score will carry no information about the original observations and consequently will be of no value in explaining anything else. If, however, the observations are correlated, then the construct score has meaning. Precisely what meaning depends upon the perceived nature of the organizing influences responsible for the correlations among observations.

A construct then is a theory which expresses how its inventor "construes" a set of interrelated observations. Construct labels (e.g., career maturity, occupational information, career decision making) serve as shorthand expressions for hypotheses regarding the nature of the predominant organizing influences responsible for correlations among observations.

What constitutes an observation? In career education measurement the most common "observation" would be a person's response to a test or rating item. Such observations provide information about a person's placement on a scale and serves as an indicant of the extent to which the subject possesses the attribute or trait being measured. A set of such indicants (items) comprises an instrument. The underlying structure or organizing influence operating on these observations is often determined by some combination of statistical structural analysis, e.g., factor analysis, and a logical analysis of the item content. Corroboration of the underlying structure is then frequently sought by confirmation via hypothesis testing and correlations with other instruments measuring conceptually similar and dissimilar constructs.

All observation whether made in service of the behavioral or physical sciences, is prone to error. Error is given more attention in behavioral sciences measurement probably because it exists in such abundance. Because of its abundance the process of construct definition must incorporate a theory of error. Various approaches to estimating the reliability of a measurement procedure rest on different assumptions about error and how it affects the observations we make.

Classical reliability theory is based on Spearman's model of an observed score (e.g., observation). Basically, an observed score is a function of two components, a true score and an error score. Within this framework, models of reliability have been formulated to assess the relative importance of each component. Campbell (1976) gives an excellent review of the historical development of reliability theory. All traditional measures of reliability (alpha, equivalent forms, retest) describe the agreement among repeated measurements of the same individuals. Although these reliability measures differ in their definition of error, they all assume a single undifferentiated source of error. Coefficient alpha attributes error to inconsistency in the extent to which individual items measure an attribute. Measures of stability such as test-retest or equivalent forms reliability coefficients, attribute error to changes in testing conditions, mood of examinee, etc.

In recent times authors such as Tryon (1957), Cronbach et al. (1963), Nunally (1967), and Lord and Novick (1968) have departed from the classic concept of true vs. error scores and have instead incorporated what has become to be known as the domain sampling theory of reliability. The notion of a true score was replaced by a "domain" or "universe" score which is an individual's score if all observations in a domain or universe could be averaged. Measurement error in this framework is the extent to which a sample value differs from the population value.

This change in focus from a "true score" to "universe score" resulted in increased importance being placed on defining the "universe" from which a particular sample of items has been drawn and to which we want to generalize. Initially the concept of universe was restricted to thinking in terms of a universe of content, e.g., sampling of reading comprehension items from a universe of possible reading comprehension items. However, the work of Cronbach, et al. has broadened this original conceptualization. His work, referred to as generalizability theory, speaks to sampling of "conditions of measurement" which include additional sources of variation to that of just variation among samples of items, or components of content. This broadened conceptualization can be viewed as a change from a focus on the reliability of an instrument to a focus on the reliability of a measurement procedure.

For example, suppose the career maturity of a group of students is rated on a number of items by a number of different teachers on several different occasions. The traditional view of a content domain would focus on the items as a sample from the universe of all such similar items. However, Cronbach's generalizability theory forces us to acknowledge that there are probably systematic differences in item scores across occasions which do not reflect true change in level of career maturity; and to recognize that there are systematic differences among students that are reflected in observed scores which are not necessarily due to difference in career maturity, e.g., socioeconomic status. Thus, from this perspective we are not only concerned with the universe of possible career maturity items, but, in addition, we need to think in terms of a universe of possible teachers and a universe of possible occasions, and a universe of possible respondents. Actually, Cronbach does not talk in terms of different universes; but rather, each of the above would be considered a "facet" in the universe of measurement conditions. The more facets one chooses to include in defining a construct, the broader the universe of generalization. Cronbach also refers to facets as either "random" or "fixed." A fixed facet would be one that would not vary, i.e., would be a constant in the universe. For example, if raters were considered to be a fixed facet in a measurement procedure, the investigator would be planning to always use the same rater(s) whenever a measurement was taken. Given this condition, there would be no systematic differences in observed scores due to idiosyncratic differences in rating behavior among raters. However, if raters were considered to be a "random" facet, the investigator would be broadening the construct definition of career maturity such that a person's "universe score" would be an average score across the universe of career maturity items judged across all possible raters. In the "fixed" case, the "universe score" would be an average score across the universe of items as judged by a particular rater or set or raters.

As discussed above, the process of construct definition begins with the recognition that observed scores (observations) are determined by some set of underlying organizing influences. In addition to "wanted" influences causing variation among scores, we must also recognize that there are "unwanted" (error) influences exercising potentially biasing or misleading effects in observed scores. Generalizability theory enables us to specify these sources of variance in observed scores in terms of characteristics of the object of measurement, characteristics of the indicants (items), characteristics of the context of measurement, and the interactions both within and across those categories.

In addition to a conceptual model, generalizability theory, using analysis of variance procedures, provides the techniques by which we can specify the sources of variance (both wanted and unwanted) in observed scores and estimate the magnitude of their effects. The procedure also yields a generalizability coefficient(s) which can be interpretated in a manner similar to traditional reliability coefficients, e.g., in estimating the standard error of measurement. However, before these analysis of variance procedures can be applied, it is necessary to design a study in which sources of variance are systematically varied.

Generalizability theory makes a distinction between G and D studies. A G study is a study in which data is collected in order to examine a wide range of sources of variance affecting a measurement procedure whereas a D study (for Decision) selects either the G study design or some modification of that design for use in estimating the generalizability coefficient that can be expected in some subsequent application of the measurement procedure. A D study does not involve the gathering of data but rather uses the variance estimates from the sources designed into the G study to estimate what the generalizability coefficient would be under alternative construct definitions and sampling specifications. For example, suppose that the authors of a career maturity scale employ a p:c × i × occ (persons nested within class crossed with items crossed with occasion) G study design. That is, the career maturity scale is administered to several classes on at least two occasions. Under this design the broadest permissible construct definition generalizes over items and occasions with either person or class as the object of measurement. Suppose that an investigator has limited time available for student testing and wanted to know what the effect would be of reducing by 50% the number of items on the scale. This scenario could be set up and a generalizability coefficient estimated given specification of the object of measurement and construct definition (which facets are considered fixed and which random). This application of generalizability theory is analogous to power analysis (Cohen, 1977) in which different sampling scenarios are evaluated to determine the probability of detecting an effect. In D studies different measurement scenarios (alternative construct definitions coupled with alternative sampling frequencies) are evaluated to determine the precision with which objects of measurement can be differentiated.

2 An Illustration of Generalizability Theory

Before proceeding with an example, it may prove useful to reflect on the meaning of a generalizability coefficient as well as its general form. A generalizability coefficient is simply the ratio of true score variance (or universe score variance) to the sum of true score variance and error variance:

$$\sum{}_{\rho}^{2} = \frac{\text{True score variance}}{\text{True score variance} + \text{error variance}} = \frac{\tau}{\tau + \delta}$$

The components that enter true score variance and error variance change as the construct definition changes, but the basic expression for a generalizability coefficient remains the same. Following are several descriptive comments about the generalizability coefficient that may help in gaining an intuitive grasp of what this ratio means:

- One task of measurement is to differentiate among objects (e.g., classrooms or children) on some scale while simultaneously generalizing over selected facets. The higher the generalizability coefficient, the better the differentiation or separation among objects.
- Children or classrooms differ on a scale for many reasons (we usually refer to these reasons as sources). Some of these reasons are important to us and represent what we want to measure, and others are not of interest and represent noise. Differences among students that arise due to reasons we are interested in, we call

true score differences whereas differences due to reasons we are not interested in, we call error differences. A generalizability coefficient is simply the ratio of average squared differences between objects that arise from wanted sources divided by the average squared differences between objects arising from wanted and unwanted sources.

- Observed score variance is the sum of true score variance and error variance. Thus the generalizability coefficient represents the proportion of observed score variance that is due to "wanted" sources of variance. If the generalizability coefficient is high, then a high proportion of the variance in observed scores is due to wanted sources of variation, whereas if the coefficient is low, it means that only a small proportion of differences among objects is due to wanted sources of variation.

- We can conceive of the generalizability coefficient as a heuristic that describes the confidence with which we can reject the null hypothesis that all objects' true scores are equal. Statisticians would use an F ratio for this purpose and, in fact, for the simple persons × items (p × i) design:

$$\sum\nolimits_\rho^2 = \frac{F-1}{F} \; or \; F = \frac{1}{1 - \Sigma\rho^2}$$

- The generalizability coefficient is the squared correlation between observed scores and true scores. The true score is the average score we would obtain if all observations across the random facets of the universe of generalization could be exhaustively sampled. Errors of measurement (unwanted reasons that objects have different scores) contribute to ordering people differently on observed scores (which are samples) than they would be ordered if their scores could be averaged over all facets of interest (e.g., items or days during a two-week period). The generalizability coefficient provides an indication of how differently people are likely to be ordered if exhaustive sampling of all relevant observations was possible.

In summary, the generalizability coefficient provides an estimate of the precision of measurement given a construct definition. It is meaningless to refer to a reliability or generalizability coefficient without reference to the governing construct definition. What construct definition is most appropriate in a given situation is a substantive question that often cannot be answered by measurement specialists. What definition to employ is a complex question which takes us back to what we want our construct to mean. Ascribing meaning to constructs and increasing our understanding of variance arising from applications of our measurement procedures is what the process of construct definition is all about.

Table 1 provides estimated G Study variance component for a p × i × m design, and Table 2 displays different D Study designs or measurement scenarios. The data used in this illustration was graciously provided by Dr. Bert Westbrook and represents a subsample of the ninth grade data used in his chapter in this volume. The sample consists of 60 students responding to the 50 attitude items of the Career Maturity Inventory (CMI).

Table 1 Illustrative example of generalizability analyses for the attitude subscale of the CMI

Source	Notation	SS	df	MS	Estimated variance component	Estimated proportion of universe variant attributable to each source
Person	P	68.87	59	1.167	0.00889	04
Item	I	270.41	49	5.519	0.04328	18
Moment	M	1.20	1	1.204	0.00030	00
Person × Item	p × i	661.62	2891	0.229	0.04111	17
Person × Moment	p × m	11.55	59	0.196	0.00098	00
Item × Moment	i × m	11.87	49	0.242	0.00159	01
Person × Item × Moment	p × i × m$_e$	423.88	2891	0.147	0.14662	60

Examination of Table 1 reveals that a large proportion (60%) of the variance on this instrument is unexplained by facets of the measurement procedures. The second and third largest components of variance are the item (i) and person × item (p × i) interaction, respectively. The person (p) component explains four percent of the universe variance. The moment (m), person × moment (p × m) interaction and item by moment (i × m) interaction explain very small proportions of the variance.

A major advantage of generalizability theory is that the theory specifies which sources of variance are to be ignored, which contribute to true score (τ) and which contribute to error (δ) in estimating the generalizability of a measurement procedure under a particular construct definition. Whether stated or not, there are two essential aspects of a measurement procedure that must be explicit before any reliability or generalizability coefficients can be interpreted. These are (1) the construct definition, i.e., which facets are to be considered random and which fixed, and (2) the sampling frequencies for each facet included in the construct definition.

Table 2 presents construct definitions and sampling specifications for five scenarios. Scenario #1 displays the generalizability coefficient under the classical reliability formulation in which moments (i.e., short-term occasions) are fixed (N_m-1) and items are random. The generalizability coefficient ($\sum_{\rho}^{2} = 0.72$) under this scenario accurately describes the precision of measurement only if our interest centers on how well students can be differentiated on a single occasion. This coefficient corresponds to the traditional coefficient alpha (or KR-20).

In scenario #2 moments are random and items are fixed. This construct definition corresponds to the traditional stability or retest coefficient. In other words, within the framework of generalizability theory the traditional retest coefficient may be computed under a generalizability design of the form p × i × m where items constitute a fixed facet and moments constitute a random facet. Note that in this case the retest coefficient is higher than the internal consistency coefficient because the p × m variance component accounts for virtually no variance whereas the p × i interaction

Table 2 Illustrative scenario table

Scenario #	Construct definition		Sampling specifications		i	m	p × i	p × m	m × i	p × i × m_e		Generalizability coefficient
	Random facets	Fixed facets	p									
1	Items	Moments	$N_i = 50$	$N_m = 1$	τ	–	–	∞	τ	–	∞	0.72
2	Moments	Items	$N_m = 1$	$N_i = 50$	τ	–	–	τ	∞	–	∞	0.78
3	Items	Moments	$N_i = 50$	$N_m = 1$	τ	–	–	∞	∞	–	∞	0.72
	Moments											
4	Items		$N_i = 50$	$N_m = 2$	τ	–	–	∞	∞	–	∞	0.81
	Moments											
5	Items		$N_i = 100$	$N_m = 3$	τ	–	–	∞	∞	–	∞	0.92
	Moments											

(which contributes to the true score when items are fixed) accounts for 17% of the universe variance.

Scenario #'s 3, 4, and 5 all employ the broadest permissible construct definition (items and moments random) but the sampling frequencies for items and/or moments differ. In Scenario #3 we estimate what the generalizability coefficient would be if 50 items were administered on one occasion (i.e., moment). Note that the generalizability coefficient under this construct is coincidentally the same as that observed under Scenario #1. As a rule when the construct definition is broadened and the sampling specifications are unchanged, the generalizability coefficient goes down. Similarly when the construct definition is narrowed and consequently the universe of generalization is narrowed, the generalizability coefficient is increased. The reasoning for this outcome is straightforward; if the universe under examination is quite broad, then a larger number of observations must be sampled to attain a specified level of precision, whereas a narrower universe permits a smaller number of observations to attain the same precision. Under Scenario #4 the item sample remains at $N_i = 50$ but the number of testing sessions is increased, $N_m = 2$, resulting in an improvement in the generalizability coefficient ($\sum_\rho^2 = 0.81$). Finally, under Scenario #5 sampling frequencies are increased for both items ($N_i = 100$) and moments ($N_m = 3$) resulting in a substantial increase in precision of measurement.

Classical reliability theory, as practiced in the field of career education measurement, is unnecessarily restrictive. Disciples of classical theory compute a number of equivalence coefficients by correlating student performance on split halves of an instrument or by computing coefficient alpha (KR-20) for an instrument administered on a single occasion. Similarly, stability or retest coefficients are computed by correlating student performance on one occasion with performance, say, two weeks later. Finally, interrater reliability coefficients are computed by correlating the ratings of two or more raters of the same behavior. Each of these forms of reliability coefficient reflect a single undifferentiated source of error and, more importantly, derive from different construct definitions. Coefficient alpha accurately reflects an instrument's reliability under the highly restricted construct definition which treats items as the only random facet and, consequently, the p × i interaction (confounded with the residual) as the only source of error. The stability coefficient properly reflects an instrument's reliability under a construct definition that treats occasions as the only random source of variance and the persons × occasion interaction as the only source of error.

It is important to recognize that traditional forms of reliability permit error variance to be confounded with true score variance because they do not differentiate among the many possible sources of error. For example, the test-retest reliability coefficient will not "break out" a p × i interaction and thus that variance will be a "hidden" component of the true score variance. Likewise, the p × occ interaction will be a hidden component of the true score variance when calculating coefficient alpha. This can result in an artificially inflated estimate of true score variance which, in turn, results in an inflated reliability coefficient. The attractiveness of generalizability theory is that it permits simultaneous consideration of items, occasions and other facets as random sources of error.

Generalizability theory provides a framework that forces us to better conceptualize the constructs we use. Unfortunately many investigators do not invest sufficient time in construct definition activities. Time and again researchers move ahead to answer substantive research questions without carefully defining the constructs that figure in their theories or program evaluations. The most common mistake found in the behavior sciences is that investigators will state a conceptually broad construct definition, but will use a reliability estimate that is based on a much narrower definition, thus yielding a coefficient which exaggerates the precision with which the construct can be measured. Elsewhere we have argued that career education will go the way of many previous fads unless, as a field, it can stake out a set of well-defined constructs and related instrumentation (Stenner et al., 1978). So far, efforts in this regard have been disappointing.

3   Some Special Applications

Many seemingly diverse issues in measurement can be accommodated within generalizability theory. Cronbach (1976) states:

What appears today to be most important in G Theory is not what the book gave greater space to. In 1972 G Theory appeared as an elaborate technical apparatus. Today the machinery looms less large than the questions that theory enables us to pose. G Theory has a protean quality. The procedures and even the issues take a new form in every contest. G Theory enables you to ask your questions better; what is most significant for you cannot be supplied from the outside (p. 199).

In the discussion to follow we attempt to illustrate the range of applications for which generalizability theory, coupled with construct definition methodology, can be useful.

4   Toward a Theory of the Indicant

A historical convention for which we can find no rational explanation has contributed to avoidance of a potentially fruitful type of construct validation study. The convention is to report person scores as number of item (or indicants) correct and item scores as proportion of respondents answering an item correctly. Thus, person scores and item scores are expressed in different metrics, leading some investigators to assume that there is some fundamental difference in the way people and items can be analyzed. For example, construct validity studies often emphasize relationships between theoretically relevant variables and the construct under study and use the person as the unit of analysis. On the other hand, little work has been done in explaining variance in item scores. Some authors, including ourselves, contend that many career development scales containing items of the multiple choice variety in fact measure verbal ability and not career maturity (see Westbrook's chapter in this volume). One approach to investigating this contention which, on the face, seems more direct than focusing on person score correlations, would involve predicting item scores using a set of item readability and syntax measures as well as theoretically derived ratings of the extent of career maturity called for by each item. If the readability and syntax measures explain a large proportion of the variance in item scores and the theory-based ratings explain little of the variance, then it is likely that the so-called career development items are really verbal reasoning items in disguise. If a construct is really well defined, then it should be possible to explain the behavior of indicants of the construct, i.e., explain variance in item or indicant scores. Unfortunately this is a test that few constructs in the behavioral sciences, let alone career education, have passed.

5 Generalizability of Ratings

Many outcomes in career education do not readily lend themselves to paper-pencil testing. For example, outcomes such as employability skills, personal work habits and job interview behavior are better measured by trained observers in either real or simulated settings. Generalizability theory provides a framework for estimating the dependability of these ratings.

Suppose a career education program sets about to improve the job interview behavior of a group of students. Five employers from the local community are called upon to interview each student and complete a rating scale. One highly informative design for examining the generalizability of these ratings would be $p \times i \times r$ (persons crossed with items crossed with raters). Thus each student would be rated by each employer on all items. Under this design the broadest permissible construct definition generalizes over items and raters.

Separate estimates of alpha or interrater agreement would overestimate the precision with which the construct as defined can be measured. The generalizability coefficient more accurately reflects our measurement precision and provides information on how the precision can be increased to an acceptable level. One excellent illustration of this type of analysis is provided by Gilmore et al. (1978). Note that this design does not include the "occasion" facet. If a sizable $p \times occ$ interaction exists, our estimate of measurement precision may be inflated if we have defined our construct to be stable over time.

6 Competency Testing

Some career education programs have objectives which state that all students will attain a particular mastery level in reading and mathematics. In assessing this objective, instrumentation is needed that has a special kind of reliability. Discussion above focused on developing instruments that would maximally differentiate among objects (e.g., students). In competency or mastery testing, the objective is to differentiate among two groups of students, those that have attained the minimal performance level and those that have not. Generalizability theory provides a framework for studying the dependability of mastery or competency decisions. The most thorough treatment of this application of generalizability theory is provided by Brennan and Kane (1977).

7 Generalizability of Class Means

Some career education evaluations employ class or school rather than student as the unit of analysis (Stenner et al., 1978; Brennan, 1975; Kane et al., 1976). Haney (1974) and Kane and Brennan (1977) have suggested that generalizability theory provides a conceptually and practically appealing approach to estimating the reliability of class means. The simplest design from which we can estimate the generalizability of class means is p:c × i (persons nested within class crossed with items). Note that this is the familiar persons × items design (from which coefficient alpha is computed) with the addition that knowledge is available on class membership. Under this design we can estimate the reliability of persons nested within classes and class means. In a more complex design such as p:c:s × i, the object of measurement might be persons (p), classes (c) or schools (s). In general, this type of split plot design can prove particularly useful in an evaluation in which multiple units of analysis (e.g., students, classes, schools) are employed (Hayman et al., 1979). As a rule, generalizability coefficients should be computed for each unit of analysis employed in a research or evaluation study.

In passing we should note that applications of generalizability theory in which class or school is the object of measurement, have focused exclusively on the mean or first moment of the distribution. Lohnes (1972), in an excellent but largely ignored paper, demonstrated that using the variance of a class or school distribution as an independent variable might also be useful in predicting outcomes. In such studies interest is centered on differentiating classrooms not in terms of their means, but rather in terms of their variances while generalizing over occasions or some other random facet.

8 Issues of Test Bias

Much attention and controversy has surrounded the issues of race and sex bias in testing. Although there are many types of bias, perhaps the most pernicious is that which gives members of particular racial, ethnic or sex groups unfair advantage in responding to certain kinds of items. It is somewhat ironical that career education has as one of its goals eradication of sex role stereotyping (Hoyt, 1975) and yet we were unable to find any studies of sex bias in career education measurement.

Some forms of item bias can be effectively studied within the framework of generalizability theory. For example, a simple p: s × i (persons nested within sex group crossed with items) will provide information on possible sex bias. In this design the component of variance related to sex bias is the s × i (sex by item) interaction. If this component of variance is large then items have a different meaning (i.e., measure something different) for males and females. Examination of the items contributing most heavily to the interaction can sometimes lead to explanations for the source of the bias (e.g., terminology unfamiliar to males or females). Students can also be nested within race groups to evaluate racial bias or nested within reading level groups to evaluate the extent to which item meaning is conditional on student reading level.

Although the literature on bias has focused almost exclusively on racial, ethnic and sex characteristics, the notion of bias is a generic concept. Any characteristic of the object of measurement which interacts with a random facet represents bias. For social and other reasons, item scores which are conditional on race and sex (i.e., interact with race and sex) have received the bulk of attention. From theoretical as well as practical perspectives, there are other types of bias that pose equally troublesome problems. For example, items that take on radically different meanings depending upon the examinee's reading level are just as invalid as indicators of career maturing as items that are conditional on sex or race of the examinee.

References

Brennan, R. L. (1975). The calculation of reliability from a split-plot factorial design. *Educational and Psychological Measurement, 35*, 779–788.

Brennan, R. L., & Kane, M. T. (1977). An index of dependability for mastery tests. *Journal of Educational Measurement, 14*(3), 227–289.

Campbell, J. P. (1976). Psychometric theory. In M. D. Dunnette (Ed.), *Handbook of industrial and organizational psychology*. Rand McNally.

Cohen, J. (1977). *Statistical power analysis for the behavioral sciences* (Rev. Ed.). Academic Press.

Cronbach, L. J. (1976). On the design of educational measures. In D. N. M. de Gruijter & L. J. T. van der Kamp (Eds.), *Advances in psychological and educational measurement* (pp. 199–208). New York: Wiley.

Cronbach, L. J., Rajaratnam, N., & Gleser, G. (1963). Theory of generalizability: A liberalization of reliability theory. *British Journal of Statistical Psychology, 16*(Part 2), 137–163.

Gilmore, G. M., Kane, M. T., & Naccarato, R. W. (1978). The generalizability of student ratings of construction: Estimation of the teacher and course components. *Journal of Educational Measurement, 15*(1), 1–13.

Green, D. R. (Ed.). (1974). *The aptitude-achievement distinction.* CTB/McGraw-Hill.

Haney, W. (1974). The dependability of group mean scores. Unpublished special qualifying paper. Harvard Graduate School of Education.

Hayman, J., Rayder, N., Stenner, A. J., & Madey, D. L. (1979). On aggregation, generalization, and utility in education evaluation. *Educational Evaluation and Policy Analysis, 1*(4).

Hoyt, K. B. (1975). An introduction to career education: A policy paper of the U.S. Office of Education, Washington, D.C.: U.S. Department of Health, Education and Welfare.

Kane, M. T., & Brennan, R. L. (1977). The generalizability of class means. *The Review of Educational Research, 47*(1), 267–292.

Kane, M. T., Gillmore, G. M., & Crooks, T. J. (1976). Student evaluations of teaching: The generalizability of class means. *Journal of Educational Measurement, 13,* 171–183.

Lohnes, P. (1972). Statistical descriptors of school classes. *American Educational Research Journal, 9,* 547–556.

Lord, F. M., & Novick, M. R. (1968). *Statistical theories of mental test scores.* Addison-Wesley.

Nunnally, J. C. (1967). *Psychometric theory.* McGraw Hill.

Stenner, A. J., Strang, E. W., & Baker, R. F. (1978). Technical assistance in evaluating career education projects: Final report. NTS Research Corporation.

Tryon, R. C. (1957). Reliability and behavior domain validity: Reformulation and historical critique. *Psychological Bulletin, 54,* 229.

Testing Construct Theories

A. Jackson Stenner and Malbert Smith III

Abstract This paper presents and illustrates a novel methodology, construct-specification equations, for examining the construct validity of a psychological instrument. Whereas traditional approaches have focused on the study of between-person variation on the construct, the suggested methodology emphasizes study of the relationships between item characteristics and item scores. The major thesis of the construct-specification-equation approach is that until developers of a psychological instrument understand what item characteristics are determining the item difficulties, the understanding of what is being measured is unsatisfyingly primitive. This method is illustrated with data from the Knox Cube Test which purports to be a measure of visual attention and short-term memory.

The process of ascribing meaning to scores produced by a measurement procedure is generally recognized as the most 'important task in evaluating a psychological measure, be it an achievement test, interest inventory, or personality scale. This process, which is commonly referred to in the literature as construct validation (Cronbach & Meehl, 1955; Cronbach, 1971; Thorndike & Hagen, 1965; Jensen, 1980), involves a family of methods and procedures for assessing the degree to which a test measures the trait or theoretical construct the test was designed to measure.

While there is general consensus among measurement theorists regarding the prominent role of construct validity, theorists and researchers have not articulated a unified and systematic approach to establishing the construct validity of score interpretations. As Buros (1977) noted in his assessment of the past fifty years in testing, there has been little progress in the field of testing except for tremendous advances in computer technology. His observations seem particularly accurate in terms of the current status of construct validity. Contemporary measurement procedures do not, in general, provide any more convincing evidence that they are measuring what

Perceptual and Motor Skills, 55. 1982.

A. J. Stenner (✉) · M. Smith III
NTS Research Corporation, Durham, NC, USA

© The Author(s) 2023
W. P. Fisher and P. J. Massengill (eds.), *Explanatory Models, Unit Standards, and Personalized Learning in Educational Measurement*,
https://doi.org/10.1007/978-981-19-3747-7_3

they purport to measure than instruments developed fifty years ago. The absence of persuasive, well-documented construct theories can be attributed, in part, to the lack of formal methods for seating and testing such theories.

A major thesis of this paper is that until the developers of a psychological instrument can adequately explain variation in item scores (i.e., difficulty), the understanding of what is being measured is unsatisfyingly primitive. Most approaches to texting and elaborating construct theories focus on explaining person score variation.[1] Cronbach (1971), for example, in his review of construct validity, emphasizes the formal testing of nomological networks and interpretation of correlations among person scores on theoretically relevant and irrelevant constructs. Except for isolated investigations, explaining variation in item scores has not been viewed by methodologists or practitioners as a particularly protean source of information about construct validity.

The rationale for giving more attention to variation in item scores is straightforward. Just as a person scoring higher than another person on an instrument is assumed to possess more of the construct in question (i.e., visual memory, reading comprehension, anxiety), an item (or task) which scores higher (in difficulty) than another item must be viewed as demanding more of the construct. The key question deals with the nature of the "something" that causes some persons to score higher than other persons and some items to score higher than other items. The process by which theories about this "something" are tested is termed "construct validation."

There are essentially two routes one can travel in testing construct theories. The traditional route is the study of between-person variation on the construct. By examining relationships between potential determinants and construct scores, progressively more sophisticated construct theories are developed. The second, much less traveled, route involves the study of relationships between item characteristics and item scores. Thurstone (1923) appears to have set the stage for a half century of neglect of this latter alternative when he argued:

> I suggest that we dethrone the stimulus. He is only nominally the ruler of psychology. The real ruler of the domain which psychology studies is the individual and his motives, desires, wants, ambitions, cravings, and aspirations. The stimulus is merely the more or less accidental fact... (p. 364).

Considering the symmetry in the person and item perspectives on construct validation, it is somewhat baffling that Thurstone, Terman, Binet, and Goodenough were so successful in focusing the attention of early correlationists exclusively on person variation. The practice, adopted early in the 1900s of expressing the row (i.e., person) as raw counts and column (i.e., item) scores as proportions may have distorted the symmetry, leading early test constructors, with few exceptions, to abandon variation in item scores as a source of knowledge about a measurement's meaning. From the perspective of construct validation, the regularity and pattern in item-score variation is no less deserving of exploration than that found in person-score variation. Only

---

[1] Throughout this paper we distinguish "construct theory" from the broader "data theory." A construct theory accounts for a regularity in a set of observations generated from a measurement procedure. A data theory interrelates a set of constructs in a nomological network.

historical accident and tradition have blinded behavioral researchers to the singularity of purpose inherent in the two forms of analysis.

Perhaps the only exception to this blindness has occurred in the field of cognitive psychology. This branch of psychology has been very diligent in analyzing items as a data source in understanding the components of behavior. As Pellegrino and Glaser (1982) explain, "performance on psychometric test tasks becomes the object of theoretical and empirical analyses" (p. 275). Similarly, Sternberg (1977, 1980), Carroll (1976), and Whitely (1981) have focused upon item variation in attempting to explicate the cognitive processes required by certain psychometric tasks. As the next section indicates, there are several advantages in adopting such an approach.

1 Advantages in Analyzing Item-Score Variation

There are advantages to studying item-score variation in the process of validating a construct theory. In this section we shall examine three such advantages.

(1) Stating theories so falsification is possible. There is an obvious need throughout the behavioral sciences for falsifiable construct theories that are broad enough to yield predictions beyond those suggested by common sense. Most verbal descriptions of constructs, as well as construct labels with their trains of connotations and surplus meaning, serve as poor first approximations to deductive theories regarding what a construct means or what an instrument measures. These verbal descriptions seldom lead to predictions and thus are not susceptible to challenge or refutation. Outside of the literature on response bias there are few studies indeed that offer testable alternative theoretical perspectives on what an instrument measures or what a construct means. Yet history shows that this type of "challenge research" is precisely what fosters intellectual revolutions (Kuhn, 1970). The behavioral sciences need more of the type of conflict such studies engender.

A suggested test interpretation is a claim that a procedure measures a con struct. Implicit in the use of any such procedure is a theory regarding the nature of the construct being measured. A major problem with the current state of behavioral science measurement is that it is not at all clear how such claims about construct meaning can be falsified. Unless and until construct interpretations are framed in ways that are in practice as well as in principle falsifiable, the behavioral sciences will not progress beyond a disjointed collection of instrumentation and methodological curios.

A major reason for emphasizing item-score variation is that theories about the meaning of a construct can be precisely stated in the form of a construct specification equation (discussed at length in the next section). Such an equation embodies a theory of item-score variation and simultaneously provides a straightforward means of confirming or falsifying alternative theories about the meaning of scores generated by a measurement procedure.

(2) Higher generalizability of dependent and independent variables is possible. Item
     scores are typically more generalizable (i.e., reliable) than are person scores. In
     a simple persons X items generalizability design with person as the object of
     measurement, the error term is divided by the number of items, whereas, if item
     is viewed as the object of measurement, the error term is divided by the number
     of people. Since most efforts to collect data involve many more people than
     items, the item scores are typically more generalizable than are person scores.
     In most studies involving psychological instruments where the number of people
     is greater than or equal to 400, the generalizability coefficient for item scores is
     greater than or equal to 0.95. When items from nationally normed instruments
     are examined, the generalizability coefficient for item scores generalizing over
     people and occasions approaches unity.

It is also true that most measures used as independent variables in construct-
specification equations can be measured with considerable precision. Io studying the
item face of the person by items matrix, it is not unusual to find one's self analyzing
a network of relationships among variables where each variable has an associated
generalizability coefficient (under a broad universe definition) approaching unity.
For those researchers accustomed to working with error ridden variables, this new
perspective can be refreshing.

(3) Experimental manipulation is possible. Items are docile and pliable subjects.
     They can be radically altered without informed consent and can be analyzed and
     reanalyzed as new theories regarding their behavior are developed. Similarly,
     controlled experiments can be conducted in which selected item characteristics
     are systematically introduced and their effects on item behavior assessed. Also,
     causal inference is strengthened when one employs items as subjects rather than
     people, due to the fact that the experimenter can control items better than people.
     Items are passive subjects, whereas people are active subjects on whom only
     minimal experimental control can be exercised; threats to the validity of a study
     are more likely to be operative when people are studied. All in all, items are
     better subjects for experimentation than people. Effects can be estimated and
     interpreted with less ambiguity, and the direct experimental manipulation is less
     costly and more efficient. Finally, this new perspective should breathe life into
     Cronbach's expressed hope that the experimental method be viewed as a proper
     and necessary means of validating interpretations of psychological scores.

2 Construct-Specification Equation

A "measurement procedure" requires (1) an "item format" which details the general
framework of fixed item characteristics within which the item con tent is to be varied,
(2) the "construct-specification equation" which operationally defines the construct
and presents the item characteristics and associated regression weights to be used
in generating items and systematically varying their difficulty, (3) a sampling plan

for selecting items from the universe bounded by the specification equation, (4) a plan for selecting persons and assigning then to items, (5) administrative procedures which govern the con text in which the measurement procedure is administered, and (6) rules for assigning scores to persons and items.

A construct-specification equation is developed by regressing item scores on selected characteristics of the items. The results from this multiple regression analysis yield two important sources of information. First, the R^2 index indicates the amount of variance in item scores accounted for by the model.[2] Second, the analysis provides a regression equation that indicates which characteristics of the items are important in predicting item scores.

The construct-specification equation expresses a theory regarding observed regularity in a set of observations generated by a measurement procedure. The equation sets forth a theory of item-score variation and simultaneously pro vides the vehicle for confirmation or falsification of the theory. Just as important, alternative theoretical formulations can be tested side by side with criteria for priority unambiguously stated.

As an example, suppose a researcher develops a measure of "career maturity" and asserts that the 100-item instrument measures "student's awareness and knowledge about the world of work and his role in it." Further suppose that another researcher examines the instrument and concludes that it is merely a poor measure of "reading comprehension." Employing the perspective encouraged in this paper, both researchers would be charged with building specification equations that embody the two competing construct theories. Both equations would be applied to the same dependent variable (i.e., same set of item scores) and the data analyzed. If the reading comprehension theory predominated, the "career maturity" theorist could revise his instrument systematically controlling reading comprehension while at the same time either modifying the specification equation to account better for observed item scores or revising items so that they conform better to predictions generated from the construct theory. Whatever the outcome of this hypothetical research, the direct confrontation between construct theories would have sharpened both theory and instrumentation.

Note that under the present formulation a construct is not operationally defined by a particular instrument but rather by a corresponding specification equation. The construct validity of a score interpretation is assessed, in part, by how closely observations generated from a measurement procedure conform to predictions made by the specification equation. If a specification equation explains a substantial proportion of variation in item scores, then person scores can be interpreted in light of the construct theory with increased confidence that such interpretations are valid. A high

---

[2] An "item difficulty" is the p value associated with an instrument and represents the proportion of respondents answering the item correctly (i.e., .68 or .74). Common sense suggests that such an index be termed "item easiness" but tradition prevails. And "item score" analogous to a "person score" is a scaled transformation of the item difficulty. Under the Rasch model the "items scores" and the "person scores" are expressed in comparable scale units. In addition, the "item scores" are not dependent on any particular sample of persons, and similarly the "person scores" are not dependent on any particular sample of items.

R^2 increases confidence in the construct theory, as well as the particular measurement procedure under examination, whereas a low R^2 raises doubts about the construct theory, the instrument, or both.

Similarly, the construct validity of a single item is assessed by the magnitude of its residual. A small residual indicates that an item conforms well to specifications and is not measuring unspecified (i.e., unwanted) influences. A large residual suggests that unspecified sources of item variability account for an item's score. Note the constant interplay between formulation of theory and development of an instrument. Given a specification equation it is possible to develop highly valid items and discard invalid items with large residuals. Conversely, examination of item residuals may suggest new variables that can be used to modify a construct theory and improve the predictive power of the specification equation. Theory and instrument are thus bound in a dynamic interplay, with a change in one implying modification of the other.

Observations are generated through applications of measurement procedures. Such observations are meaningless unless interpreted by a construct theory. A good construct theory explains the observed regularity in a set of observations and thus imparts meaning to a set of person or item scores. Construct-specification equations aid us in making explicit our theories about what is being measured by a particular collection of items. As more comprehensive construct theories emerge, predictions about observations (on both the person and item face) become more precise, and broader generalizations are possible.

One pretender to legitimate explanation is the practice of describing the same thing in a new language and passing the translation off as a contribution to understanding. As, for example, when judges rate the complexity of a group of test items and this rating is then employed in accounting for variance in item scores (Kirsch & Guthrie, 1980). Unless the reasoning underlying the judgments is explicated, such activity contributes little to construct validation. An individual is not incapable of tenderness because he is unloving and an item is not difficult because it is complex. Such pseudo explanations are not only theoretically sterile but intellectually quite unsatisfying.

Construct-specification equations provide a useful, objective methodology for explicating the components which account for the complexity in a set of items. We turn now to an illustration of how a construct-specification equation can be built and used.

3   Illustration

To illustrate the development of a "construct-specification equation," we have conducted an analysis of data from the Knox Cube Test which purports to be a measure of visual attention and short-term memory (Arthur, 1947). The test uses five 1-in. cubes; four of the cubes are fixed 2 in. apart on a board, and the fifth cube is used to tap a series on the other four. To avoid confusion we will arbitrarily number the cubes from left to right as "1", "2", "3", and "4". The easiest item is a two-tap item in which the examiner taps cube 2, then cube 3. The most difficult item is a seven-tap item in which the sequence is 4, 1, 3, 4, 2, 1, 4.

Theoreticians have conceptualized short-term memory as an individual's "working" memory span. It is the immediate memory in a person's cognitive system which appears to have a very limited capacity in terms of the amount of material that can be retained and the length of time it can be retained. The limited nature of short-term memory, in terms of volume of storage, tends to be around $7 + 2$ bits of information (Miller, 1956), while the limit in terms of time appears to be 18 to 20 s. (Gagne, 1977).

Most psychologists have postulated that material is lost (forgotten) from short-term memory due to two factors, interference and/or time decay. Due to limited working capacity, the strength of memory traces is diminished as more bits of information are presented. As more demands are placed upon short-term memory, the recall of any particular memory trace becomes more difficult. That is, as the background complexity increases, the probability of retrieving a particular trace (signal) from the background (noise) is reduced.

The second determinant of loss from short-term memory is time. The probability of retrieving an item from short-term memory is negatively correlated with the passage of time. The longer an item stays in memory, the weaker it becomes, until at some point it is completely lost.

To assess the extent to which the Knox Cube Test measures the construct, "short-term memory," item scores computed on a sample of 101 subjects ages 3–16 year. were subjected to the previously described analysis. First, the items were arranged in order of difficulty from the easiest to the most difficult. Second, each item was carefully examined in an attempt to ascertain whether items differed from one another in ways consistent with a "short-term memory" construct interpretation. In trying to identify characteristics of the Knox Cube Test items which accounted for their difficulty, we selected the following attributes: (1) number of taps, (2) number of reversals, and (3) total distance covered. Table 1 presents difficulty indices for the items along with the tapping sequence and data on the three identified facets of each item.

The descriptive statistics for these variables are presented in Tables 2 and 3. Table 2 presents the mean, standard deviation, and range for each variable, while Table 3 presents the correlation matrix.

An examination of these two tables provides several characteristics of the Knox Cube Test. First, the three identified facets (number of taps, number of reversals, and distance covered) seem to be important determinants of item difficulty. As the zero-order correlations indicate, items become harder as the distance covered increases ($r = 0.95$), as the number of taps increases ($r = 0.94$ and as the number of reversals increases ($r = 0.87$). Second, there is substantial multicollinearity among the independent variables. For example, as the number of taps increases, there are concomitant increases in the distance covered ($r = 0.90$) and the number of reversals ($r = 0.82$).

To develop and refine the construct specification equation, a series of multiple regression analyses were conducted. First, hierarchical set-wise multiple regression was utilized in which the three main effects were entered as a first set, and the

Table 1 Distributions of 106 children's responses to the Knox Cube Test (taken from best test design (Wright & Stone, 1979))

Item no	Tapping sequence	Rescaled Rasch item difficulties	No. of taps	No. of reversals	Distance covered
2	2-3	1.0	2	0	2
3	1-2-4	1.4	3	0	4
7	1-2-3-4	3.0	4	0	4
4	1-3-4	3.2	3	0	4
6	3-4-1	4.5	3	1	5
5	2-1-4	4.7	3	1	5
8	1-4-3-2	4.7	4	1	6
9	1-4-2-3	5.2	4	2	7
10	1-3-2-4	5.2	4	2	6
15	1-2-3-4-3	5.7	5	1	5
11	2-4-3-1	6.2	4	1	6
13	2-1-4-3	6.4	4	2	6
16	1-2-3-4-2	65	5	1	6
14	4-2-1-3	6.7	4		6
12	3-1-4-2	7.1	4	2	8
18	1-3-2-4-3	8.6	5	3	7
17	1-3-1-2-4	9.2	5	2	8
19	1-4-3-2-4	9.2	5	2	8
20	1-4-2-3-4-1	10.1	6	3	11
21	1-3-2-4-1-3	10.6	6	4	11
22	1-4-2-3-1-4	10.6	6	4	12
23	1-4-3-1-2-4	11.7	6	2	10
25	3-2-4-1-3-4-2	13.0	7	4	12
24	4-1-3-4-2-1-4	13.5	7	3	13

Table 2 Means, standard deviations, and range of four item facets of the Knox Cube Test

Variable	M	SD	Range
Item difficulty	7.00	3.44	1–13.5
Number of taps	4.54	1.32	2–7
Number of reversals	1.75	1.26	0–4
Distance covered	7.17	2.94	2–13

Table 3 Correlations among the four item facets of the Knox Cube Test

	No. of taps	No. of reversals	Distance covered
Item difficulty	0.94	0.87	0.95
Number of tabs		0.82	0.90
Number of reversals			0.90

Table 4 Results of multiple regressional analysis of item difficulties on item characteristics

	df	SS	R2	F	p	Predictors	Beta	F
Regression	2	253.23	0.93	139.8	0.01	Distance covered	0.56	14.8
Res, Jual	21	19.03				Number of taps	0.43	8.8
Total	23	272.56						

three two-way interactions were entered as a second set. The first set of main effects accounted for 93% of the variance in item difficulties. Although the addition of the set of interactions increased the variance explained to 96%, the incremental variance accounted for was not statistically significant ($F = 4.7$). In the second analysis, the three independent variables were entered in a stepwise fashion. The results of this analysis are presented in Table 4.

An inspection of Table 4 indicates that the construct-specification equation requires only two variables, distance covered and number of taps. With this two-variable model, we can account for 93% of the variance in item difficulties; we have a firm understanding of what is governing item difficulties on the Knox Cube Test. This model predicts that items become more difficult as the distance covered is increased and the number of taps is increased. Since there is a high degree of multicollinearity among the predictors, the regression results were subjected to commonality analysis. The commonality analysis indicated that each variable accounted for only a small amount of unique variance. The unique contribution of distance covered and the number of taps was 0.05 and 0.03 respectively. The variance shared in common between the two variables was 0.85.

The empirically generated two-variable model of the Knox Cube Test is consistent with our theoretical understanding of the processes which account for loss of information from short-term memory. As previously discussed, the loss of information can be attributed to two factors, interference and time decay, which are highly interrelated. Distance covered corresponds directly with interference, and the number of taps serves as a proxy for time decay. Interference, which can be conceptualized as the ratio of signal to noise, is represented by the distance covered in a sequence. A two-tap sequence such as "2–3" is easier than a two-tap sequence which is spread out over more blocks, such as "1–4." In the latter case, there is more background noise which the individual must process and filter out.

The second factor, time decay, also has its analogue in the item components in terms of the number of taps. As the number of taps increases, the string of blocks to be recalled increases the memory load not only in terms of number but also in terms of time. That is, the second individual block tapped in a four-tap sequence must be held in memory twice as long as the second block in a two-tap sequence.

The construct-specification equation derived from the Knox Cube Test provides satisfying evidence that it is measuring what it purports to measure. From this illustration, we now have a more clear and precise understanding of what characteristics govern the item difficulties. Difficulty is determined by two item characteristics (distance covered and number of taps) which conform nicely to the previously discussed construct theory. In addition to providing confirming evidence for the construct theory, the specification equation has an immediate practical application. New items, with predictable item difficulties, can be quickly and easily generated by manipulating the specification equation.

The construct-specification equation provides a means for testing different conceptualizations of a hypothesized construct. For example, if a researcher suspects that a self-concept instrument is actually measuring verbal reasoning and social desirability, then this hypothesis can be empirically tested by translating the hypothesis into a specification equation. If the new equation reflecting the parameters of verbal reasoning and social desirability of the items ac counts for a large percentage of the variance, then the test is not measuring the construct its authors intended. Second, the development and utilization of con struct-specification equations permits the test designer to control systematically unwanted sources of variation. If a wide range vocabulary scale was contaminated by items requiring fine discriminations among word meanings, the resulting construct-specification equation could be employed to create items which systematically controlled such influences.

Note that issues of item bias and culture-fairness are easily addressed under this procedure. LISREL (Joreskog, 1976) can be used to test the hypothesis that the same item-specification equation fits black, white, and Spanish populations. If it does not, then items with high or low residuals can be identified and either modified or discarded. Through an iterative procedure, an equation that fits all language and cultural groups might be identified, ensuring that the measurement procedure is valid (i.e., measures the same thing) for each group.

An admittedly untested intuitive leap is that for each component identified as an important determinant of item scores, there exists an analogous person characteristic. Such an inference could be tested by developing relatively pure component tests and determining how well person scores on these components predict person scores on the more componentially complex instrument. Profiles of person scores on these components might prove more useful to teachers and diagnosticians than current score profiles computed from scores of unknown composition.[3]

---

[3] While conceptually very similar to componential analysis (Sternberg, 1977, 1980), the construct-specification equation approach differs in terms of its orientation and level of inference. Although both methodologies employ linear regression models, componential analysis focuses upon the cognitive process (components) that occur in a task, whereas construct-specification equations focus on concrete observable characteristics of the task (i.e., item).

4 Conclusion

As Cronbach (1957) has noted, a test interpretation is a claim that a test measures a construct:

> To decide how well a purported test of anxiety measures anxiety, construct validation is necessary, i.e., we must find out whether scores on the test behave in accordance with the theory that defines anxiety. This theory predicts differences in anxiety between certain groups, and traditional correlational methods can test those predictions. But the theory also predicts variation in anxiety, hence in the test score, as a function of experiences or situations, and only an experimental approach can test those predictions (p. 767).

We believe that much can be gained from a more balanced perspective than that suggested by Cronbach and others with its nearly exclusive emphasis on person score variation. Let us return the stimulus (i.e., item) to its rightful place on the throne and let it share equally in the load of testing construct theories. Item-score variation is far more than an "accidental fact" and much can be learned about how and why individuals vary from an examination of how and why items vary.

References

Arthur, G. (1947). *A point scale of performance tests.* Psychological Corp.

Buros, O. K. (1977). Fifty years in testing: Some reminiscences, criticisms, and suggestions. *Educational Researcher, 6,* 9–15.

Carroll, J. B. (1976). Psychometric tests as cognitive tasks: A new structure of intellect. In L. B. Resnick (Ed.), *The nature of intelligence* (pp. 27–56). Erlbaum.

Cronbach, L. J. (1957). The two disciplines of scientific psychology. *American Psychologist, 12,* 671–684.

Cronbach, L. J. (1971). Test validation. In R. L. Thorndike (Ed.), *Educational measurement* (2nd ed., pp. 443–507). American Council on Education.

Cronbach, L. J., & Meehl, P. E. (1955). Construct validity in psychological tests. *Psychological Bulletin, 52,* 281–302.

Gagne, R. M. (1977). *The conditions of learning.* Holt, Rinehart & Winston.

Jensen, A. R. (1980). *Bias in mental testing.* Free Press.

Joreskog, K. G. (1976). *Structural equation models in the social sciences: Specification, estimation and testing.* Uppsala Univer., (Research Report 76–9).

Kirsch, I. L., & Guthrie, J. T. (1980). Construct validity of functional reading tests. *Journal of Educational Measurement, 17,* 81–93.

Kuhn, T. S. (1970). *The structure of scientific revolutions* (2nd ed.). Univer. of Chicago Press.

Miller, G. A. (1956). The magical number seven, plus or minus two: Some limits on our capacity for processing information. *Psychological Review, 63,* 81–97.

Pellegrino, J. W., & Glaser, R. (1982). Analyzing aptitudes for learning: Inductive reasoning. In R. Glaser (Ed.), *Advances in instructional psychology* (pp. 269–345). Erlbaum.

Sternberg, R. J. (1977). *Intelligence, information processing, and analogical reasoning: The componential analysis of human abilities.* Erlbaum.

Sternberg, R. J. (1980). Factor theories of intelligence are all right-almost. *Educational Researcher, 9,* 6–13.

Thorndike, R. L., & Hagen, E. P. (1965). *Measurement and evaluation in psychology and education.* Wiley.

Thurstone, L. L. (1923). The stimulus-response fallacy in psychology. *Psychological Review, 30*, 354–369.
Whitely, S. E. (1981). Measuring aptitude processes with multicomponent latent trait models. *Journal of Educational Measurement, 18*, 67–84.
Wright, B. D., & Stone, M. H. (1982). *Best test design*. MESA Press. Accepted 31 July 1982.

Toward a Theory of Construct Definition

A. Jackson Stenner, Malbert Smith III, and Donald S. Burdick

The process of ascribing meaning to scores produced by a measurement procedure is generally recognized as the most important task in developing an educational or psychological measure, be it an achievement test, interest inventory, or personality scale. This process, which is commonly referred to as construct validation (Cronbach, 1971; Cronbach & Meehl, 1955; ETS, 1979; Messick, 1975, 1980), involves a family of methods and procedures for assessing the degree to which a test measures a trait or theoretical construct.

Theorists and practitioners have not yet articulated a unified and systematic approach to establishing the construct validity of score interpretations. Contemporary measurement procedures are not, in general, supported by more convincing evidence that they are "measuring what they purport to measure" than instruments developed 50 years ago. The absence of persuasive, well documented construct theories can be attributed, in part, to the lack of formal methods for stating and testing such theories.

An expanded version of this paper was presented at the Fifth International Conference on Educational Testing, Sterling, Scotland, June 30, 1982. We wish to thank the participants of the Scotland symposium, along with the anonymous reviewers of JEM for helpful comments and suggestions. We also thank E. B. Page, J. B. Carroll, and W. G. Katzenmeyer for their comments and suggestions.

A. J. Stenner (✉) · M. Smith III
Computerland, Durham, NC, USA

D. S. Burdick
Department of Mathematics, Duke University, Durham, NC, USA

© The Author(s) 2023   43
W. P. Fisher and P. J. Massengill (eds.), *Explanatory Models, Unit Standards, and Personalized Learning in Educational Measurement*,
https://doi.org/10.1007/978-981-19-3747-7_4

Until the developers of educational and psychological instruments can adequately explain variation in item scale values[1] (i.e., item difficulty). The understanding of what is being measured will remain unsatisfyingly primitive. In contrast, most approaches to testing and elaborating construct theories focus on explaining person score variation. Cronbach (1971), for example, in his review of construct validity, emphasizes interpretation of correlations among person scores on theoretically relevant and irrelevant constructs. A related approach is the use of multimethod-multitrait matrices of correlations between person scores (Campbell & Fiske, 1959). Explaining variation in item scale values has rarely been used as a source of information about construct validity.

The rationale for giving more attention to variation in item scale values is straightforward. Just as a person scoring higher than another person on an instrument is assumed to possess more of the construct in question (e.g., visual memory, reading comprehension, anxiety), an item (or task) that scores higher in difficulty than another item presumably demands more of the construct. The key question deals with the nature of the "something" that causes some persons to score higher than other persons and some items to score higher than other items. The process by which theories about this "something" are tested is termed "construct definition," where "definition" denotes specification of meaning (Kaplan, 1964).

Stenner and Smith (1982) illustrate construct definition theory using data from the Knox Cube Test (KCT). The KCT is purported to measure visual attention and short-term memory (Arthur, 1947). Five cubes are placed in a row before the examinee, who is to repeat various tap sequences, ranging from two-tap to seven-tap sequences.

To assess the extent to which the KCT measures short-term memory, item scale values based on a sample of 101 subjects were analyzed. A two-variable equation accounted for 93% of the variance in item scale values. The two variables, number of taps and distance covered between taps, have direct analogues in several theories about how information is lost from short-term memory (Gagne, 1977; Miller, 1956). That is, as the number of taps increases and the distance (in number of blocks) between tapped blocks increases, the combined effects of decay and interference serve to make the tap sequence more difficult.

Traditional approaches to testing construct theories analyze between-person variation on a construct. By examining relationships of other variables to construct scores, progressively more sophisticated construct theories are developed. As noted above, the study of relationships between item characteristics and item scale values is much less developed. Thurstone (1923) appears to have set the stage for a half century of neglect of this alternative when he argued:

---

[1] "Item difficulty" is the p value associated with an item and represents the proportion of respondents answering that item correctly (e.g., 0.38 or 0.74). Common sense would suggest that such an index be termed "item easiness," but tradition prevails. Under the Rasch model, the "item scale values" and "person scores" are expressed in comparable scale units. In addition, the "item scale values" are not dependent on any particular sample of persons, and similarly, the "person scores" are not dependent on any particular sample of items.

I suggest that we dethrone the stimulus. He is only nominally the ruler of psychology. The real ruler of the domain which psychology studies is the individual and his motives, desires, wants, ambitions, cravings and aspirations. The stimulus is merely the more or less accidental fact (p. 364).

Considering the symmetry of the person and item perspectives, it is somewhat baffling that early correlationists were so successful in focusing the attention of psychometricians on person variation. The practice, adopted early in this century, of expressing the row (i.e., person) scores as raw counts and column (i.e., item) scores as proportions may have distorted the symmetry, leading early test constructors, with few exceptions, to ignore variation in item scores as a source of knowledge about a measurement's meaning. In terms of construct definition theory, the regularity and pattern in item-scale-value variation is no less deserving of exploration than that found in person-score variation. Only historical accident and tradition have blinded educators and psychologists to the common purpose in the two forms of analysis.

1 Terminology

Constructs are the means by which science orders observations. Educational and psychological constructs are generally attributes of people, situations, or treatments presumed to be reflected in test performance, ratings, or other observations. We take it on faith that the universe of our observations can be ordered and subsequently explained with comparatively small number of constructs. Thurstone (1947) states:

> Without this faith, no science could ever have any motivation. To deny this faith is to affirm the primary chaos of nature and the consequent futility of scientific effort. The constructs in terms of which natural phenomena are comprehended are man-made inventions. To discover a scientific law is merely to discover that a man-made scheme serves to unify, and thereby to simplify, comprehension of a certain class of natural phenomena (p. 51).

The task of behavioral science is to recast the seemingly unlimited number of variously correlated observations in terms of a reduced set of constructs capable of explaining variation among observations. For example, "Why does person A have a larger receptive vocabulary than person B?", and "Why is the meaning of the word 'altruism' known by fewer people than the word 'compassion'?" The term "receptive vocabulary" is a construct label that refers simultaneously to the attribute and the construct theory which offers a provisional answer to these two questions. *Construct definition* is the process whereby the meaning of a construct, such as "receptive vocabulary," is specified. We impart meaning to a construct by testing hypotheses suggested by construct theories.

The meaning of a measurement depends on a construct theory. The simple fact that numbers are assigned to observations in a systematic manner implies some hypothesis about what is being measured. *Instruments* (e.g., tests) are the link between theory and observation, and scores are the readings or values generated by instruments. The validity of any given construct theory is a matter of degree, dependent, in part, on how well it predicts variation in item scale values and person scores and the breadth of these predictions.

Educational and psychological instruments are generally collections of stimuli hypothesized to be *indicators* of a particular attribute. An instrument (or individual item) may be characterized as being more or less well-specified depending upon how well a *construct specification equation* explains observed variation in item scale values. The specification equation links a construct theory to observations. As Margenau (1978) stated:

> If observation denotes what is coercively given in sensation, that which forms the last instance of appeal in every scientific explanation or prediction, and if theory is the constructive rationale serving to understand and regularize observations, then measurement is the process that mediates between the two, the conversion of the immediate into constructs via number or, viewed the other way, the contact of reason with nature (p. 199).

The construct specification equation affords a test of fit between instrument-generated observations and theory. Failure of a theory to account for variation in a set of item scale values invalidates the instrument as an operationalization of the construct theory and limits the applicability of that theory. Competing construct theories and associated specification equations may be suggested to account for observed regularity in item scale values. Which construct theory emerges (for the time) victorious, depends upon essentially the same norms of validation that govern theory evaluation in other sciences.

Often, a construct theory begins as no more than a hunch about why, after exposure to one another, items and people order themselves in a consistent manner. The hunch evolves as hypotheses about item and person variation, suggested by a construct theory, are systematically tested. The specification equation provides a direct means for hypotheses to interact with measurement such that construct theory and measurement each force the other into line (Kuhn, 1961).

Several terms to be used later require brief discussion. A *response model* is a mathematical function that relates the probability of a correct response on an item to characteristics of the person (e.g., ability) and to characteristics of the item (e.g., difficulty). Several such models are discussed by Lord (1980). Two of the most popular are the one-parameter Rasch Model (Wright & Stone, 1979) and the three-parameter logistic model (Birnbaum, 1968). In the former, person ability and item difficulty are viewed as sufficient to estimate the probability of a correct response. In the latter, two additional item characteristics, item discrimination and a guessing parameter, are used.

The response model does not embody a construct theory. The fit of data to a particular response model can be tested without knowledge of where the data came from, whether the objects are people, nations, or laboratory animals, or whether the indicants (i.e., items) are bar presses, scored responses to verbal analogy questions, or attitudinal responses. Nothing in the fit between response model and observation contributes to an understanding of what the regularity means. In this sense, the response model is atheoretical. Once a set of observations has been shown to fit a response model, the important task remains of ascribing meaning to scaled responses. In a way, the distinction is similar to that between classical reliability and validity. Like a well-fitting response model, high reliability suggests that "something" is being measured; but what that "something" is remains to be specified.

Under the present formulation, a construct can, in large part, be *defined* by the specification equation, whereas a test is simply a set of items from the structured universe bounded by the equation. This position might be interpreted as a neo-operationalist perspective on the relationship between construct and test. A more classical operationalism is exemplified in Bechtoldt's (1959) critique of construct validity: "Each test defines a separate ability; several tests involving the 'same content' but different methods of presenting the stimuli or of responding likewise define different abilities" (p. 677). Our position is that two tests measure the same construct if essentially the same specification equation can be shown to fit item scale values from both tests. Note that if the equation fits, it is irrelevant that these two instruments differ radically in name, method of presentation, scoring, scaling, or superficial appearance of the items. Similarly, two nominally similar tests that appear to the naked eye to measure the same construct may found to require different theories and construct labels. Such has been found to be the case in our research on certain norm-referenced reading comprehension tests. One publisher varies item difficulties by manipulating vocabulary, sentence length, and syntactic complexity of the reading passages. Another publisher, however, standardizes these variables over passages within a test level, and manipulates the logical complexity of the questions asked about each passage. Substantially different specification equations are needed to explain variation in item scale values on the two kinds of tests. As the example indicates, even careful, expert examination of educational and psychological tests may lead to erroneous conclusions about what a test does or does not measure.

2 Advantages of Focusing on Variation in Item Scale Values

Construct definition theory assigns considerable importance to explaining variation in item scale values. Availability of a good fitting specification equation is viewed from both theoretical and practical perspectives as an absolutely essential feature of a measurement procedure. There are at least four advantages to focusing on item-scale-value variation as well as person-score variation in ascribing meaning to a construct.

1. Stating theories so that falsification is possible. There is an obvious need throughout the behavioral sciences for falsifiable construct theories that are broad enough to yield predictions beyond those suggested by common sense. Most verbal descriptions of constructs, or construct labels (with their trains of connotations and surplus meaning), are poor first approximations to theories regarding what a construct means or what an instrument measures. These verbal descriptions seldom lead to definite predictions and us are not susceptible to challenge or refutation. Outside of the response-bias literature (Edwards, 1953, 1970), few studies offer testable alternative theoretical perspectives on what an instrument measures or what a construct means. Yet, history reveals that this type of "challenge research" is precisely what fosters intellectual revolutions (Kuhn, 1970). The behavioral sciences need more of the type of conflict such studies engender.

A suggested test interpretation is generally a claim that the measurement procedure measures a certain construct. Implicit in the use of any such procedure is a theory regarding the construct being measured. A major problem with the current state of educational and psychological measurement is that it is not at all clear how such claims about construct meaning can be falsified. Construct theories should be regarded as tentative, transient, and inherently doomed to falsification. A major reason for emphasizing item scale values is that theories about the meaning of a construct can be precisely stated in the form of a specification equation. Such an equation embodies a theory about the universe from which items have been sampled and simultaneously provides an objective means of confirming or falsifying theories about the meaning of scores.

2. Generalizability of dependent and independent variables. Estimated item scale values are typically more generalizable (i.e., reliable) than are person scores. In a simple person by items (p. xi) generalizability design with person as the object of measurement, the error variance is divided by the number of items, whereas, if item is viewed as the object of measurement, the error variance is divided by the number of people (Cardinet et al., 1976). Because most data-collection efforts involve many more people than items, the item scale values are typically more generalizable than are person scores. In most studies involving psychological instruments where the number of people is equal to or greater than 400, the generalizability coefficient for item scale values is equal to or greater than 0.90. When items from nationally normed instruments are examined, the generalizability coefficient for item scale values generalizing over people and occasions approaches unity.

Many independent variables in construct specification equations can be measured with a high degree of precision. Thus, in studying the item face of the person by items matrix, it not unusual to analyze a network of relationships among variables where each variable has an associated generalizability coefficient (under a broad universe definition) approaching unity. For those researchers accustomed to working with error-ridden variables, this new perspective can be refreshing.

3. Ease of experimental manipulation.[2] Items are docile and pliable subjects. They can be radically altered without informed consent and can be analyzed and

---

[2] Although a correlational perspective is adopted in this paper, it should be noted that any number of other procedures including experimental designs are easily incorporated under the proposed methodology. Typically, a construct definition study of an existing instrument or instruments will employ regression or path analytic techniques, whereas a construct definition study of an instrument under development might employ more traditional experimental methods. The objective in either case is to understand the features of items responsible for the lawful behavior of item scale values.

reanalyzed as new theories regarding their behavior are developed. Controlled experiments can be conducted in which selected item characteristics are systematically introduced and their effects on item behavior assessed. Because the experimenter can control items better than people, causal inference is simpler when the focus is on items. Items are passive; people are active subjects on whom only minimal experimental control can be exercised. All in all, items are better subjects for experimentation than people. Effects can be estimated and interpreted with less ambiguity, and the direct experimental manipulation is less costly and more efficient. Finally, this new perspective should breathe life into Cronbach's (1957) expressed hope that the experimental method will become a proper and necessary means of validating score interpretations.

4. Intentions can be explicitly tested. The notion that a test is valid if it measures what it purports to measure implies that we have some independent means of making this determination. Of course, we usually do not have an independent standard; consequently, validation efforts devolve into circular reasoning where the circle generally possesses an uncomfortably small circumference. Take, for example, Nunnally's (1967) statement, "A measuring instrument is valid if it does what it is intended to do" (p. 75). How are we to represent intention independent of the test itself? In the past, educators and psychologists have been content to represent intentions very loosely, in many cases letting the construct label and its fuzzy connotations reconstruct the intentions of the test developers. Unfortunately, when intentions are loosely formulated, it is difficult to compare attainment with intention. This is the essence of the problem faced by classical approaches to validity. Until intentions can be stated in such a way that attainment can be explicitly tested, efforts to assess the adequacy of a measurement procedure will be necessarily characterized by post hoc procedures and interpretations. The next section shows that construct specification equations offer a straightforward means of explicitly stating intentions and testing attainment.

3   Building a Theory of Receptive Vocabulary

In our daily lives, we use language to communicate with ourselves and others, to express emotions, to instruct, to abuse, to convey insight and intensity of feeling. Language is useful in enabling us to remember, in facilitating thought, in creating, in framing abstractions, and in adopting fresh perspectives.

To illustrate construct definition theory, we have started on a theory for receptive vocabulary. The theory attributes difficulty of items to three characteristics of stimulus words: (1) common logarithm of the frequency of a word's appearance in large samples of written material; (2) the degree to which the word is likely to be encountered across different content areas (e.g., mathematics, social studies, science, etc.); and (3) whether the word is abstract or concrete. The theory is restricted in application to pictorial representations of primary word meanings in the English language, and to nouns, verbs, adjectives, and adverbs. The construct theory emphasizes the importance of frequency of contact with words in context and thus is an exposure theory of receptive vocabulary. Words that appear frequently in writing or in speech are more likely to be learned, and thus are less difficult than are words that appear less frequently. Similarly, words that are less dispersed over a range of content areas (e.g., mathematics, cooking, music) are more difficult than words that are more widely distributed. Finally, for words of equal frequency and comparable dispersion, abstract words (i.e., referent cannot be touched, tasted, smelled, seen, or heard) are more difficult than concrete words.

The construct theory predicts that: (a) words can be scaled from easy to hard; (b) location on the scale will be highly predictable from a specification equation combining the three variables described above: and (c) person scores will correlate with any variable that reflects exposure to language in the environment. Research support for the last prediction can be marshalled from several suggestive studies that have found relationships between a child's verbal ability and the verbal ability of the teacher (Hanushek, 1972) and parents (Jordan, 1978; Kelly, 1963; Shipman & Hess, 1965; Wolbert et al., 1975). An illustration of how a construct specification equation is built and tested follows.

The instrument used in the illustration is the Peabody Picture Vocabulary Test-Revised (PPVT-R), Forms Land M (Dunn & Dunn, 1981).[3] The authors state: "The PPVT-R is designed primarily to measure a subject's receptive (learning) vocabulary for Standard American English. In this sense, it is an achievement test, since it shows the extent of English vocabulary acquisition" (p. 2). Each item has four high quality, black-and-white illustrations arranged in a multiple-choice format. The subject's task is to select the picture that best illustrates the meaning of a word spoken by the examiner.

The Rasch item scale values for the 350 words of the PPVT-R were used as the dependent variable in the analysis. The three predictor variables were word frequency, dispersion, and abstractness. Word frequency and dispersion of each word were obtained by referencing each keyed response word in the extensive word sample of Carroll et al. (1971). The Carroll et al., sample is based upon words selected from schoolbooks used by third- to ninth-grade students in the United States. The third measure, abstraction, was based upon independent ratings of each stimulus picture by two educational psychologists. Each of these predictors is described in more detail below.

1. *Log frequency.* Our early work used the common log of the frequency with which a word appeared in a running count of 5,088,721 words sampled by Carroll et al. (1971) from a broad range of written materials. We subsequently discovered that the log frequency of the "word family" was more highly correlated with word difficulty. Consequently, this variable was used in the latter stages of our work. A "word family" included: (1) the stimulus word; (2) all plurals (adding *s* or changing *y* to *ies*); (3) adverbial forms; (4) comparatives and superlatives; (5) verb forms (*s, d, ed, ing*); (6) past participles; and (7) adjective forms. The

---

[3] The authors thank Dr. Gary Robertson and American Guidance Service for providing norming data for the PPVT-R.

frequencies were summed for all words in the family and the log of that sum was entered in the specification equation.

2. *Dispersion.* This variable is a measure of dispersion of word frequencies over 17 subject categories specified by Carroll et al. (e.g., literature, mathematics, music, art, shop). It appears to be highly useful in differentiating words that are equally likely to appear in different kinds of verbal materials from words that tend to appear only in specialized subject matter. The values for dispersion range from 0 to 1.0. Dispersion approaches the value of 0 for words that are concentrated in only a few content areas, whereas dispersion approaches the value of 1.0 when the word is distributed fairly evenly over the entire range of contents.

3. *Abstract/concrete.* As noted above, the abstract/concrete ratings were made independently by two educational psychologists. Although several investigators have employed a Likert-type scale for the measurement of this variable (e.g., Paivio et al., 1968; Rubin, 1980), the operational definition of abstraction lends itself to a dichotomous classification. The dichotomy was created to differentiate words that refer to tangible objects (e.g., car, coin, ice cream) from words denoting concepts (e.g., transportation, money, sweet). Abstract words were assigned a value of 1, and concrete words were assigned a value of 0. Inter-observer agreement for the two forms was 92% on Form L and 95% on Form M.

Table 1 presents the means, standard deviations, intercorrelations, raw score regression equation, and R^2 for the PPVT-L and PPVT-M. The construct specification equation for Form L explains 72% of the observed score variance in item scale values. Frequency and dispersion are highly correlated ($r = 0.848$) indicating that less frequently appearing words are likely to be concentrated in a small number of content areas, whereas more frequently appearing words tend to be more uniformly distributed over content areas. The abstract/concrete dichotomy is moderately correlated with item scale values ($r = 0.352$) but negligibly correlated with frequency ($r = -0.033$) and dispersion ($r = -0.081$). All relationships are consistent with the exposure theory: Less frequently appearing words are more difficult than more frequently appearing words, and abstract words are, on average, more difficult than concrete words.

A comparison of Forms L and M reveals several similarities. First, there is less than 0.15 standard deviation difference in the means of item scale values, frequencies, and dispersions. The largest difference appears on the abstract/concrete dichotomy where Form M includes 67% abstract words and Form L only 60%. Second, the variances are remarkably similar. Third, the pattern of intercorrelations appears to be largely invariant over forms of the PPVT. Last, the proportions of variance explained are similar (Form L, $R^2 = 0.722$; Form M, $R^2 = 0.712$.

The combined Form L and M analysis is presented in Table 2. Item scale values for 350 items were regressed on the three variables in the construct specification equation. As might be expected, the results are very similar to those obtained in the individual Form L and M analysis. The R^2 on the aggregate data set is 0.713 and the pattern of intercorrelations remains consistent with the previous results. The

Table 1 Descriptive data and construct specification for forms L and M

	Rasch item scale value	Log of frequency	Dispersion	Abstract/Concrete	Rasch item scale value	Log of frequency	Dispersion	Abstract/Concrete
Rasch item scale value		-0.776	-0.736	0.352		-0.783	-0.791	0.359
Log of frequency			0.848	-0.033			0.870	-0.146
Dispersion				-0.081				-0.188
Mean	%0.94	1.72	0.49	0.60	98.82	1.80	0.52	0.67
S.D.	26.05	0.86	0.26	0.49	25.95	0.91	0.27	0.47
Regression coefficients		-17.53	-21.84	16.71		-11.59	-38.65	12.48
Standard error of coefficients		2.30	7.68	2.15		2.38	8.18	2.31
	$R''' = 0.722$				$Rl = 0.712$			

Table 2 Construct specification equation for combined forms PPVT-R

	Rasch item scale value	Log of frequency	Dispersion	Abstract/Concrete
Rasch item scale value		-0.779	-0.763	0.354
Log of frequency			0.859	-0.086
Dispersion				-0.130
Mean	98.88	1.76	0.51	0.64
S.D.	25.96	0.89	0.26	0.48
Regression coefficients		-14.57	-29.73	14.70
Standard error of coefficients		1.65	5.61	1.57
	$R' = 0.713$			

only discernible difference is in the expected reduction of the standard errors of the coefficients with the larger data set.[4]

The amount of shrinkage was examined by cross-validating each equation on data from the opposite form. When the Form L equation was applied to Form M data, the

---

[4] It should be noted that heterogeneity of the item set is not a satisfactory explanation for the explanatory powers of the specification equation. We have examined the fit of the equation to items with drastically restricted ranges and found the R^2 to be smaller (as we might expect), but not dramatically so. Furthermore, the form of the equation remained invariant.

R^2 fell slightly from 0.722 to 0.702. Likewise, application of the Form M equation to Form L data resulted in a decrement from 0.712 to 0.709. As the cross-validation studies indicate, the variables in the construct specification equation are stable and reliable.

In an attempt to improve the amount of variance explained, 50 additional variables were examined for inclusion in the specification equation. Among the variables considered were: (1) part of speech; (2) phonetic complexity of the word; (3) number of letters; (4) modal grade level at which the word appears in school materials; (5) distractor characteristics; (6) experiential frequency; (7) content classification of word; (8) semantic features; (9) frequencies based upon the Kucera and Francis (1967) corpus; and (10) numerous interactions and polynomials. The results of this search were disappointing in that only negligible improvement in fit was obtained when the number of variables in the equation was increased to 8, 10, or 12.

4 Conclusion

Throughout the history of educational and psychological measurement, there has been an abundance of attention to the statistical characteristics of measurement procedures and a disturbing lack of concern for the role that theory plays or should play in measurement. We have confused quality of measurement with statistical sophistication and high internal consistency coefficients. The related notions of construct theories and specification equations may provide a vehicle for restoring theory as the foundation of measurement in psychology and education. One conclusion seems certain: Much can be learned from attempts at building construct specification equations for the major instruments used in education and psychology. Our expectation is that there will be no shortage of surprises once such work begins.

References

Arthur, G. (1947). *A point scale of performance tests*. Psychological Corp.

Bechtoldt, H. P. (1959). Construct validity: A critique. *American Psychologist, 14*, 619–629.

Birnbaum, A. (1968). Some latent trait models and their use in inferring an examinee's ability. In F. Lord & M. Novick (Eds.), *Statistical theories of mental test scores*. Addison-Wesley.

Campbell, D. T., & Fiske, D. W. (1959). Convergent and discriminant validation by the multitrait multimethod matrix. *Psychological Bulletin, 56*, 81–105.

Cardinet, J., Tourneur, Y., & Allal, L. (1976). The symmetry of generalizability theory: Applications to educational measurement. *Journal of Educational Measurement, 13*, 119–134.

Carroll, J. B., Davies, P., & Richman, B. (1971). *Word frequency book*. Houghton Mifflin.

Cronbach, L. J. (1957). The two disciplines of scientific psychology. *American Psychologist, 12*, 671–684.

Cronbach, L. J. (1971). Test validation. In R. L. Thorndike (Ed.), *Educational measurement* (2nd ed.). American Council on Education.

Cronbach, L. J., & Meehl, P. E. (1955). Construct validity in psychological tests. *Psychological Bulletin, 52,* 281–302.

Dunn, L. M., & Dunn, L. M. (1981). *Manual for forms land M of the Peabody picture vocabulary test-revised.* American Guidance Service.

Edwards, A. L. (1953). *Edwards personal preference schedule.* Psychological Corp.

Edwards, A. L. (1970). *The measurement of personality traits by scales and inventories.* Holt, Rinehart, & Winston.

Educational Testing Service. (1979, October). *Construct validity in psychological measurement: Proceedings of a colloquium in theory and applications in education and employment.* Henry Chauncey Conference Center.

Gagne, R. M. (1977). *The conditions of/earning.* Holt, Rinehart, & Winston.

Hanushek, E. A. (1972). *Education and race: An analysis of the educational production process.* Lexington Books.

Jordan, T. E. (1978). Influences on vocabulary attainment: A five year prospective study. *Child Development, 49,* 1096–1106.

Kaplan, A. (1964). *The conduct of inquiry.* Chandler.

Kelly, S. (1963). The social world of the urban slum child: Some early findings. *American Journal of Orthopsychiatry, 33,* 823–831.

Kucera, H., & Francis, W. N. (1967). *Computational analysis of present day American English.* Brown University Press.

Kuhn, T. S. (1961). The function of measurement in modem physical science. In H. Woolf (Ed.), *Quantification: A history of the meaning of measurement in the natural and social sciences.* Bobbs-Merrill.

Kuhn, T. S. (1970). *The structure of scientific revolutions* (2nd ed.). University of Chicago Press.

Lord, F. M. (1980). *Applications of item response theory to practical testing problems.* Erlbaum.

Margenau, H. U. (1978). *Physics and philosophy.* Reidel.

Messick, S. (1975). The standard problem: Meaning and values in measurement and evaluation. *American Psychologist, 30,* 955–966.

Messick, S. (1980). Test validity and the ethics of assessment. *American Psychologist, 35,* 1012–1027.

Miller, G. A. (1956). The magical number seven, plus or minus two: Some limits on our capacity for processing information. *Psychological Review, 63,* 81–97.

Nunnally, J. C. (1967). *Psychometric theory.* McGraw-Hill.

Paivio, A., Yuille. J. C., & Madigan, S. A. (1968). Concreteness, imagery and meaningfulness values for 925 nouns. *Journal of Experimental Psychology, 76,* 1–25.

Rubin, D. C. (1980). 51 properties of 125 words: A unit of analysis of verbal behavior. *Journal of Verbal Learning and Verbal Behavior, 17,* 736–755.

Shipman, V., & Hess, R. D. (1965). *Social class and sex differences in the utilization of language and the consequences for cognitive development.* Paper presented at the meeting of the Midwestern Psychological Association.

Stenner, A. J., & Smith, M. (1982). Testing construct theories. *Perceptual and Motor Skills, 55,* 415–426.

Thurstone, L. L. (1923). The stimulus-response fallacy in psychology. *Psychology Review, 30,* 354–369.

Thurstone, L. L. (1947). *Multiple factor analysis.* University of Chicago Press.

Wolbert, M., Inglis, S., Kriegsmann, E., & Mills, B. (1975). Language delay and associated mother-child interactions. *Developmental Psychology, 11,* 61–70.

Wright, B. D., & Stone, M. H. (1979). *Best test design.* MESA Press.

Most Comprehension Tests Do Measure Reading Comprehension: A Response to McLean and Goldstein

A. Jackson Stenner, Ivan Horabin, Dean R. Smith, and Malbert Smith III

Abstract There is nothing wrong with the NAEP reading exercises, the sampling design, or the NAEP Reading Proficiency Scale, these authors maintain. But adding a rich criterion-based frame of reference to the scale should yield an even more useful tool for shaping U.S. educational policy.

In the January 1988 *Kappan*, Leslie McLean and Harvey Goldstein fired off a vitriolic attack on the design and interpretation of the National Assessment of Educational Progress.[1] In particular, they attacked the "fictional" properties of the NAEP Reading Proficiency Scale "RPS" and concluded with the bold assertion, "In reality, reading achievement is not unidimensional."[2]

Remarkably, this conclusion followed a summary of the NAEP's study on the dimensionality of reading, which concluded: "Overall, the four dimensionality analyses of the NAEP reading items indicate that it is not unreasonable to treat the data as unidimensional."[3] In addition it should be pointed out that the NAEP spent $150,000 trying to reject the unidimensionality hypothesis for reading,[4] and this work stands

---

[1] Leslie D. McLean and Harvey Goldstein, "The U.S. National Assessments in Reading: Reading Too Much into the Findings," *Phi Delta Kappan*, January 1988, pp. 369–372.

[2] McLean and Goldstein, p. 371.

[3] Albert E. Beaton, *Implementing the New Design: The NAEP 1983–84 Technical Report* (Princeton, N.J.: National Assessment of Educational Progress/Educational Testing Service, Report No. 15-TR20, 1987), p. 273. See also Rebecca Zwick, "Assessing the Dimensionality of NAEP Reading Data," *Journal of Educational Measurement*, Vol. 24, 1987, pp. 293–308.

[4] Albert Beaton, personal communication, May 1987.

Phi Delta Kappan. 69. 1988

A. J. Stenner (✉) · I. Horabin · D. R. Smith
MetaMetrics, Inc., Durham, NC, USA

M. Smith III
Computerland, Durham, NC, USA

© The Author(s) 2023
W. P. Fisher and P. J. Massengill (eds.), *Explanatory Models, Unit Standards, and Personalized Learning in Educational Measurement*,
https://doi.org/10.1007/978-981-19-3747-7_5

as probably the most carefully designed and skillfully implemented study of its kind in the literature.

For most of this century, debate over the nature of reading comprehension and over the behaviors measured by reading comprehension tests has waxed and waned. Why should such questions be so important? The answers to these questions directly affect the way we measure reading comprehension and the way we teach children to read.

Comprehension testing and reading instruction in this nation have been fragmented into hundreds of objectives, skills, and taxonomies that may well be misdirecting the efforts of reading teachers. The dissection of reading comprehension into a myriad of subskills has been fostered to a great extent by the way reading tests have been constructed. When we examined the reading comprehension tests in widespread use today, we are immediately struck by their diversity. The tests vary greatly in the subskills each claims to measure. By the same token, item formats vary from single sentences or paragraphs too lengthy passages. Some involve pictures; some, cloze responses; and some, separate questions about each passage. The kinds of questions that are asked of a student also appeared to vary, but some of this diversity may be the result of the taxonomies used to catalog these differences.

With the sponsorship of the National Institute of Child Health and Human Development "NICHHD", we conducted a study of 3,000 reading comprehension items appearing on 14 reading comprehension tests, one of which was the NAEP test. Our study asked the question, is there a common dimension being measured by these tests? To examine this question, we used the Lexile theory of reading comprehension,[5] which states that the difficulty of text can be predicted from knowledge of the familiarity/rarity of the vocabulary used and the demands the text places on short term memory.

For our study we correlated the difficulties of the test items (provided by published norms) with the difficulties of the items (as measured by a computer analysis using the Lexile theory). Table 1 shows the results of our computations, which produced an average correlation of 0.93.

It seems reasonable to conclude from these results that most attempts to measure reading comprehension—no matter what the item form, type of prose, or response requirement—end up measuring a common comprehension factor captured by the Lexile theory. In short, the majority of reading comprehension tests in use today measure a construct labeled "reading comprehension." Furthermore, it should be evident from Table 1 that the RPS is no more a fiction than our scales derived from the other 13 tests we examined.

The source of McLean and Goldstein's confusion regarding the NAEP can be traced to a failure to separate what a test measures from the usefulness of a score. By failing to distinguish between what the NAEP Reading Proficiency Scale measures and how useful it is as a communication device, McLean and Goldstein become their

---

[5] A. Jackson Stenner, Dean R. Smith, Ivan Horabin, and Malbert Smith, "Fit of the Lexile Theory to Item Difficulties on Fourteen Standardized Reading Comprehension Tests," paper presented at the National Reading Conference, St. Petersburg, FL, 1987.

Table 1 Correlations between the Lexile measure of difficulty and the standard scores for items from 14 reading comprehension Tests

Test	Correlation
California achievement test (Form C)	1.00
SRA achievement series	0.99
Comprehensive test of basic skills	0.98
Iowa tests of basic skills	0.98
Lexile test of reading comprehension	0.97
Peabody individual achievement test	0.97
California achievement test (Form E)	0.97
Metropolitan achievement test	0.95
NAEP	0.94
Gates-MacGinitie (Form 2)	0.91
Stanford achievement test	0.90
Gates-MacGinitie (Form 1)	0.89
Woodcock-Johnson battery	0.84
Woodcock-Reading mastery test	0.77

own biggest detractors. The important point they raise about the utili-ty of a score is obscured and eventually lost amid a welter of non sequiturs.

Bringing meaning to test scores is the single most important activity that test designers engage in. Two frames of reference for interpreting test scores are generally distinguished: norm referenced and criterion referenced.

A norm referenced framework for score interpretation involves locating a score in a normal distribution; usually that distribution is normed on a national sample, and the result is often expressed as a percentile. This is by far the most common means of imparting meaning to a score, but it has serious defects if used exclusively. Foremost among these defects is that knowing that a student reads at the 80th percentile for fourth-graders nationally does not help a teacher select instructional materials nor does it tell us what the student can or cannot read. The NAEP has done a splendid job of providing a normative frame of reference for reading test scores.

It is on the criterion-referenced side that improvement is both necessary and achievable. The approach to criterion referencing adopted by the NAE P is to equate points along the Reading Proficiency Scale and test items used in the assessment. Thus a score of 250 on the RPS is matched with the test items that students who scored 250 can answer correctly 80% of the time.

Such a strategy for imparting meaning to scores is clearly better than an exclusive reliance on norm referencing. But it is context-bound in that only items used in the assessment can be used in describing what a score at any given level means. What is needed is a method for building a criterion-based frame of reference that is free from the local context of the assessment. Such a frame of reference would enable us to ask, for example, what RPS ability score is required for a student to read a local newspaper with a 75% comprehension rate.

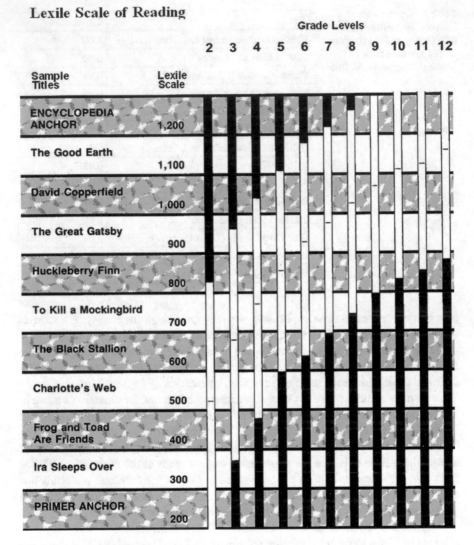

Fig. 1 Lexile scale of reading

Using Lexile theory, it is possible to construct a scale of the difficulty of "real world" prose encountered by fourth-grade students, high school seniors, or young adults trying to find a job. Test scores can be expressed as the expected comprehension rate for an individual encountering a text with a given predicted "theoretical" difficulty. An example of such a scale is found in Fig. 1.

The works listed in the far right column of Fig. 1 contain representative sample passages that differ in difficulty by 100 Lexile increments. (There are also two anchor titles that are used to establish the extremes of the scale.) These works represent the kinds of prose that individuals with corresponding Lexile scores can read with a

75% comprehension rate. To the right of the scale are bands represent-ing national percentiles for grades 2 through 12.

By juxtaposing the normative and criterion-based frames of reference, it is possible to see what a student with a given ability is able to read. This knowledge pro-vides the teacher with a way to operationalize the meaning of a test score. It can al-so help the teacher place that student in a basal series or to select appropriate supplemental materials. Finally, the criterion reference can provide policy makers with a tool that can be used to answer such questions as the following:

- What proportion of our nation's fourth-graders can read their fourth-grade basal readers with at least a 75% comprehension rate?
- What proportion of our nation's high school graduates can read USA Today with at least a 75% comprehension rate?
- What ability level is required to comprehend warning labels, tax forms, insurance policies, etc.?

In Literacy: Profiles of America's Young Adults, Irwin Kirsch and Ann Jungeblut note that "the important question facing our society today is not 'How many illiterates are there?' but rather "What are the nature and levels of literacy skills demonstrated by various groups in the population?" The question of the levels of skill can be answered by means of the existing norm-referenced frame-work. The question of the nature of those skills, however, requires a rich criterion-based frame of reference.

Thus the policy analyst in Washington, D. C., and the fourth-grade teacher in Omaha both want to know what students can read. A satisfactory answer to this question requires a criterion-based frame of reference that is independent of the as-sessment context. That Lexile framework can yield interpretations of scores in terms of what students can read at various levels of development, and this frame-work can easily be imposed on the existing RPS.

McLean and Goldstein argued that "to have relevance for policy, however, such assessments [as the NAEP] must use measures that are connected to teaching and learning." We hope that we have made it clear that it is through a criterion-based frame of reference for score interpretation that we link test scores to teaching, learning, and so-called "real-world" tasks.

There is nothing wrong with the NAEP reading exercises and sampling design- or with the RPS. Adding a rich criterion-based frame of reference to the RPS should address the objections of McLean and Goldstein and fashion an even more useful tool for shaping our nation's educational policy.

Measuring Reading Comprehension with the Lexile Framework

A. Jackson Stenner

Abstract Implicit in the idea of measurement is the concept of objectivity. When we measure the temperature using a thermometer, we assume that the measurement we obtain is not dependent on the conditions of measurement, such as which thermometer we use. Any functioning thermometer should give us the same reading of, for example, 75 °F. If one thermometer measured 40 °, another 250 and a third 150, then the lack of objectivity would invalidate the very idea of accurately measuring temperature.

1 Objectivity and the Idea of Measurement

It is this general objectivity that distinguishes physical science measurement from behavioral science measurement. General objectivity requires a construct theory embodied in a specification equation that is capable of estimating indicant calibrations. When these theory-based calibrations are employed in the Rasch model, observations (i.e., raw scores) can be converted into measures without relying on individual or group data on indicants (e.g., items) or objects of measurement (e.g., persons). The benefits of these methods include: (1) the construct theory is exposed to falsification, (2) it is possible to build correspondence tables between observations and measures with recourse only to theory, (3) a generalized linking solution is available for placing observations of all kinds on a common scale, (4) a reproducible unit of measurement can be developed, (5) the framework for fit statistics that is sample-dependent under the Rasch model becomes sample-independent, and (6) a complete frame of reference for measure interpretation can be constructed.

This paper shows how the concept of general objectivity can be used to improve behavioral science measurement, particularly as it applies to the Lexile Framework a

---

Paper Presented at the Fourth North American Conference on Adolescent/Adult Literacy, Washington, D.C. 1996.

A. J. Stenner (✉)
MetaMetrics, Inc., Durham, NC, USA

W. P. Fisher and P. J. Massengill (eds.), *Explanatory Models, Unit Standards, and Personalized Learning in Educational Measurement*, https://doi.org/10.1007/978-981-19-3747-7_6

63

tool for objectively measuring reading comprehension. We begin with a dialogue between a physicist and a psychometrician that details some of the differences between physical science and behavioral science measurement. Building on these distinctions, we offer a definition of measurement that describes what goes on in the physical sciences and that represents an attainable ideal of what should go on in the behavioral sciences. This definition of measurement is formalized in an equation that turns out to be the Rasch model, with the important difference that indicant calibrations are obtained via theory, not data. Through the use of theory-based calibrations, we achieve a generally objective estimation of the measure parameter in the Rasch model. The next section of the paper examines the differences between local objectivity obtained with the Rasch model and general objectivity obtained with a theory-enhanced version of that model. Next, we report on a 10-year study of reading comprehension measurement that implemented the concept of general objectivity through the development of the Lexile Framework. Finally, we summarize several of the benefits of objective measurement and general objectivity as they might be realized in the measurement of constructs other than reading comprehension.

2   The Problem of General Objectivity: A Dialog Between the Behavioral and Physical Sciences

2.1   Psychometrician

As I understand our purpose here today, each of us will solve a measurement problem specific to our discipline, and then discuss our respective methods. Since I opened the dialog, I will present you with your problem first.

In the box in front of you is a long thin glass tube open at one end, some mercury, and a ruler. Your task is to answer the following question: 'What change in temperature is indicated by an increase of five centimeters in column height of the mercury?'.

2.2   Physicist

I have completed my analysis and found that the answer is 18 ° Fahrenheit.

My method was straightforward. I calculated the volume of mercury by first calculating the capacity of the glass tube and then estimating the fraction of that capacity occupied by the mercury. Next, I consulted the table of expansion coefficients and found the equation for mercury. Given the observation (column height of mercury), I set up an equation and solved for the measure (i.e., temperature) corresponding to each of two volumes of mercury separated by the indicated five centimeters. For those of us trained in physics and chemistry, this is a standard measurement problem.

Through theory, a context is created for expressing an observation as a measure. The theory enables me to extract the relevant information from the context surrounding an observation (what some call an observation model) and use this information to convert observations to measures. The theory specifies what is essential to record about the context of an observation. All other features are either ignored as irrelevant (e.g., the time of day the thermometer is used) or are considered during the process of observation (e.g., removing the potential contaminating effects of barometric pressure by sealing the top of the thermometer).

Now here is your problem: In front of you is a newspaper, USA Today. Your task is to construct a 50-item reading comprehension test and compute the increment in reading comprehension that would be reflected in an increase from thirty to forty correct responses on the test. You may consult any other references you choose.

2.3 Psychometrician

I spent the first few hours generating the 50-item measure and the last two hours trying to understand how you were able to solve the temperature problem while I cannot begin to imagine how to solve the reading comprehension problem. If given more time, I would administer the test to, say, 200 high school seniors, analyze the data with an appropriate measurement model (e.g., the Rasch model) and report the difference between 60% correct and 80% correct on some appropriately transformed logit scale. I can see, however, that this is a very different process from the one you used in solving the temperature problem.

What I concluded is that I lack a theory and associated equations for transforming observations into measures. Where possible, I attempt to make up for this lack of theory by basing instrument calibrations on data instead of theory. Once I frame the problem in this way, it is clear that all I would need to accomplish what you did with the temperature problem is a good reading comprehension theory and associated equations for calibrating the items. Then I could solve the reading comprehension problem in a manner identical to the one you used in solving the temperature problem. What is your view?

2.4 Physicist

First of all, I don't think that most of my colleagues would agree that your data-based calibration procedure is "measurement". If you don't know enough about the items you use to specify their calibrations independent of data, how in the world can you be sure that you end up measuring what you intended to measure?

Furthermore, this data-based calibration procedure-used in lieu of theory-seems to impose a local boundedness on the measures you compute. It seems clear that a new 50-item reading comprehension test administered to a different group of high school seniors would yield measures on a different scale than did the first test. Each set of measures is bound to the particulars of the instrument and local context of measurement. As an aside, might this lack of objectivity in your measurement procedures partly explain the behavioral sciences' reliance on significance testing, correlation coefficients, meta-analytic methods and other metric-insensitive procedures? Could the excessive use of these methods be the inevitable consequence of working with measures that lack complete objectivity?

2.5 Psychometrician

In response to your first observation, we have elaborate procedures for validating inferences from behavioral science measurements. In general, the most persuasive evidence that we know what we are measuring comes from high correlations among instruments purporting to measure the same construct.

I will grant you that a construct theory capable of supporting theory-based item calibrations might prove more persuasive, but at present, we have few such theories. As for the second observation about lack of objectivity, we attempt to solve this problem, first, by employing the Rasch model which yields "local objectivity" and second, by using either a common-persons or common-items linking design, thus bringing all measures onto one scale. These procedures may appear primitive and cumbersome, but they do yield a kind of objectivity that is superior to what our measures have enjoyed in the past.

Finally, from the new perspective you have given me, I think it likely that, since we have no standard metrics for major behavioral science constructs, it is not surprising that we would gravitate towards measures of association and inference that ignore this failing.

2.6 Physicist

Let me try to end on a positive note, for I do believe there is hope.

It should be clear from the temperature problem that measures of temperature possess a special kind of objectivity (i.e., absolute measures are separated from the conditions of measurement). The unavoidable consequence of using data-based calibrations rather than theory- based calibrations when converting observations to measures is that the resulting measures are expressed in units which, although of equal intervals, possess a kind of location indeterminacy. The only means I know of to remove this indeterminacy is to use theory rather than data to calibrate instruments. If you can develop construct theories and associated equations capable of calibrating your instruments, it seems possible that your measures can aspire to the kind of objectivity enjoyed by temperature measurement.

3 Measurement Defined

Measurement is the process of converting observations into quantities through theory. Measurement as a "process" implies an "act of ascertainment of finding out" (Leonard 1962, p. 4). The term "observation" refers to the qualitative observation or count, such as the height of a column of mercury in a thermometer. The "quantity", or measure, is the number assigned to the attribute of the object being measured (e.g., a person). Quantities, unlike other collections of numbers, possess an additive conjoint property (Luce & Tukey, 1964). The term "theory" in this definition makes clear that "every instance of measurement presupposes an extensive background of explicitly confirmed, scientific theory" (Leonard 1962, p. 4).

A construct theory, which, in its more colloquial form, is just a story about what it means to move up and down a scale, is used to calibrate indicants. Examples of calibration include the placement of lines on the tube of a liquid-in-a-glass thermometer or the assignment of difficulty calibrations to a series of vocabulary test items. The theory creates a context in which the observation can be understood as an estimator for the measure. In the case of the attribute "reading comprehension", the "process" is the act of ascertaining the level of reading comprehension attained by a person.

Measurement is a process of which the product is a quantity. The "observation" is often a raw score or count correct on some set of items. The "quantity" is the amount of reading comprehension ability that a person possesses expressed in some metric. The actual conversion of observations into measures through theory is accomplished using the Rasch (1980) model, which states a requirement for the way that theory (expressed as item calibrations) and observations (count of correct items) interact in a probability model to make measures.

4 Local and General Objectivity

Objectivity is the foundation of valid measurement. Indeed, it is central to the whole idea of measurement. When I report a number to be used as a measure, the underlying assumption is that the measure is objective, that is, it has been sufficiently well-separated from the conditions of measurement that I can ignore these conditions when I report the measure. As noted earlier, if I report that it is 75 ° Fahrenheit, the inherent assumption is that the validity of this measure does not depend on any conditions of measurement, such as which thermometer was used.

As seen in the previous dialog, this objectivity, which is the foundation of physical science measurement, has not been achieved to the same degree in the behavioral sciences, although it has been sought after beginning with Thurstone:

> It should be possible to omit several test questions at different levels of the scale without affecting the individual score.

It should not be required to submit every subject to the whole range of the scale. The
starting point and the terminal point being selected by the examiner should not directly affect
the individual score. (Thurstone, 1926, p. 446)

It is clear that Thurstone believed that person measures should be independent of
the particular items used in the measurement instrument. What is not clear is whether
or not Thurstone intended that relative scores (i.e., differences) or absolute scores
(i.e., point locations) should be free of effects due to indicants. Two years later, he
stated:

The scale must transcend the group measured. One crucial experimental test must be applied
to our method of measuring attitudes before it can be accepted as valid. A measurement instrument must not be seriously affected in its measuring function by the object of measurement.
To the extent that its measuring function is so affected, the validity of the instrument is
impaired or limited. If a yardstick measured differently because of the fact that it was a rug,
a picture or a piece of paper that was being measured, then to that extent the trustworthiness
of that yardstick as a measuring device would be impaired. Within the range of objects for
which the measuring instrument is intended, its function must be independent of the object
of measurement. (Thurstone, 1928, p. 547)

From 1926 to 1931, Thurstone published examples of "objectivity", including
weight and height, that suggest an interest in sample-free, absolute measurement,
although all his models result in an approximation of specific objectivity only. Thus
Thurstone's philosophizing on the attributes of "good" measures focused on general
objectivity (absolute scale locations are independent of the instrument), whereas
his mathematical models and research applications realized only local objectivity
(differences among persons are independent of the measuring instrument) (Stenner,
1994).

Although Thurstone did not use the word objectivity, he clearly had this concept at
the forefront of his thinking. Georg Rasch, however, made objectivity the centerpiece
of a new psychometric model. Rasch (1960) states:

Individual-centered statistical techniques require models in which each individual is characterized separately and from which, given adequate data, the individual parameters can be
estimated. It is further essential that comparisons between individuals become independent
of which particular instruments-tests or items or other stimuli-within the class considered
have been used. Symmetrically, it ought to be possible to compare stimuli belonging to the
same class-'measuring the same thing'-independent of which particular individuals within
a class considered were instrumental for the comparisons. (Rasch, 1980, p. x)

Where this law can be applied, it provides a principle of measurement on a ratio scale of
both stimulus parameters and object parameters, the conceptual status of which is comparable
to that of measuring mass and force. Thus, by way of an example, the reading accuracy of a
child-as ascertained by means of any of the oral reading tests catalogued in the appendix-can
be measured with the same kind of objectivity as we may tell its weight, though not with the
same degree of precision, to be sure, but that is a different matter. (Rasch, 1980, p. 115)

**Thus, if a set of empirical data cannot be described by the [Rasch] model, then
complete specifically objective statements cannot be derived from them**. Firstly, the
failing of specific objectivity means that the conclusions about, say, any set of person parameters will depend on which other persons also are compared. As a parody we might think of
the comparison of the volumes of a glass and a bottle as being influenced by the heights of
some books on a shelf.

Secondly, the conclusions about the persons would depend on just which terms were chosen for the comparison, a situation to which a parallel would be that the relative height of two persons would depend on whether the measuring stick was calibrated in inches or in centimeters." (Rasch, 1968, p. 7)

Thus, in principle, the [Mp's] stand for properties of the objects per se, irrespective of which [Ci's] might be used for locating them. Therefore, they really ought to be appraised without any reference to the [Ci's] actually employed for this purpose, just like reading the temperature of an object should give essentially the same result whichever adequate thermometer was used. (Rasch, no date, p. 5)

Rasch (1960) coined the term "specific objectivity" and realized that his model achieved a separation of instrument and measure long sought after, but never achieved. Whether or not Rasch used a distinction between specific or local and general objectivity is not clear.

On the one hand, he was always careful to point out that it was comparisons (i.e., relative measures) that were independent of the instrument, suggesting that he clearly understood the distinction. On the other hand, his favorite physical science examples (mass and temperature) clearly possess a more general and complete objectivity not shared by the reading comprehension tests he developed.

Finally, Wright (1968) offered an accessible and complete statement on specific objectivity. He wrote:

Let us call measurement that possesses this property 'objective'. Two conditions are necessary to achieve it. First, the calibration of measuring instruments must be independent of those objects that happen to be used for calibration. Second, the measurement of objects must be independent of the instrument that happens to be used for measuring. In practice, these conditions can only be approximated, but their approximation is what makes measurement objective.

Object-free instrument calibration and instrument-free object measurement are the conditions that make it possible to generalize measurement beyond the particular instrument used, to compare objects measured on similar but not identical instruments, and to conceive or partition instruments to suit new measurement requirements." (Wright, 1968).

Wright (1991) continues a quarter-century of exploration of "objectivity" as a fundamental requirement of measurement. He states:

Objectivity is the expectation and, hence, requirement that the amount and meaning of a measure has been well enough separated from the measuring instrument and the occasion of measurement that the measure can be used as a quantity without qualification as to which was the particular instrument or what was the specific occasion.

Although a measuring occasion is necessary for a measure to result, the utility of the measure depends on the specifics of the occasion disappearing from consideration. It must be possible to take the occasion for granted and, for the time being, to forget about it. Were such a separation of meaning from the circumstances of its occasion not possible, not only science, but also commerce, and even communication, would become impossible. (Wright, 1991, p. 1)

Thus, objectivity is clearly the cornerstone of all measurement. Measures must be completely independent of the particular instruments used and the particular conditions of measurement surrounding their use. A critical distinction exists between specific or local objectivity, as achieved by Rasch, and the general objectivity that is inherent in the concept of measurement used in the physical sciences.

"Local objectivity" is a consequence of a set of data fitting the Rasch model. When the data fit, differences between object measures and indicant calibrations are sample-independent. This means that two indicants must be found to differ by the same amount no matter which sample of objects actually responds to the indicants. Similarly, two objects must be found to differ by the same amount no matter which samples of indicants (from the relevant universe) are used to implement the measurement procedure. Consequently, if the data fit the Rasch model then the relative locations of objects and indicants on the underlying continuum for a construct are sample-independent.

An ideal, approximated by measures in physics and chemistry (e.g., thermometers), is that absolute location of an object on, for example, the Fahrenheit scale, is independent of the instruments and conditions of measurement. Temperature theory is well enough developed that thermometers can be constructed without reference to any data. In fact, routine manufacture of thermometers occurs without even checking the calibrations against data with known values prior to shipping the instruments to customers. Such is our collective confidence in temperature theory. We know enough about liquid expansion coefficients, gas laws, glass conductivity and fluid viscosity to construct a usefully precise measurement device with recourse only to theory.

By operating with a construct theory and associated calibration equations, we achieve general objectivity. Measurement of the temperature of two objects results in not just sample independence for the difference between their temperatures but sample independence for the point estimate of each object's temperature reading.

Under a generally objective measurement framework however, thanks to the sufficiency of raw scores, object measures are also entirely free of any reference to individual or group data. In short, specific or local objectivity as achieved with the Rasch model ensures only that relative measures, that is, the differences between people, are independent of the conditions of measurement. In contrast, general objectivity ensures that absolute measures, the amounts themselves, are similarly independent.

The fundamental requirement of a measurement procedure, therefore, is that it be capable of converting an observation (raw count) into a measure without recourse to individual or group data on indicants or objects. We call this feature of a measurement procedure "general objectivity":

> The difference between local and general objectivity is seen not to be a consequence of the fundamental natures of the social and physical sciences, nor to be a necessary outcome of the method of making observations, but to be entirely a matter of the level of sophistication of the theory underlying the construction of the particular measurement instruments. (Stenner, 1990, p. 111)

We turn now to an application of these methods to the measurement of reading comprehension.

5   An Application to Reading Comprehension

Reading comprehension is the most tested construct in education. Among students aged six to 18, reading comprehension ability probably is measured more frequently than temperature, height, or weight. It is widely recognized as the best predictor of success in higher education and on the job performance. Economists and educators have joined in identifying low literacy rates as a causal factor in the United States' dwindling economic prowess.

The importance of reading comprehension is underscored in today's "information age", in which the ability to read easily and well has become a survival skill. Even in production jobs, workers must be able to read complex operational and safety manuals in order to run the computerized equipment on which the modern factory depends. Strong reading skills also are necessary for the continuing education that rapidly changing technology and economic conditions demand, as well as for the requirements of citizenship, such as keeping up with political issues and current events.

Sadly, as many as one-third of Americans are functionally illiterate, unable to read standard adult-level text, such as employee manuals and newspaper articles, with reasonable comprehension. As the need for strong reading skills for work continuing education and citizenship rises, these Americans are at an increasing disadvantage. Hence, the importance that educators and society attach to the construct of reading comprehension.

The Ninth Mental Measurements Yearbook (Mitchell, 1985) reviews 97 reading comprehension tests. Associated with each of these tests is a conceptual rationale (however primitive) and a scale. With no unifying theory for reading comprehension, it is impossible to convert a raw score (i.e., count correct) on one test into a scale score on another test. The current status of reading comprehension measurement is reminiscent of late seventeenth-century temperature measurement, in which the absence of a unifying temperature theory resulted in some 30 different scales competing for favor throughout Europe. The consequence for science and commerce was chaos. In a similar fashion, the presence of dozens of competing reading comprehension scales results in confusion for educators, researchers, policy makers, and parents.

6   The Lexile Theory

People communicate using various symbol systems, including mathematics, music, and language. All symbol systems share two features: a semantic component and a syntactic component. In mathematics, the semantic units are numbers and operators that are combined according to rules of syntax into mathematical expressions. In music, the semantic unit is the note, arranged according to rules of syntax to form chords and phrases. In language, the semantic units are words. Words are organized according to rules of syntax into thought units and sentences (Carver, 1974). In all cases, the semantic units vary in familiarity and the syntactic structures vary in complexity. The comprehensibility or difficulty of a message is governed largely by the familiarity of the semantic units and by the complexity of the syntactic structures used in constructing the message.

7   The Semantic Component

As far as the semantic component is concerned, it is clear that most operationalizations are proxies for the probability that a person will encounter a word in context and thus infer its meaning (Bormuth, 1966). This is the basis of exposure theory, which explains the way receptive or hearing vocabulary develops (Miller & Gildea, 1987; Stenner et al., 1983). Klare (1963) builds the case for the semantic component varying along a familiarity-to-rarity continuum, a concept that is further developed by Carroll et al., (1971), whose word-frequency study examined the reoccurrence of words in a five- million-word corpus of running text. Knowing the frequency of words as they are used in written and oral communication provides the best means of inferring the likelihood that a word will be encountered and thus become a part of an individual's receptive vocabulary.

Variables such as the average number of letters or syllables per word are actually proxies for word frequency. They capitalize on the high negative correlation between the length of words and the frequency of word usage. Polysyllabic words are used less frequently than monosyllabic words, making word length a good proxy for the likelihood of an individual being exposed to them.

Stenner et al., (1983) analyzed more than 50 semantic variables in hopes of identifying those elements that contributed to the difficulty of the vocabulary items on Forms L and M of the Peabody Picture Vocabulary Test-Revised (Dunn & Dunn, 1981). Variables included were part of speech, number of letters, number of syllables, the modal grade at which the word appeared in school materials, content classification of the word, the frequency of the word from two different word counts, and numerous algebraic transformations of these measures. We then ran correlations between the logit difficulties of the test items and each predictor variable. We found that the best operationalization of the semantic component of reading was word frequency.

The word frequency measure used was the raw count of how often a given word appeared in a corpus of 5,088,721 words sampled from a broad range of school materials (Carroll et al., 1971). Through exploratory data analysis, we tested the explanatory power of this variable. This analysis involved calculating the mean word frequency for each of 66 reading comprehension test passages from the *Peabody Individual Achievement Test* (Dunn & Markwardt, 1970). Correlations were then run between algebraic transformations of these means and the rank order of the test items. Since the items were ordered according to increasing difficulty, the rank order was used as the observed item difficulty. The mean of the log word frequencies provided the highest correlation with item rank order.

8 The Syntactic Component

Sentence length is a powerful proxy for the syntactic complexity of a passage. One important caveat is that sentence length is not the underlying causal influence (Chall, 1988). Researchers sometimes incorrectly assume that manipulation of sentence length will have a predictable effect on passage difficulty. Davidson and Kantor (1982), for example, illustrate rather clearly that sentence length can be reduced and difficulty increased and visa versa.

Klare (1963) provides a possible interpretation for how sentence length works in predicting passage difficulty. He speculates that the syntactic component varies in the load placed on short-term memory. This explanation also is supported by Crain and Shankweiler (1988), Shankweiler and Crain (1986), and Liberman, Mann, Shankweiler, and Westelman (1982), whose work has provided evidence that sentence length is a good proxy for the demands that structural complexity places upon verbal short-term memory.

Again, we correlated algebraic transformations of the mean sentence length for the 66 *Peabody Individual Achievement Test* (PIAT) reading comprehension items with item rank order. We found that the log of the mean sentence length was the best predictor of passage difficulty.

9 The Calibration Equation

We then combined the word-frequency and sentence-length measures in hopes of producing a regression equation that could explain most of the variance found in any set of reading comprehension task difficulties. A provisional equation was developed from a regression analysis of the PIAT reading comprehension items. The log of the mean sentence length and the mean of the log word frequencies combined to explain 85% of the variance ($r = 0.92$) in PIAT item rank order.

Using the regression equation produced by this analysis, we assigned theoretical difficulties to 400 pilot test items (see Fig. 1). The pilot items were ordered by difficulty and administered to approximately 3,000 students ranging from grades two to 12. Misfitting items were removed, leaving a total of 262 test items for which observed logit difficulties were computed using **M-scale** (Wright, Rossner, and Congdon, 1985).

The mesa plain had an appearance of great antiquity, and of incompleteness; as if, with all the materials of world-making assembled, the Creator had desisted, gone away and left everything on the point of being brought together, on the eve of being arranged into mountain, plain, plateau. The country was still waiting to be made into a landscape. **It looked** _____

A. arid
B. deserted

C. fertile
D. unfinished

The final specification equation was based upon the observed logit difficulties for the remaining 262 pilot test items. Again, the sentence length and word frequency variables were entered into a regression analysis of these logit difficulties. The resulting correlation between the observed logit difficulties and the theoretical difficulties was 0.97 after correction for range restriction and measurement error. The respective weights produced by the regression run formulated the following equation:

$$(9.82247 * LMSL) - (2.14634 * MLWF) - constant = Theoretical Logit \quad (1)$$

where: LMSL = Log of the Mean Sentence Length
  MLWF = Mean of the Log Word.

10 The Lexile Scale

Once we established this equation, we re-scaled the theoretical logit difficulties. The logit scale is limited in that it has no fixed zero and, therefore, comparisons among different items or different populations are difficult (i.e., measures lack general objectivity). The method of imposing such a scale is quite simple.

For example, when a set of test items from a generic achievement test is given to fifth- graders from Podunk Primary, item difficulties will range from -4 to + 4 logits, centered around the average item difficulty. When the same items are given to fifth-graders from Excel Elementary, the item difficulties also will be in logits from -4 to + 4, again centered around the average difficulty. The average zero, however, floats depending upon the population taking the items. The students from Excel have, on average, a higher ability, and so the logit values will be lower (i.e., the items will appear to be easier). The logit values from the Podunk students will be higher (i.e., the items will appear to be more difficult) because the students have less ability.

Relative to a construct theory, though, test items have a fixed difficulty. The observed variation in difficulty occurs when the test item is given to people of different ability. Unless the logit scores obtained from a test administration are tied to a fixed zero, there is no way to compare the results of these test items given to two different populations.

The method of imposing a scale with a fixed zero point is simple. First, identify two anchor points for the scale. They should be intuitive, easily reproduced, and widely recognized. For thermometers, the anchor points are the freezing and boiling points of water. For the Lexile Scale, the anchor points are the text from seven basal primers for the low end and text from the *Electronic Encyclopedia* (Grolier, 1986) for the high end.

Second, using the regression equation, obtain the logit difficulty of the two anchors. For the Lexile scale, the mean logit difficulty of the primer material was -3.3 and the mean logit difficulty for the encyclopedia samples was +2.3.

Third, decide what the unit size should be. For the Celsius thermometer, the unit size (a degree) is 1/100 of the difference between freezing (0 °) and boiling (100 °) water. For the Lexile scale the unit size was defined as 1/1000. Therefore, a Lexile by definition equals 1/1000th of the difference between the comprehensibility of the primers and the encyclopedia.

Fourth, assign a value to the lower anchor. To minimize the occurrence of negative Lexile values, we did not use zero as the low-end value, but instead assigned a value of 200.

Finally, we developed an equation that converts logit difficulties to Lexile calibrations. When the regression equation is used to analyze the anchors, the resulting difficulties are -3.3 logits for the primers and 2.3 logits for the encyclopedia. In order to set the -3.3 logits for the primer anchor equal to 200, we used the following equation:

$$(-3.3 + 3.3) + 200 = 200 \text{ Lexiles} \tag{2}$$

The 3.3 that offsets the negative difficulty of the primer now becomes one of the two constants in the final formula. The second constant is determined when this equation is made to equal 1200 Lexiles, which is where the encyclopedia has been located.

$$[(2.26 + 3.3) * \text{Constant}] + 200 = 1200 \text{ Lexiles} \tag{3}$$

The second constant turns out to be 180, which is the amount needed to convert the logit difficulty of the encyclopedia to 1200 Lexile units. The final equation that converts theoretical logit difficulties produced by the Lexile equation into Lexile units is as follows:

$$[(\text{Logit} + 3.3) * 180] + 200 = \text{Lexile calibration} \tag{4}$$

Measurements for persons and text are now reportable in Lexiles, which are similar to the degree calibrations on a thermometer. Essentially, the higher the Lexile measure for a text, the more difficult the material and the more ability a student must possess to comprehend the text.

Text measures are located on the Lexile map at the point corresponding to a person with the ability to achieve 75% comprehension. People are located on the scale by analyzing their performance on calibrated reading tasks. They are located on the map at the point where they are forecasted to achieve 75% comprehension. A person with a Lexile measure of 1000L is expected to answer correctly 75% of native Lexile items sampled from a text with a 1000L measure. This provides the means for directly matching a person's measure with a measure for a text. The difference between these two measures is used to forecast the comprehension that the person will have with that text. Comprehension is always relative to the difference between person measure and text measure.

The Lexile map can be used to bring meaning to these Lexile measures of text and persons. This richly annotated four color, poster-sized graphic provides an extensive list of texts, from novels and non-fiction books to newspapers and magazines, at various levels of Lexile measurement. The map makes it easy to "see" how reading develops and to select other reading materials as students progress in their reading development.

11   Testing the Lexile Equation

A computer program incorporating the Lexile equation is available that analyzes continuous prose and reports the difficulty in Lexiles (MetaMetrics 1995). In order to test the power of the theory, we analyzed 1,780 reading-comprehension test items appearing on nine nationally normed tests (Stenner et al., 1987). The study involved correlating Rasch item difficulties provided by the publisher with the Lexile calibrations generated from the computer analysis of the text of each item. In those cases where multiple questions were asked about a single passage, we averaged the reported item difficulties to yield a single observed difficulty for the passage.

We obtained the observed difficulties in one of three ways. Three of the tests included observed logit difficulties from either a Rasch or three-parameter analysis (e.g., NAEP). For four others, logit difficulties were estimated based upon item p-values and raw score means and standard deviations (e.g., CAT). To obtain these logit difficulties, we used TestCalc (), a computer program for analyzing test data. Two of the tests provided no item parameters, but in each case items were ordered on the test in terms of difficulty (e.g., PIAT). For those tests, the observed difficulty was approximated by the difficulty rank order of the item.

Once theory-based calibrations and data-based item difficulties were computed, we correlated the two arrays and plotted them separately for each test. The plots were checked for unusual residual distributions and curvature, and we discovered that the equation did not fit poetry items or non-continuous prose items (e.g., recipes, menus, or shopping lists). This indicated that the universe to which the Lexile equation could be generalized was limited to continuous prose, which still accounts for a majority of reading material. The poetry and non-continuous prose items were removed and correlations were again obtained and used to describe the fit of observation to theory.

Two major influences other than model misspecification operate to artificially deflate the relationship between theory and observation. The first is range restriction in the item difficulties. Some tests purposely do not cover the full developmental continuum for reading comprehension. The NAEP (1983), for example, is administered to grades four, eight, and 11. As might be expected, the resulting restriction in the range of item difficulties tends to attenuate the relationship between theory and data. Thorndike (1949) gives the procedure for correcting a correlation for restriction in range where the range of the variable in the unrestricted group is known.

A second influence that operates to reduce the correlation between theory and data is unreliability in the theory-based item calibrations. Theories are rarely perfectly operationalized. As we have noted, the Lexile equation contains two terms that are both proxies for the presumed underlying causes of item difficulty. Proxies are imperfect substitutes for the theoretical causes and, as such, act to attenuate correlations. The data-based difficulties, on the other hand, are so well estimated that the reliabilities are typically near 0.99. Stanley (1971) gives the procedure for disattenuating a correlation for unreliability in one of the variables.

Finally, note that the Lexile analysis was applied only to the passages and did not include the questions and their respective answers. This decision most likely introduced error, since it has long been recognized that the questions themselves add to the overall difficulty of a test item. The magnitude of these influences is difficult to estimate, but we can safely assume that some of the remaining differences between theoretical calibrations and data-based difficulties are due to these factors.

Table 1 presents the results of correlating the theoretical calibrations and observed difficulties. The last three columns of the table show the raw correlation between observed (O) item difficulties and theoretical (T) item calibrations, with the correlations corrected for restriction in range and measurement error. The mean of the raw correlations is $r_{(OT)} = 0.84$. When corrections are made for range restriction and measurement error, the average disattenuated correlation between theory-based calibration and data-based difficulty in an unrestricted group of reading comprehension items is $R'_{(OT)} = 0.93$.

$r_{(OT)}$ = raw correlation between observed difficulties (O) and theory-based calibrations (T).

$R_{(OT)}$ = correlation between observed difficulties (O) and theory-based calibrations (T) corrected for range restriction.

$R'_{(OT)}$ = correlation between observed difficulties (O) and theory-based calibrations (T) corrected for range restriction and measurement error.

Table 1 Correlation Between Theory-Based Calibrations Produced by the Lexile Equation and Data-Based Item Difficulties

Test	Number of questions	Number of passages	r(OT)	R(OT)	R'(OT)
SRA	235	46	0.95	0.97	1.00
CAT-E	418	74	0.91	0.95	0.98
Lexile	262	262	0.93	0.95	0.97
PIAT	66	66	0.93	0.94	0.97
CAT-C	253	43	0.83	0.93	0.96
CTBS-U	246	50	0.74	0.92	0.95
NAEP	189	70	0.65	0.92	0.94
Battery	26	26	0.88	0.84	0.87
Mastery	85	85	0.74	0.75	0.77
Total grand means	1780	722	0.84	0.91	0.93

It seems reasonable to conclude from these results that most attempts to measure reading comprehension, no matter what the item form, type of skill objectives purportedly being measured, or response requirement used, all end up measuring a common comprehension factor specified by the Lexile theory (Stenner et al., 1988).

In a second study, we obtained Lexile measures for units within 11 basal series. It was assumed that each basal series was sequenced by difficulty. So, for example, the latter portion of a third-grade reader is presumably more difficult than the first portion of that same book.

Likewise, a fourth-grade reader is presumed to be more difficult than a third-grade reader. We estimated observed difficulties for each unit in a basal series by the rank order of the unit in the series. Thus, the first unit in the first book of the first-grade was assigned a rank order of one and the last unit of the eighth-grade reader was assigned the highest rank order number. Correlations were computed between the ranked order and the Lexile calibration of each unit. After correction for range restriction and measurement error, the average correlation between the Lexile calibration of text comprehensibility and the rank order of the basal units was 0.99 (see Table 2).

$r(OT)$ = raw correlation between observed rank order (O) and theory-based calibrations (T).

$R(OT)$ = correlation between observed rank order (O) and theory-based calibrations (T) corrected for range restriction.

$R'(OT)$ = correlation between observed rank order (O) and theory-based calibrations (T) corrected for range restriction and measurement error.

*Means are computed on Z transformed correlations.

The fact that the Lexile theory accounted for the unit rank ordering of 11 basal series is all the more noteworthy when we recognize that the series differ in prose selections, the developmental range addressed, the types of prose introduced (i.e.,

Table 2 Correlations between the Lexile Calibration and The Rank Order of Units

Basal series	Number of units	r(OT)	R(OT)	R'(OT)
Ginn rainbow series (1985)	53	0.93	0.98	1
HBJ eagle series (1983)	70	0.93	0.98	1
Scott foresman focus series (1985)	92	0.84	0.99	1
Riverside reading series (1986)	67	0.87	0.97	1
Houghton-Mifflin reading series (1983)	33	0.88	0.96	0.99
Economy reading series (1986)	67	0.86	0.96	0.99
Scott foresman American tradition (1987)	88	0.85	0.97	0.99
HBJ odyssey series (1986)	38	0.79	0.97	0.99
Holt basic reading series (1986)	54	0.87	0.96	0.98
Houghton-Mifflin reading series (1986)	46	0.81	0.95	0.98
Open court headway program (1985)	52	0.54	0.94	0.97
Totals means	660	0.86	0.97	0.99

narrative versus expository), and the purported skills and objectives they emphasize. The theory works throughout the full developmental range, from pre-primer (-200 Lexiles to 200 Lexiles) through advanced graduate school material (1400 Lexiles to 1800 Lexiles).

12   Lexile Measures and Probable Error

A fundamental concept in the use of a measure is "probable error", or how much error is attributable to that measure. A well-known feature of all measurement procedures is that repeated measurement of the same thing results in a series of non-identical amounts. The root mean square of the differences among repetitions is the standard deviation of the distribution, also called the standard error of measurement (SEM). This statistic describes how close together are the repeated measurements. The SEM also defines the prediction interval for the question, "What would happen if we measured again?".

There are many ways of defining what we mean by "measuring again". Definitions of a "repetition" differ in what facets of the measurement procedure vary from one repetition to another. As more facets vary, more variation is observed in the replicate measures and the SEM increases. A highly restricted recipe for replication may result in a smaller SEM, but may overstate the precision of measurement that results from "normal use" of the measurement procedure.

With regard to the Lexile Framework, if we were to process 20 pages of text again and again, we would see a standard deviation of 40 Lexiles. Similarly, if we were to test individuals according to a broad recipe-different items on different days-we would end up with a standard deviation of 70 Lexiles.

13   Application of the Lexile Scale

One of the major weaknesses of current testing procedures is the limited usefulness of the normative interpretation of a score. A normative interpretation only expresses how a student did on the test compared to other students of the same age or grade. A student's performance is typically reported as a percentile. A percentile of 65 for a third-grade girl indicates that she scored better than 65% of all third-grade students involved in the norming study. Percentile scores on standardized reading tests, however, do not provide any information about what a student can or cannot read. What does a teacher or parent actually do with a percentile score? What kind of instruction can a teacher give a student when the only information provided is that a particular child is reading at the 65th percentile of all third-graders across the nation?

The Lexile scale is designed to provide both a normative and a criterion-referenced interpretation of a measure. Since the Lexile scale is based upon the Rasch model, the probability of a person answering a reading item correctly is governed only by the difference between the person's measure and the task's calibration. This relationship is captured in the following equation:

$$O = \sum_i \frac{e^{(M_o - C_i)}}{1 + e^{(M_o - C_i)}}$$

O = observation (count correct)

P_{ni} = the probability of a correct response

M_o = the person's measure

C_i = item calibration

If a person's measure is equal to the task's calibration, then the Lexile scale forecasts that the individual has a 75% comprehension rate on that task. If 20 such tasks were given to this person, one would expect three-fourths of the responses to be correct. If the task is more difficult than the person is able, then the probability is less than 75% that the response of the person to the task will be correct; similarly, if the task is easier compared to a person's measure, then the probability is greater that the response will be correct.

A person with a Lexile ability of 600L who is given a text measured at 600L will have a 75% comprehension rate. If the same student is given a text measured at 350L, the forecast comprehension rate improves to 90%. Give the same student a 100L title, and comprehension improves to 96%. The more a person's Lexile measure surpasses the Lexile measure for a task the higher the forecasted comprehension rate. The more the Lexile measure for a task surpasses a person's Lexile measure, the lower the forecasted comprehension rate. See Tables 3 and 4 illustrate the relationship between person measure, text measure and the forecasted comprehension rate.

Empirical evidence supporting a 75% target comprehension rate, as opposed to, say, a 50% or 90% rate, is limited. Squires et al., (1983) did find that reading achievement for second-graders peaked when the success rate reached 75%. A 75% success rate also is supported by the findings of Crawford et al., (1975). It may be, however, that there is no one optimal rate, but rather a range in which individuals can operate

Table 3 Comprehension Rates for the Same Individual with Materials of Varying Comprehensibility

Person measure	Text calibration	Sample titles	Forecast comprehension
1000	500	Are You There God? It's Me, Marqaret - Blume	96%
1000	750	The Martian Chronicles - Bradbury	90%
1000	1000	Reader's Digest	75%
1000	1250	The Call Of The Wild - London	50%
1000	1500	On the Equality Among Mankind—Rousseau	25%

Table 4 Comprehension Rates of Different Ability Person with the Same Material

Person measure	Calibrated for Sports Illustrated	Forecast comprehension rate
500	1000	25%
750	1000	50%
1000	1000	75%
1250	1000	90%
1500	1000	96%

Note that it is the difference in Lexiles between person and text that governs comprehension. The difference between a 200L text and a 450L reader results in the same success rate as with a 600L text and an 850L reader. Each case produces a 90% comprehension rate

to improve optimally their reading ability. We have found, however, that te subjective report of readers reading at 50% comprehension is frustration whereas readers reading at 75% comprehension report comfort and confidence with the text.

Since the Lexile theory provides complementary procedures for measuring people and text, the scale can be used to match a person's level of comprehension with books that the person is forecast to read with a high comprehension rate. Up to this time, trying to identify possible supplemental reading for students has, for the most part, relied on a teacher's familiarity with the tittles. For example, an eighth-grade girl who is interested in sports but is not reading at grade level might be able to handle a biography on a famous athlete. The teacher may not know, however, whether that biography is too difficult or too easy for the student. The Lexile Framework can provide a person measure and text measure on the same scale. Armed with this information, a teacher or parent can plan for success. This ability to provide strong linkages between test results and reading material, among students, teachers and parents, and between schools and workplaces is one of the primary advantages of the Lexile Framework.

Improving student's progress in reading requires that they read properly targeted prose accompanied by frequent response requirements. Response requirements range from having a more competent reader ask occasional questions as the reader progresses through the prose to embedding questions in the text, much as is done with Lexile test items. The reason for requiring that readers do more than simply read is that unless there is some evaluation, there can be no assurance that the reader is properly targeted and comprehending the material. Students should be given text on which they can practice being a competent reader (Smith, 1973). The above approach does not represent a fully articulated instructional theory, but its prescription is straightforward. Students should read more targeted prose and teachers should monitor this reading with some efficient response requirement. One implication of these notions is that some of the time spent on skill sheets might be better spent reading targeted prose with attendant response requirements (Anderson, Hiebert, Scott, and Wilkinson, 1984).

As the reader improves, new titles with higher text measures can be chosen to match the growing person measure, thus keeping the comprehension rate at the chosen level. In essence, we need to locate a reader's "edge" and then systematically expose the reader to text that plays on that edge. When this approach is followed in any domain of human experience, the edge moves and the capacities of the individual are enhanced.

What happens when the "edge" is over-estimated and repeatedly exceeded? In any kind of physical exertion, if you push beyond the edge you feel pain; if you demand even more performance on the part of the muscle, you will experience severe muscle strain or ligament damage. In reading, playing on the edge is a satisfying and confidence-building activity, but exceeding that edge by over-challenging readers with materials well out of their reach reduces self-confidence, stunts growth and eventually results in the individual "tuning out". With the tremendous emphasis placed on reading in daily activities, virtually every encounter with written text becomes a reconfirmation of a poor reader's inadequacy. Is it any wonder that 15 to 20% of U. S. high school students decide to find some other way to spend their days (Hahn, 1987)?

For students to become competent readers, they need to be exposed to text that results in a comprehension rate of 75% or better. If an 850L reader is faced with an 1100L text (resulting in a 50% comprehension rate), there will be too much unfamiliar vocabulary and too much of a load placed on the reader's tolerance for syntactical complexity for that reader to attend to meaning. The rhythm and flow of familiar sentence structures will be interrupted by frequent unfamiliar vocabulary, resulting in inefficient chunking and short-term memory overload. When readers are properly targeted, they read fluently with comprehension; when improperly targeted, they struggle both with the material and with maintaining motivation and their self-esteem. In reality, there are no poor readers-only mistargeted readers who are being challenged inappropriately.

Leading researcher Jeanne S. Chall states that the Lexile Framework, while under-girded by highly sophisticated statistical procedures, stands firmly in the tradition of classic readability formulas with its emphasis on comprehension as a function of semantic and syntactic components. In contrast with cognitive-structural readability researchers, she notes that the developers of the Lexile Framework view comprehension as a unidimensional ability that subsumes different types of comprehension. Under the Lexile theory, no special consideration is given to prior or special subject knowledge.

Chall also points out several unique features of the Lexile Framework, including the use of calibrated scores to represent equal units of difficulty and the fact that the same scores measure both the difficulty of text and the ability of readers. These Lexile scores correlate well with nine classic readability formulas and with item difficulties on several tests of reading comprehension (Chall, 1995).

14 Benefits of Objective Measurement

What has been proposed in this paper can be characterized as a minor adjustment to a well-known model that results in profound implications for what we can do with behavioral science measures. Truly objective measurement can be achieved by simply replacing the data-based difficulties of the Rasch model with theory-based calibrations. When this is done, several benefits accrue: (1) the construct theory is exposed to falsification, (2) it is possible to build correspondence tables between observations and measures with recourse only to theory, (3) a generalized linking solution is available for placing observations of all kinds on a common scale, (4) a reproducible unit of measurement can be developed, (5) the framework for fit statistics that is sample-dependent under the Rasch model becomes sample-independent, and (6) a complete frame of reference for measure interpretation can be constructed.

15 A Refutable Construct Theory

A construct theory is a story about what it means to move up and down a scale. A specification equation is a regression model, based on the theory, that forecasts indicant calibrations on the scale. To the extent that the theory is adequate and the specification equation forecasts accurately, it is possible to achieve generally objective parameter estimation. Construct theories embodied in specification equations are exposed to test and falsification. Through such attempts at falsification, construct theories are sharpened, specification equations are made more accurate and understanding of what our measures, in fact, do measure is enhanced.

16 Correspondence Tables via Theory

Measurement is the process of converting observations into quantities through theory. The graphic that represents the essential correspondence between observation and measure (i.e., quantity) is termed a "correspondence table". This table is constructed via theory in physical science measurement and via data in behavioral science measurement. General objectivity means that two observers will arrive at the same numbers to assign as a measure, given that they begin with the same observation. Objectivity refers to the process of converting observations into measures and is not meant to imply that two applications of the same measurement procedure will arrive at identical measures.

17   A Generalized Linking Solution

Rasch measurement recognizes two linking designs: common objects and common indicants. Under the common-objects design, different instruments (sets of indicants) are administered to the same sample of objects, thus establishing calibration differences between the various measurement procedures. The common-indicants design involves administering a sub-set of indicants taken from two or more instruments to a sample of persons and using the differences in mean calibrations to equate scales. Unfortunately, the most frequently encountered situation that requires a linking procedure involves two tests that share no indicants and that have not been administered to the same group of persons. This situation has been considered intractable.

Objective measurement provides a means for linking or equating instruments in the case of uncommon objects and uncommon items. The procedure uses a common theory embodied in a common specification equation to link instruments. The specification equation provides theory- based calibrations for each indicant, thus equating all measures of a construct to a common theory-referenced scale. Under this new procedure, all reading comprehension tests developed in the last century can now be placed on a common scale (Stenner et al., 1988).

18   A Reproducible Unit of Measurement

Most units of measurement used in the behavioral sciences are fractions of some sample standard deviation. The unit, as such, cannot be reproduced in the same way that a degree Celsius can be reproduced, but an approximation can be gained through a common-objects or common-indicants design.

A sounder strategy is to select as a unit some fraction of the difference between two publicly reproducible states or artifacts. In the case of the Celsius scale, the unit is 1/100 of the distance between freezing and boiling "normal" water at sea level. In the case of the Lexile framework, the unit is 1/1000 of the distance between the comprehensibility of first-grade primers and encyclopedias. This unit also is reproduced easily from the Lexile theory and associated specification equations.

19   Model Fit

Objective measurement shifts the focus away from fit to the model to theory and measure validation. Theory validation is tested by checking the correspondence between theory-based and data-based estimates of indicant parameters. The analysis of interest is the plot between theory-based calibrations of the indicant and data-based difficulties. The construct theory and associated specification equation is well confirmed to the extent that the above correlation approaches r = 1.0 upon repeated application of the specification equation to indicants from different instruments.

Measure validation corresponds to person fit in the Rasch model and uses the identical statistics and interpretive framework. The difference is that Rasch measurement does not require an a priori statement of intention. Rather, "intention" is arrived at a posteriori in the form of estimated difficulties. Fit is then checked against this data-dependent statement of intention. Surely, intention must be stated prior to data analysis and its argument must be framed in language that moves beyond the immediate materials and moment of measurement. Objective measurement through the specification equation and theory-based indicant calibrations makes an explicit a priori statement of intention. Whether our intention is realized is examined at the theory or specification equation level and at the individual object measure level. Thus, we speak of construct theory validation and of measure validation.

Elsewhere, we have argued that the conventional approach to validity confuses the question of whether an instrument measures what it is intended to measure with the question of how useful the obtained measures are in the description and prediction of phenomena (Stenner et al., 1988). Objective measurement clearly distinguishes between these two questions. It is quite possible for an instrument to generate measures that are valid in the sense discussed above, but have no known use. The absence of measure utility, however, does not mean the construct theory is invalid or that the measures produced are invalid. On the other hand, low theory validity and or measure validity does reduce the expected utility of a measure.

20 Frames of Reference

For most of this century, the dominant frame of reference for score interpretation has been a normative one. Early attempts at criterion-reference score interpretation either emphasized objectives or proposed so-called mastery or cut-scores. The former generally had little to do with what was actually being measured by an instrument and the latter proved unsatisfyingly primitive as a means of attaching meaning to various score levels.

Perhaps the best-known break with this tradition was the scale annotation procedure illustrated by Thurstone (1925), used to good advantage by Woodcock (1974) and now employed by NAEP (1987). The Keymath test manual (Connolly, Nachtman & Pritchett, 1971) provided a graphic that annotated selected points along the scale with items taken from the test. The test user could examine these items and "see" the progression in math ability as scale scores increased.

This criterion-referencing procedure represented a major advance over objective-based and mastery-level procedures. The disadvantage is that the annotation procedure is context-bound. Only items used in the test or linked to the test through a common-objects or common-indicants design can be used in the annotation process. With objective measurement, the specification equation can be used to build a rich criterion frame-of-reference that is independent of any particular set of indicants. In the reading comprehension domain, for example, a specification equation is used to generate Lexile measures for books and periodicals. These titles are then arranged along the scale to describe what readers at various levels of reading comprehension can read at selected comprehension rates. We refer to this process as scale specification. It is distinguished from scale annotation by the use of specification equations that can compute theory-based calibrations (in a common metric) for any task from the construct universe governed by the construct theory. Just as we don't need data to know what a particular hash mark on a thermometer means, we don't need data to know what a given raw count or raw score means on a theory-referenced task or measure.

Most, if not all, norms in use today reflect norm-group performance on a particular test. Rasch measurement practitioners and theoreticians have written on the prospect of norming a reference scale (Wright & Stone, 1979), but, in the absence of general objectivity, it has not been worth the effort. Objective measurement changes this state of affairs. Since absolute scale locations for persons are sample-independent, when measures are generally objective, it makes sense to norm-reference the measurement scale and not just scores produced by a single instrument. Once this has been done and a norm-maintenance program has been established, all past and future measures of the construct in question are norm-referenced.

References

Anderson, R. C., & Davison, A. (1988). Conceptual and empirical bases of readability formulas. In Davison, A. & Green, G. M. (Eds.), *Linguistic complexity and text comprehension?: Readability issues reconsidered*. Hillsdale, NJ: Erlbaum.

Anderson, R. C., Hiebert, E. H., Scott, J. A., & Wilkinson, I. (1985). Becoming a nation of readers: The report of the commission on reading. U. S. Department of Education.

Bormuth, J. R. (1966). Readability: New approach. *Reading Research Quarterly, 7*, 79–132.

California achievement test: Form C (1977). New York: McGraw-Hill.

California achievement test: Form E (1985). New York: McGraw-Hill.

Carroll, J. B. (1980). Measurement of abilities constructs. In U. S. office of Personnel Management, *Construct Validity in Psychological Measurement*. Princeton, NJ: Educational Testing Service.

Carroll, J. B., Davies, P., & Richman, B. (1971). *Word frequency book*. Houghton Mifflin.

Carver, R. P. (1974). Measuring the primary effect of reading: Reading storage technique, understanding judgments and cloze. *Journal of Reading Behavior, 6*, 249–274.

Chall, J. S (1988). The beginning years. In Zakaluk, B. L., & Samuels, S. J. (Eds.), Readability: Its past. present. and future. Newark, DE: International Reading Association.

Chall, J. S., & Dale, E. (1995). Readability revisited: The New Dale-Chall readability formula. Brookline Books.

Comprehensive test of basic skills: Form U (1981). New York: McGraw-Hill.

Connolly, Nachtman, & Pritchett (1971). Key math diagnostic arithmetic test. American Guidance Service, Inc.

Crain, S., & Shankweiler, D. (1988). Syntactic complexity and reading acquisition. In Davidson, A., & Green, G. M. (Eds.), Linguistic complexity and text comprehension: Readability issues reconsidered. Hillsdale, NJ: Erlbaum Associates.

Crawford, W. J., King, C. E., Brophy, J. E., & C. M. (1975, March). Error rates and question difficulty related to elementary children's learning. Paper presented at the annual meeting of the American Educational Research Association, Washington, D.C.

Gick, J. E., & Brerman, R. L. (1982). GENOVA: A generalized analysis of variance system. [computer program]. Dorchester, MA: University of massachusetts at Boston.

Davidson, A., & Kantor, R. N. (1982). On the failure of readability formulas to define readable text: A case study from adaptations. *Reading Research Quarterly, 17*, 187–209.

Dunn, L. M., & Dunn, L. M. (1981). Peabody picture vocabulary Test-Revised: Forms L and M. American Guidance Service.

Dunn, L. M., & Markwardt, F. C. (1970). Peabody individual achievement test. American Guidance Service.

Electronic encyclopedia (1986). Danbury, CT: Grolier.

Hahn, A. (1987). Reaching out to America's dropouts: What to do? *Phi Delta Kappan, 67*, 256–263.

Hitch, G. J., & Baddeley, A. D. (1974). Verbal reasoning and working memory. *Journal of Experimental Psychiatry, 28*, 603–621.

Horabin, I. (1989a). TestCalc [computer program]. Durham, NC: Ivan Horabin.

Horabin, I. (1989b). TestCalc [computer program]. Durham, NC: MetaMetrics.

Horabin, I. (1987). PC-LEX: A computer program for rating the difficulty of continuous prose in Lexiles [computer program]. Durham, NC: MetaMetrics.

Klare, G. R. (1974). Assessing readability. *Reading Research Quarterly, 1*, 63–102.

Klare, G. R. (1963), The measurement of readability. Iowa State University Press.

Liberman, I. Y., Mann, V. A., Shankweiler, D., & Werfelman, M. (1982). Children's memory for recurring linguistic and non-linguistic material in relation to reading ability. *Cortex, 18*, 367–375.

Lord, F. M. (1980). Applications of item response theory to practical testing problems. Erlbaum.

Luce, R. D., & Tukey, J. W. (1964). Simultaneous conjoint measurement: A new type O fundamental measurement. *Journal of Mathematical Psychology, 1*, 1–27.

Miller, G. A., & Gildea, P. M. (1987). How children learn words. *Scientific American, 257*, 94–99.

Mitchell, J. V. (1985). The ninth mental measurements yearbook. University of Nebraska Press.

National Assessment of Educational Progress. (1984). Princeton. Educational Testing Service.

Rasch, G. On Objectivity and Specificity of the Probalistic Basis for Testing, mimeographed, no date, 1–19.

Rasch, G. A. (1968). Mathematical theory of objectivity and its consequences for model construction. In report from *European Meeting on Statistics, Economics, and Management Sciences*. Amsterdam.

Rasch, G. (1960). *Probabilistic models for some intelligence and attainment tests* (Reprint, with Foreword and Afterword by B. D. Wright, Chicago: University of Chicago Press, 1980). Copenhagen, Denmark: Danmarks Paedogogiske Institut.

Readability Calculations [computer program] (1984). Dallas, TX: Micro Power and Light Company.

Thorndike, R. L. (1949). Personnel selection. Wiley.

Thurstone, L. L. (1925). A method of scaling psychological and educational tests. *Journal of Psychology*, October 1925.

Thurstone, L. L. (1928). Attitudes can be measured. *American Journal of Sociology, 33*, 529–544.

Shankweiler, D., & Crain, S. (1986). Language mechanisms and reading disorder: A modular approach. *Cognition, 14*, 139–168.

Stanley, J. C. (1971). Reliability in R. C. Thorndike (Ed.) Educational Measurement: 2nd edition. Washington D. C.: American Council on Education.

Stenner, A. J. (1994). Specific objectivity - local and general. *Rasch Measurement Transactions, 8(3)*, 374 (http://www.rasch.org/rmt/rmt83e.htm).

Stenner, A. J., & Smith, M. (1982). Testing construct theories. *Perceptual and Motor Skills, 55*, 415–426.

Stenner, A. J., Smith, M., & Burdick, D. S. (1983). Toward a theory of construct definition. *Journal of Educational Measurement. 20*, 305–315.

Stenner, A. J., Smith, D. R., Horabin, I., & Smith, M. (1987). Fit of the Lexile theory to item difficulties on fourteen standardized reading comprehension Tests. Durham, NC: MetaMetrics.

Stenner, A.J., Horabin, I., Smith. D.R. & Smith, M. (1988). Most comprehension tests do measure reading comprehension: A response to McLean and Goldstein. Phi Delta Kappan, June 1988, 765–769.

Squires, D. A., Huitt, W. G., & Segars, J. K. (1983). Effective schools and classrooms. Alexandria. Association for Supervisor and Curricular Development.

Woodcock, R. W. (1974). Woodcock reading mastery tests. American Guidance Service.

White, E. B. (1952). Charlotte's web. New York: Harper and Row.

Wright, B. D. (1968). Sample-free test calibration and person measurement. In proceedings of the 1967 *Invitational Conference on Testing Problems*. Princeton, NJ: Educational Testing Service.

Wright, B. D., & Stone, M. H. (1979). Best test design. MESA Press.

Wright, B. D. & Stone, M. (1991). Objectivity: measurement primer No. 2. Wilmington, DE: Jastak Associates.

Readability and Reading Ability

Benjamin D. Wright and A. Jackson Stenner

1 Uniform Measures

The world of education has long been waiting for a sunrise. Believe it or not, a popular compilation of educational tests lists 97 different reading tests (Mitchell, The Ninth Mental Measurements Yearbook, University of Nebraska Press, 1985). This situation produces 97 different "reading ability measures." What a mess! But now, with the dawn of uniform educational measures, the sun is rising here in Melbourne.

When I was a physicist, I came to appreciate the essential part uniform measures play in science. In the seventeenth century, there were many ways to observe the effects of heat. It was thought, therefore, that there were many kinds of heat. That was a brutal barrier to progress. Nevertheless, seventeenth century scientists thought they were observing "57 varieties." After all, is not bathtub heat different from teacup heat, different from cauldron heat, different from fireplace heat—all of which are different from the heat of the sun? Eventually, it was discovered that it was not only desirable but also necessary to have just one kind of heat. Today, for science and commerce, we do our thinking about heat in terms of one entirely abstract unit, the "degree." Whether it's a Kelvin, Celsius, or Fahrenheit degree does not matter. We know exactly how to get from one to another. They all measure, what we insist is, the same one kind of heat. Measures are older than talking. Birds measure. So do bees. Our own measures evolved from our bodies—our feet, our arms, our hands, our fingers. An inch is the distance from thumb to knuckle. A span is the distance between thumb tip and little finger. A cubit is the length of a forearm. A fathom is

Paper Presented to the Australian Council on Education Research (ACER). 1998.

B. D. Wright
Departments of Education and Psychology, University of Chicago, Chicago, USA

A. J. Stenner (✉)
MetaMetrics, Inc., Durham, NC, USA

2. The length of one curtain shall be eight and twenty cubits, and the breadth of one curtain four cubits; and every one of the curtains shall have one measure.

Fig. 1 Exodus 26

13. Thou shalt not have in thy bag diverse weights, a great and a small. 14. Thou shalt not have in thine house diverse measures, a great and a small. 15. Thou shalt have a perfect and just weight, a perfect and just measure.

Fig. 2 Deuteronomy 25

the distance between outstretched arms. A pace is two steps. A furlong is 200 paces. A mile 1,000.

Abstractly equal units of length were counted on before the oldest fragments of writing. Figure 1 is Moses' plan for the Tabernacle.

Without approximations to equal units, Babylonians, Egyptians and Hebrews could not have imagined, let alone built, their towers.

Fair measurement is embedded in Judeo-Christian morality. But the "perfect and just measure" demanded in Deuteronomy 25, Fig. 2, is an ideal that can only be approximated in practice.

The "weight" referred to is a shekel stone, understood to weigh 11.4 oz. However, archeologists have never found two shekel stones that weighed exactly the same. No technology, no matter how advanced, can fabricate perfect weights. Nevertheless, even when Deuteronomy was written, we already understood the essential necessity and justice of fair units.

The necessity of uniformity in the representation of quantity appears again in King John's Magna Carta, Fig. 3. Without the ideal of uniform measures, there would be no money. There would be no fitted clothes, because there would be no way to fit them. Imagine what life would be like if there were no abstract unit of length like the inch.

Suppose that taking an inch were complex—differing with every situation and material. Imagine that wood inches were different from brick inches, were different from steel inches. We would not have civilization. We would have a mess—a mess like the mess that permeates most of what we misleadingly refer to as "educational tests and measurements."

35. There is to be one measure of wine and ale and corn within the realm, namely the London Quarter, and one breadth of cloth, and it shall be the same with weights.

Fig. 3 The Magna Carta

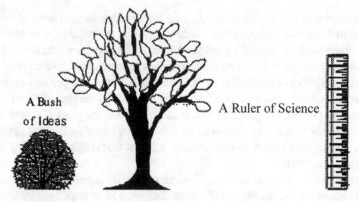

Fig. 4 A tree of knowledge

2 The Evolution of Science

The study of any subject begins with tangles of speculations. Ideas branch in all directions. But as we work through the tangle, we connect what we experience with what we see. We coax our ideas into shape, form unities, develop lines of inquiry. We fit our ideas together and make them something. We evolve our bush of ideas into a tree of knowledge, Fig. 4.

The bush was a tangle. The tree has direction. Our final step in wrestling a useful abstract assertion from a complex concrete confusion is to carve a ruler out of our tree. The ruler does not exist until we imagine it and carve it. The carving is not perfect. It is just an approximation. But what it approximates—a perfectly straight line—enables us to use it as though it were marked off in perfectly equal intervals.

We can pace off land in somewhat equal steps. But steps inevitably vary according to conditions. To produce reliable measurements, we need something more reproducible than pacing. The scientific measurement of length was born as we connected our experience of stride with man made marks on straight pieces of wood extracted from tree trunks. A piece of tree is more stable than anyone's paces. A ruler does not change its bench marks. When we grow a confusing bush of tangled ideas into a tree of useful knowledge and make a ruler, then we can plan and build a pyramid, a temple, a house—and also measure the height of a child (Rasch, 1980).

3 The Imaginary Inch

An inch is pure, abstract and without content. It has no meaning of its own. It is an imaginary unit of length. A height of inches, however, has meaning. As we grow, we learn the advantages to growing taller. Brick size has meaning. As we build, we learn the advantages of same-sized bricks. What makes bricks useful is that their interchangeability is maintained by approximations to the fiction we call an inch.

It is essential that our idea of an abstract inch is always the same. If we let our idea of an inch change each time we made a measure, we would not produce useful bricks or keep track of a child's growth. As our child grew, we would not know by how much they had grown. But with a uniform unit of measure, like an inch, we can measure the height of our children, we can refer to last year—or perhaps to the height of an average second grader because, as it turns out, child height is related to school grade. We can guess what grade a child is in by how tall they are–and how tall they are by what grade they are in. That is an understanding based entirely on applications of rulers. The applications would be useless without that single, unvarying inch that our rulers approximate.

No metric has content of its own. The ruler, with its equal measurement units is merely an approximate realization of a pure idea—an ideal which we invented from tangled experiences of length—invented to make uniform measures available for any application we care to undertake.

4   One Kind of Reading Ability

Let's turn to the measure of reading. We can think of reading as the tree in Fig. 5. It has roots like oral comprehension and phonological awareness. As reading ability grows, a trunk extends through grade school, high school and college branching at the top into specialized vocabularies. That single trunk is longer than many realize. It grows quite straight and singular from first grade through college.

Reading has always been the most researched topic in education (Thorndike, 1965). There have been many studies of reading ability, large and small, local, and national. When the results of these studies are reviewed, one clear picture emerges. Despite the 97 ways to test reading ability (Mitchell, 1985), many decades of empirical data document definitively that no researcher has been able to measure more than one kind of reading ability. This has proven true in spite of intense interest in discovering diversity. Consider three examples: the 1940s Davis Study, the 1970s Anchor Study and six 1980s and 1990s ETS studies.

4.1   Davis—1940s

Fred Davis went to a great deal of trouble to define and operationalize nine kinds of reading ability (1944). He made up nine different reading tests to prove the separate identities of his nine kinds. He gave his nine tests to hundreds of students, analyzed their responses to prove his thesis, and reported that he had established nine kinds of reading. But when Louis Thurstone reanalyzed Davis' data (1946), Thurstone showed conclusively that Davis had no evidence of more than one dimension of reading.

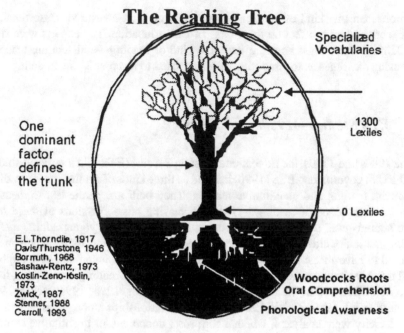

Fig. 5 A tree of reading

4.2   Anchor Study—1970s

In the 1970s, worry about national literacy moved the U.S. government to finance a national Anchor Study (Jaeger, 1973). 14 different reading tests were administered to a great many children in order to uncover the relationships among the 14 different test scores. Millions of dollars were spent. Thousands of responses were analyzed. The final report required 15,000 pages in 30 volumes—just the kind of document one reads overnight, takes to school the next day and applies to teaching (Loret et al., 1974). In reaction to this futility, and against a great deal of proprietary resistance, Bashaw and Rentz were able to obtain a small grant to reanalyze the Anchor Study data (Rentz & Bashaw, 1975, 1977). By applying new methods for constructing objective measurement (Rasch, 1980; Wright & Stone, 1979), Bashaw and Rentz were able to show that all 14 tests used in the Anchor Study—with all their different kinds of items, item authors, and publishers—could all be calibrated onto one linear "National Reference Scale" of reading ability.

The essence of the Bashaw and Rentz results can be summarized on one easy-to-read page (1977)—a bit more useful than 15,000 pages. Their one page summary shows how every raw score from the 14 Anchor Study reading tests can be equated to one linear National Reference Scale. Their page also shows that the scores of all 14 tests can be understood as measuring the same kind of reading on one common scale. The Bashaw and Rentz National Reference Scale is additional evidence that, so far, no more than one kind of reading ability has ever been measured. Unfortunately, their work had little effect on the course of U.S. education. The experts went right on claiming there must be more than one kind of reading—and sending teachers confusing messages as to what they were supposed to teach and how to do it.

4.3 ETS Studies—1980s and 1990s

In the 1980s and 1990s, the Educational Testing Service (ETS) did a series of studies for the U.S. government. ETS (1990) insisted on three kinds of reading: prose reading, document reading and quantitative reading. They built a separate test to measure each of these kinds of reading—greatly increasing costs. Versions of these tests were administered to samples of school children, prisoners, young adults, mature adults, and senior citizens. ETS reported three reading measures for each person and claimed to have measured three kinds of reading (Kirsch, Jungeblut, & Campbell, 1991). But reviewers noted that, no matter which kind of reading was chosen, there were no differences in the results (Kirsch & Jungeblut, 1993, 1994; Reder, 1996; Salganik & Tal, 1989; Zwick, 1987). When the relationships among reading and age and ethnicity were analyzed, whether for prose, document, or quantitative reading, all conclusions came out the same.

Later, when the various sets of ETS data were reanalyzed by independent researchers, no evidence of three kinds of reading measures could be found (Bernstein & Teng, 1989; Reder, Rock & Yamamoto, 1994; 1996; Salganik & Tai, 1989; Zwick, 1987). The correlations among ETS prose, document and quantitative reading measures ranged from 0.89 to 0.96. Thus, once again and in spite of strong proprietary and theoretical interests in proving otherwise, nobody had succeeded in measuring more than one kind of reading ability.

5 Lexiles

Figure 6 is a reading ruler. Its Lexile units work just like the inches. The Lexile ruler is built out of readability theory, school practice, and educational science. The Lexile scale is an interval scale. It comes from a theoretical specification of a readability unit that corresponds to the empirical calibrations of reading test items (Rasch, 1980; Stenner, 1997). It is a readability ruler. And it is a reading ability ruler.

Readability formulas are built out of abstract characteristics of language. No attempt is made to identify what a word or a sentence means. This idea is not new. The Athenian Bar Association used readability calculations to teach lawyers to write briefs in 400 B.C. (Chall, 1988; Zakaluk & Samuels, 1988). According to the Athenians, the ability to read a passage was not the ability to interpret what the passage was about. The ability to read was just the ability to read. Talmudic teachers who wanted to regularize their students' studies, used readability measures to divide the Torah

Fig. 6 Educational status by average Lexile

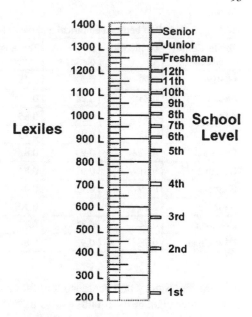

readings into equal portions of reading difficulty in 700 A.D. (Lorge, 1939). Like the Athenians, their concern in doing this was not with what a particular Torah passage was about, but rather the extent to which passage readability burdened readers.

In the twentieth century, every imaginable structural characteristic of a passage has been tested as a potential source for a readability measure: the number of letters and syllables in a word; the number of sentences in a passage; sentence length; balances between pronouns and nouns, verbs and prepositions (Stenner, 1997). The Lexile readability measure uses word familiarity and sentence length.

5.1 Lexile Accuracies

Table 1 lists the correlations between readability measures from the ten most studied readability equations and student responses to different types of reading test items. The columns of Table 1 report on five item types: Lexile Slices; SRA Passages; Battery Test Sentences; Mastery Test Cloze Gaps; Peabody Test Pictures.

The item types span the range of reading comprehension items. The numbers in the table show the correlations between theoretical readability measures of item text and empirical item calibrations calculated from students' test responses. Consider the top row. The Lexile readability equation predicted how difficult Lexile slices would be for persons taking a Lexile reading test at a correlation of 0.90, the SRA passage at 0.92, the Battery Sentence at 0.85, the Mastery Cloze at 0.74 and the Peabody Picture at 0.94 (Stenner, 1996, 1997). With the exception of the cloze items, these predictions are nearly perfect. Also note that the simple Lexile equation, based only

Table 1 Correlations between empirical and theoretical item difficulties

Ten readability equations	Lexile slice	Five SRA passage	Test item types battery sentence	Mastery cloze	Peabody picture
Lexile	0.90	0.92	0.85	0.74	0.94
Flesch	0.85	0.94	0.85	0.70	0.85
ARI	0.85	0.93	0.85	0.71	0.85
FOG	0.85	0.92	0.73	0.75	0.85
Powers	0.82	0.93	0.83	0.65	0.74
Holquist	0.81	0.91	0.81	0.84	0.86
Flesch-1	0.79	0.92	0.81	0.61	0.69
Flesch-2	0.75	0.87	0.70	0.52	0.71
Coleman	0.74	0.87	0.75	0.75	0.83
Dale-Chall	0.76	0.92	0.82	0.73	0.67

on word familiarity and sentence length, predicts empirical item responses as well as any other readability equation—no matter how complex. Table 1 documents, yet again that one, and only one, kind of reading is measured by these reading tests. Were that not so, the array of nearly perfect correlations could not occur. Table 1 also shows that we can have a useful measurement of text readability and reader reading ability on a single reading ruler!

An important tool in reading education is the basal reader. The teaching sequence of basal readers records generations of practical experience with text readability and its bearing on student reading ability. Table 2 lists the correlations between Lexile readability and basal reading order for the 11 basal readers most used in the United States Each series is built to mark out successive units of increasing reading difficulty. Ginn has 53 units—from book 1 at the easiest to book 53 at the hardest. HBJ Eagle has 70 units.

Teachers work their students through these series from start to finish. Table 2 shows that the correlations between Lexile measures of the texts of these basal readers and their sequential positions from easy to hard are extraordinarily high. In fact, when corrected for attenuation and range restriction, these correlations become approach perfection (Stenner et al., 1992, 1996, 1997, 1998).

Each designer of a basal reader series used their own ideas, consultants and theory to decide what was easy and what was hard. Nevertheless, when the texts of these basal units are Lexiled, these Lexiles predict exactly where each book stands on its own reading ladder—more evidence that, despite differences among publishers and authors, all units end up benchmarking the same single dimension of reading ability.

Finally there are the ubiquitous reading ability tests administered annually to assess each student's reading ability. Table 3 shows how well theoretical item text Lexiles predict actual readers' test performances on eight of the most popular reading tests. The second column shows how many passages from each test were Lexiled. The third column lists the item type.

Table 2 Correlations between basal reader order and Lexile readability

Basal reader Series	Basal units	r	R	R'
Ginn	53	0.93	0.98	1.00
HB JE Agle	70	0.93	0.98	1.00
SF FOCUS	92	0.84	0.99	1.00
Riverside	67	0.87	0.97	1.00
HM (1983)	33	0.88	0.96	0.99
Economy	67	0.86	0.96	0.99
SF Amer Trad	88	0.85	0.97	0.99
HB J Odyssey	38	0.79	0.97	0.99
Holt	64	0.87	0.96	0.98
HM (1986)	46	0.81	0.95	0.97
Open court	52	0.54	0.94	0.97

Table 3 Correlations between passage Lexiles and item readabilities

Tests	Passages analyzed	Item type	r	R	R'
SRA	46	Passage	0.95	0.97	1.00
CAT-E	74	Passage	0.91	0.95	0.98
CAT-C	43	Passage	0.83	0.93	0.96
CTBS	50	Passage	0.74	0.92	0.95
NAEP	70	Passage	0.65	0.92	0.94
Lexile	262	Slice	0.93	0.95	0.97
PIAT	66	Picture	0.93	0.94	0.97
Mastery	85	Cloze	0.74	0.75	0.77

Once again there is a very high correlation between the difficulty of these items as calculated by the entirely abstract Lexile specification equation and the live data produced by students answering these items on reading tests. When we correct for attenuation and range restriction, the correlations are just about perfect. Only the Mastery Cloze test, well-known to be idiosyncratic, fails to conform fully (Stenner, 1997).

What does this mean? Not only is only one reading ability being measured by all of these reading comprehension tests, but we can replace all the expensive data used to calibrate these tests empirically with one formula—the abstract specification equation. We can calculate the reading difficulty of test items by Lexiling their text without administering them to a single student!

Putting the relationship between theoretical Lexiles and observed item difficulties into perspective, the uncorrected correlation of 0.93, when disattenuated for error and corrected for range restrictions, approaches 1.00. The Lexile equation produces an almost perfect correlation between theory and practice (Stenner, 1997).

Variation comes from item response options that have to compete with each other or they do not work. But there has to be one correct answer. Irregularity in the composition of multiple choice options, even when they are reduced to choosing one word to fill a blank, is unavoidable. What the item writer chooses to ask about a passage and the options they offer the test taker to choose among are not only about reading ability. They are also about personal differences among test writers.

There are also variations among test takers in alertness and motivation that disturb their performances. In view of these unavoidable contingencies, it is surprising that the correlation between Lexile theory and actual practice is so high.

How does this affect the measurement of reading ability? The root mean square measurement error for a one item test would be about 172 Lexiles (Stenner, 1997). What are the implications of that much error? The distance from First Grade school books to Second Grade school books is 200 Lexiles. So we would undoubtedly be uneasy with measurement errors as large as 172 Lexiles. However, when we combine the responses to a test of 25 Lexile items, the measurement error drops to 35 Lexiles. And when we use a test of 50 Lexile items, the measurement error drops to 25 Lexiles—one eighth of the 200 Lexile difference between First and Second Grade books. Thus, when we combine a few of Lexile items into a test, we get a measure of where a reader is on the Lexile reading ability ruler, precise enough for all practical purposes. We do not plumb their depths of understanding. But we do measure their reading ability.

5.2 Lexile Items

One might now ask, how hard is it to write a Lexile test item? Fig. 7 describes a study to find out whether Lexile items written by different authors produce usefully equivalent results (Stenner, 1998).

Five apprentice item authors were each asked to choose their own text passages and to write their own response illustrated missing word options (Fig. 8) each author

5 different item authors compose
5 different sets of 60 Lexile items evenly sequenced from 900 to 1300 Lexiles

5 different 60 item tests are assembled
Each test containing 12 items selected at random from each author's set of 60

7 grade school students take a different 60 item test each day for 5 days

This produces, for each student
5 measures across 5 days balanced over authors
and
5 measures across 5 authors balanced over days

Fig. 7 Stability study

Wilbur likes Charlotte better and better each day. Her campaign against insects seemed sensible and useful. Hardly anybody around the farm had a good word to say for a fly. Flies spent their time pestering others. The cows hated them. The horses hated them. The sheep loathed them. Mr. and Mrs. Zuckerman were always complaining about them, and putting up screens.

Everyone _____ about them.
a) agreed
b) gathered
c) laughed
d) learned

Fig. 8 An 800 Lexile slice test item

wrote 60 items spanning 900 to 1300 Lexiles. From these ($5 \times 60 = 300$) items, five 60 item tests were constructed by drawing 12 items at random from each author. Then seven grade school students were given a different test each day for five days. This produced five measures for each student over the five days. And, by pooling days, five measures for each student over the five authors.

The question becomes "Is the variation by author in a student's reading ability measure any larger than the variation by day?" If not, that would imply that writing useful Lexile test items, as in Fig. 8, was not a problem, since even apprentice authors can do it well enough to obtain measures as stable as the differences in a person's reading performance from day to day.

Findings indicate that no more noise is introduced into the Lexile way of making a reading measure by a difference among item authors than by the difference a day makes.

These five Lexile item authors were not experts. They were just well educated persons, instructed in Lexile item writing for four hours. Courtney, 27, is a psychology student. John, 23, is a math student. Gail, 35, is a law student. Chris, 22, is a football player. Gayle, 45, is a teacher.

5.3 Calculating Lexiles

Lexile measures of reading are easy to understand and easy to use. Lexile readability—measured by word familiarity and sentence length—establishes how difficult a text is to read. Lexile reading ability—measured by how well a reader is able to recognize words and connect them into sentences—establishes how able a reader is to read a text (Stenner, 1982, 1983, 1987).

Readability is passage reading difficulty.

Reading Ability is ability to read passages.

Lexile reading ability is measured by finding out what Lexile passage readability a person can read with 75 percent success.

Success is defined as recognizing what words are needed to mend gaps inserted in passages.

Divide the BOOK into natural SLICES of 125-140 words.
For each TEXT SLICE i, determine:

Log mean sentence length $=$ SL_i

Mean log word frequency $=$ WF_i

Then   .

CALIBRATE SLICE i
at
Readability $= 582 + 1768* SL_i - 386*WF_i$ Lexiles

The Lexile Measure of a Book is equal to the Lexile Level of a Reader
Who succeeds at 75% of that Book's Slices

Fig. 9 How to Lexile a book

The Lexile formula is based on two axioms.

The *semantic axiom*: the more familiar the words, the easier the passage is to read;

The more unfamiliar the words, the harder.

The *syntactic axiom*: the shorter the sentences, the easier the passage is to read; the longer the sentences, the harder.

These axioms apply to whatever is read, quite apart from content. They apply whether we like what we are reading or not, whether it is prose, document or quantitative.

The Lexile system calculates passage readability from just these two characteristics—both of which are explicit in the passage. Sentence lengths are there to see. We count and average them. Word familiarities are obtained from compilations of word usage. The Lexile Analyzer uses John Carroll's sample of 5 million words (Carroll, Davies, & Richmond, 1971).[1]

If readers do not know the words, they cannot read the passage. If they do know the words, they can begin to make the passage take shape by stringing its words into sentences. If they can make the sentences, they can read the passage and then, and only then, begin to think about what the passage has to say. Knowing the words and making the sentences sets the threshold for reading (Hitch & Baddeley, 1974; Lieberman et al., 1982; Shankweiler & Crain, 1986; Miller & Gildea, 1987).

To Lexile a passage, we look up the occurrence frequency of each word. The Lexile Analyzer uses the average log word frequency and the logarithm of average sentence length. The final Lexile measure for the passage is a weighted sum of these two logarithms. Figure 9 shows how to Lexile a book.

---

[1] The familiarity of the words used in a passage can be estimated from any comprehensive word usage compilation—A Basic Vocabulary of Elementary School Children (Rinsland, 1945); The Teacher's Word Book of 30,000 Words, Thorndike and Lorge (1952); The Word Frequency Book, Carroll, Davies and Richman (1971); The Educators' Word Frequency Guide, Zeno, Ivens, Millard and Davvuri (1995).

Test the READER
with L Response IllustratedLexile Calibrated SLICES
of Average slice Lexile $=$ H
and slice Lexile Standard Deviation $=$ S

Then

Count the READER'S right answers for Score= R
This Reader's Lexile MEASURE is
Reading Ability= H $+$ $(180 + S^2/1040)$ log [R/L-R)]

The Lexile Measure of a Reader is equal to the Lexile Level of a Text
For which a Reader succeeds on 75% of the Slices

Fig. 10 How to Lexile a reader

The coefficients in the formula are set to provide the most efficient balance between log word familiarity and log sentence length and to define a metric that reaches 1000 Lexiles from the books used in First Grade at 200 Lexiles to the books used in Twelfth Grade at 1200 Lexiles. The full Lexile range of readability goes from zero to 1800. The equation is simple. Word familiarity and sentence length are all there is to it. Figure 10 shows how to Lexile a reader.

5.4 Lexile Relationships

When a reader with a Lexile ability of 1000L is given a 1000L text, we expect them to experience a 75% success rate (Stenner, 1992). If the same reader is given a 750L text, then we expect their success rate to improve to 90%. If the text is at 500L, their success rate should improve to 96%. The more readers' Lexile abilities surpass the Lexile readability of a text, the higher their expected success rates. However, the more a text Lexile readability surpasses readers' Lexile reading abilities, the lower their expected success rates.

Success rates are relative. They are the results of Lexile differences between readers and texts. The 250L difference between a 750L text and a 1000L reader results is the same success rate as the 250L difference between a 1000L text and a 1250L reader. Each reader-text combination produces 90% reading success. Success rates are centered at 75% because readers forced to read at 50% success report frustration, while readers reading at 75% report comfort, confidence and interest.[2]

Each reader has their own range of reading comfort. As a result, there is a natural range of text readability that most motivates each reader to improve their reading ability. Some readers are challenged by a success rate as low as 60%. Others find

---

[2] Squires, Huitt and Segars (1983) found that reading achievement for second-graders peaked when their success rate reached 75%. A 75% rate is also supported by the findings of Crawford, King, Brophy and Evertson (1975).

that burdensome. Once a reader places themselves and their books in the Lexile Framework, they can discover what Lexile difference between their reading ability and text readability challenges them in the most productive way.

Book readability varies from page to page. Some books have a narrow range, their passages cluster around a common level. As we read these books, the reading challenge stays level. There are no hills or valleys. Other books have a wide range of readability. There are easy passages and hard passages. These books can enable us to use the momentum we gain from the easier passages to surmount the challenge of the harder ones. Overcoming this kind of resistance improves reading ability.

When we want to help a student read, we can Lexile them and then offer them books with a readability that matches their reading ability. It is also helpful to know the book's passage difficulty variation. If we want our students to learn to read by reading, then we want to give them material that fascinates, motivates, absorbs and also challenges them.

We do that best by giving them books they want to read that are a little too hard for them, with passages that vary in passage difficulty. Then as they read along, they speed up and slow down. The speed-ups give them the energy and confidence needed to work through the slow-downs.

5.5 Using Lexiles

Books are brought into the Lexile Framework by Lexiling the books. Tests are brought into the Framework by Lexiling their items and using these Lexile calibrations as the basis for estimating readers and reading ability.

To write a Lexile test item, we can use any natural piece of text. If we wish to write an item at 1000 Lexiles, we select books that contain passages at that level. We select a 1000 Lexile passage and add a relevant continuation sentence at the end with a crucial word missing. This is the "response illustration." Then we compose four one word completions, all of which fit the sentence but only one of which makes sense. Thus, the only technical problem is to make sure all choices complete a perfectly good sentence, but that only one choice fits the passage.

The aim of a Lexile item is to find out whether the student can read the passage well enough to complete the response illustrated sentence with the word that fits the passage. Lexiled items like this are available at the Lexile website (www.lexile.com).

The Lexile Slice is a simple, easy to write item type. But in practice, we may not even need the slice to determine how well a person reads. Instead, we may proceed as we do when we take a child's temperature. Since, the Lexile Framework provides a ruler that measures readers and books on the same scale, we can estimate any person's reading ability by learning the Lexile level of the books they enjoy.

6 The One Minute Self-report

When our child says, "I feel hot!" we infer they have a fever. When a person says, "I like these books," and we know the books' Lexile levels, we can infer that the person reads at least that well.

7 The Three Minute Observation

To find out more about our child, we feel their forehead. The three minute way to measure a person's reading is to pick a book with a known Lexile level and ask the person to "Read me a page." If they read without hesitation, we know they read at least that well. If they stumble, we pick an easier book. With two or three choices, we can locate the Lexile level at which the person is competent, just by having them read a few pages out loud.

With a workbook of Lexile calibrated passages, we can implement the three minute observation this simply by opening the workbook and turning the pages to give them successive passages to read.

8 The Fifteen Minute Measurement

To find out more, we use a thermometer to take our child's temperature, perhaps several times. For reading, we give the person some Lexiled passages ended with an incomplete sentence. To measure their reading ability, we find the level of Lexiled passages at which that person correctly recognizes what words are needed to replace the missing words 75% of the time.

The Lexile reading ruler connects reading, writing, speaking, listening with books, manuals, memos and instructions. This stable network of reproducible connections empowers a world of opportunities of the kind that the inch makes available to scientists, architects, carpenters and tailors (Luce & Tukey, 1964).

In school, we can measure which teaching method works best and manage our reading curriculae more efficiently and easily. In business, we can Lexile job materials and use the results to make sure that job and employee match. When a candidate applies for a position, we can know ahead of time what level of reading ability is needed for the job and evaluate the applicant's reading ability by finding out what books they are reading and asking them to read a few sentences of job text out loud. This quick evaluation of an applicant's reading ability will show us whether the applicant is up to the job. When an applicant is not ready, we can counsel them, "You read at 800 Lexiles. The job you want requires 1000 Lexiles. To succeed at the job you want, you need to improve your reading 200 Lexiles. When you get your reading ability up to 1000, come back so that we can reconsider your application."

8.1 Lexile Perspectives

Jobs—Twenty-five thousand adults reporting their jobs to the 1992 National Adult Literacy Study (Campbell et al., 1992; Kirsch et al., 1993, 1994). Their reading ability was also measured. In 1992, the average laborer read at 1000 Lexiles. The average secretary at 1200. The average teacher at 1400. The average scientist at 1500.

When we can see so easily how much increasing our reading ability can improve our lives, we cannot help but be motivated to improve, especially when what we must do is so obvious. If we want to be a teacher at 1400 Lexiles but read at only 1000, it is clear that we have 400 Lexiles to grow to reach our goal. If we are serious about teaching, the Lexile Framework shows us exactly what to do. As soon as we can take 1400 Lexile books off the shelf and read them easily, we know we can read well enough to be a teacher. But if we find that we are still at 1000 Lexiles, then we cannot avoid the fact that we are not ready to qualify for teaching, not yet, not until we teach ourselves how to read more difficult text.

School—Reading is learned in school. The 1992 National Adult Reading Study shows that there is a strong relationship between the last school grade completed and subsequent adult reading ability. On average, we are never more literate than the day we left school. The average 7th grade student reads at 800 Lexiles. The average high school graduate reads at 1150 Lexiles. College graduates can reach 1400 Lexiles. For many of us the last grade of school we successfully complete defines our reading ability for the rest of our lives. Once we leave school—and we no longer benefit from the reading challenge that school provides—we tend to stop learning.

Income—Reading ability also limits how much we can expect to earn. The average incomes of readers in the 1992 National Adult Literacy Study indicated that, on average, an adult reading at 950 Lexiles made $10,000, at 1200 Lexiles, $30,000, at 1400 Lexiles, $60,000 and at 1500 Lexiles, $100,000. From 1000 to 1300 Lexiles, each reading ability increases of 150 Lexiles doubles our earning expectations. If we read at 1000 Lexiles and want to double our potential, then we have to improve are reading to 1150 Lexiles.

When students can see the financial consequences of reading ability on an easy to understand scale that connects reading ability and income, then they have a persuasive reason to spend more time improving their reading abilities. No need to berate students, "Do your homework!" Instead, we can show them, "You want more money? You want to be a doctor? Here is the road. Learn to read better. It's up to you. But we'll help you learn."

8.2 Reading Education

Education can only succeed if we connect learning to each learner's selfish motives. We need to involve our students individually, to engage their desires and arouse their drives. When we do that, student education will drive itself. Then, all we need do is to add support and guidance. Otherwise, we will continue to deceive ourselves into running a penitentiary system that keeps some troublesome kids off the street, but only for a while.

Remember, when we know text readability, all we need do to learn how well a student reads is to ask them to read a page or two aloud. If they succeed, we can give them a harder page. If not, we know their reading ability is below the readability of the text we asked them to read. No need for debate. No need for guesswork. No need for confession or reproach. The student's status is plain to us and plain to them. We have not tricked them with a mysterious test score. All we have done is to help them see for themselves how high they can read.

References

Bernstein, I. H., & Teng, G. (1989). Factoring items and factoring scales are different: Spurious evidence for multidimensionality due to item categorization. *Psychological Bulletin, 105*(3), 467–477.

Campbell, A., Kirsch, I. S., & Kolstad. (1992). *Assessing literacy: The framework for the national adult literacy survey.* National Center for Education Statistics, U.S. Department of Education.

Carroll, J. B., Davies, P., & Richmond, B. (1971). *The word frequency book.* Houghton Mifflin.

Chall, J. S. (1988). The beginning years. In B. L. Zakaluk & S. J. Samuels (Eds.), *Readability: Its past, present and future.* Newark, DE: International Reading Association.

Crawford, W. J., King, C. E., Brophy, J. E., & Evertson, C. M. (1975). Error rates and question difficulty related to elementary children's learning. In *Paper Presented at the annual meeting of the American Educational Research Association.* Washington, D.C.

Davis, F. (1944). Fundamental factors of comprehension in reading. *Psychometrika, 9,* 185–197.

Educational Testing Service. (1990). *ETS tests of applied literacy skills.* Simon & Schuster Workplace Resources.

Jaeger, R. M. (1973). The national test equating study in reading (the anchor test study). *Measurement in Education, 4,* 1–8.

Hitch, G. J., & Baddeley, A. D. (1974). Verbal reasoning and working memory. *Journal of Experimental Psychiatry, 28,* 603–621.

Kirsch, I. S., Jungeblut, A., & Campbell, A. (1991). *The ETS tests of applied literacy.* Educational Testing Service.

Kirsch, I. S., Jungeblut, A., Jenkins, L., & Kolstad, A. (1993). *Adult literacy in America: A first look at the results of the national adult literacy survey.* National Center for Education Statistics, U.S. Department of Education.

Kirsch, I. S., Jungeblut, A., & Mosenthal, P.B. (1994). Moving toward the measurement of adult literacy. In *Paper Presented at the March NCES Meeting.* Washington, DC

Liberman, I. Y., Mann, V. A., Shankweiler, D., & Werfelman, M. (1982). Children's memory for recurring linguistic and non-linguistic material in relation to reading ability. *Cortex, 18,* 367–375.

Lorge, I. (1939). Predicting reading difficulty of selections for school children. *Elementary English Review, 16,* 229–233.

Loret, P. G., Seder, A., Bianchini, J. C., & Vale, C. A. (1974). *Anchor test study final report: Project report* and vols. 1–30. Berkeley, CA: Educational Testing Service (ERIC Document Nos. Ed 092 601 - ED 092 631).

Luce, R. D., & Tukey, J. W. (1964). Simultaneous conjoint measurement: A new type of fundamental measurement. *Journal of Mathematical Psychology, 1,* 1–27.

Miller, G. A., & Gildea, P. M. (1987). How children learn words. *Scientific American., 257,* 94–99.

Mitchell, J. V. (1985). *The ninth mental measurements yearbook.* University of Nebraska Press.

Rasch, G. A. (1980). The University of Chicago Press (first published in 1960).

Reder, S. (1996). Dimensionality and construct validity of the NALS assessment. In M. C. Smith (Ed.) *Literacy for the 21st century: Research, policy and practice.* Greenwood Publishing (in Press).

Rentz, R. R., & Bashaw, W. L. (1975). Equating reading tests with the Rasch model, vl: *Final Report & vol 1 & vol2: Technical reference tables. Final Report of US. Department of Health, Education, and Welfare* Contract OEC-O72-5237. Athens, GA: The University of Georgia. (ERIC Document Reproduction Nos. ED 127 330 & ED 127–331.

Rentz, R. R., & Bashaw, W. L. (1977). The national reference scale for reading: An application of the Rasch model. *Journal of Educational Measurement, 14,* 161–179.

Rinsland, H. D. (1945). *A basic vocabulary of elementary school children.*

Rock, D. A., & Yamamoto, K. (1994). *Construct validity of the adult literacy subscales.* Educational Testing Service.

Salganik, L. H., & Tai, J. (1989). *A review and reanalysis of the ETS/NAEP young adult literacy survey.* Pelavin Associates.

Shankweiler, D., & Crain, S. (1986). Language mechanisms and reading disorder: A modular approach. Cognition, 14, 139–168.

Squires, D. A., Huitt, W. G., & Segars, J. K. (1983). *Effective schools and classrooms.* Alexandria, VA: Association for Supervisor and Curricular Development.

Stenner, A. J., Smith, M., & Burdick, D. S. (1987). *Fit of the Lexile theory to item difficulties on fourteen standardized reading comprehension tests.* Durham, NC: Metametri cs.

Stenner, A. J. (1992). Meaning and method in reading comprehension. In *Paper Presented at AERA, Division,* D. Rasch Special Interest Group, San Francisco, California.

Stenner, A. J. (1996). Measuring reading comprehension with the Lexile framework. In *Paper Presented at the Fourth North American Conference on Adolescent/Adult Literacy.* Washington, D.C.

Stenner, A. J. & Burdick, D. S. (1997). *The objective measurement of reading comprehension: In response to technical questions raised by the California Department of Education Technical Study Group.* Durham, NC: Metametrics.

Stenner, A. J. Writing Lexile Items (1998). *Paper Presented at the American Educational Research Association Annual Meeting.* University of Chicago.

Thorndike, E. L. & Lorge, I. (1952). *The teacher's word book of 30,000 word.*

Thorndike, E. L., & Hagen, E. P. (1965). *Measurement and evaluation in psychology and education.* Wiley.

Thurstone, L. L. (1946). *Note on a reanalysis of Davis' reading tests. Psychometrika, v11*(n2), 185ff.

Wright, B. D., & Stone, M. H. (1979). *Best test design.* MESA Press.

Zeno, S. M., Ivens, S. H., Millard, R. T. & Davvuri, R. (1995). *The educators word frequency guide.* Touchstone.

Zakaluk, & Samuels, S. J. (Eds.) (1988). *Readability: Its past, present and future, Newark.* DE: International Reading Association.

Zwick, R. (1987). Assessing the dimensionality of the NAEP reading data. *Journal of Educational Measurement, 24,* 293.

Mapping Variables

Mark H. Stone, Benjamin D. Wright, and A. Jackson Stenner

Abstract This paper describes Mapping Variables, the principal technique for planning and constructing a test or rating instrument. A variable map is also useful for interpreting results. Modest reference is made to the history of mapping leading to its importance in psychometrics. Several maps are given to show the importance and value of mapping a variable by person and item data. The need for critical appraisal of maps is also stressed.

The **Map of a Variable** is the beginning and end of assessment. But we must immediately add that variable construction never ends because it is never complete; it is ever continuing. Variables require continuous attention for their development and maintenance. The map of a variable is a visual representation of the current status of variable construction. It is a pictorial representation of the "state of the art" in constructing a variable.

1 The Origin of Mapping

Maps are visual guides. They ground us in a stable frame of reference and give a sense of direction. How frequently we use expressions of belief implying vision, "Do you see?", "Now I see.", "Show me what you mean.", and "Put me in the picture."

Journal of Outcome Measurement, 3(4). 1999.

M. H. Stone
Adler School of Professional Psychology, Chicago, IL, USA

B. D. Wright
Departments of Education and Psychology, University of Chicago, Chicago, IL, USA

A. J. Stenner (✉)
MetaMetrics, Inc., Durham, NC, USA

© The Author(s) 2023
W. P. Fisher and P. J. Massengill (eds.), *Explanatory Models, Unit Standards, and Personalized Learning in Educational Measurement,*
https://doi.org/10.1007/978-981-19-3747-7_8

109

These expressions testify to the visual power inherent in pictorial representations and conveyed in speech and writing. Mapping visualizes the extent of our knowledge.

Maps are indispensable to planning and traveling. Map making has great utility. The inability to understand or make use of maps is a handicap to understanding the world.

The earliest maps used naturally occurring phenomena—celestial and terrestrial—to identify features. If we look at the sky on a starry night we can use the "pointers" of the Big Dipper to locate Polaris, the pole star. Although both dippers move, they rotate around Polaris which appears fixed and determines north. From this "fixed" star we orient ourselves to the points of the compass. More comprehensive maps of the heavens include the popular constellations of the Zodiac, lesser known constellations and other celestial features. The more celestial features we know, the better oriented we become to a starry night.

Terrestrial markers also serve to orient. A lake, a river, or a mountain may be used to anchor locations. Celestial and terrestrial maps have been used for centuries and are sometimes brought into relationship with one another. Today's roadmaps are but a current update of the state of knowledge in local geography. A map is an analogy, an idea that pictures an abstraction. While the map may initially seem superficial, incomplete, or even inaccurate, it still serves a purpose. The map shows the current status of what is known about a domain.

Maps by their very nature, invite improvement. Every edition of a map calls attention to its accuracy and inaccuracy. Each new edition incorporates changes from a previous one resulting in a new and more accurate version.

Consider a map with the lines of longitude and latitude. This illustrates the benefits of superimposing an abstraction upon the natural contours of land and sea. Abstractions enhance maps by expediting generalization.

Natural reference points also explain by serving as markers to ground our observations. The more natural reference points we can employ, the fewer the resulting errors. The wider apart the natural markers, the greater the possibility of error. Brown (1949) provides a comprehensive history of map building with numerous illustrations that record how maps have become increasingly more accurate. Wilford (1981) has produced a similar, but more recent history. Edward Tufte's recent publications (1979, 1983) offer a panorama of useful visual strategies together with his critique of how visual displays can facilitate the interpretation of data or mislead.

The use of maps illustrates several important aspects:

1. Maps are useful pictures of experience.
2. Inaccuracies are successively and inevitably corrected.
3. Abstractions, such as longitude and latitude, enhance mapping.
4. More knowledge produces greater accuracy.

2 Maps of Variables

Map topography is a useful application to psychometrics because a map is an abstraction of a variable. The variable implied by a test can first be pictured as a line (Wright & Stone, 1979, pp. 1–6). It is a line with direction illustrated by an arrow. The variable is defined by items and persons, but other useful characteristics can also be incorporated on the map. Continuous improvement is irresistible. Maps invite further corrections. The more information we gather about the variable, the more accurate our representation becomes. Finally, this pictorial representation of the variable invites yet further abstractions that generalize understanding.

Rudolph Carnap wrote:

> The nineteenth-century model was not a model in this abstract sense (i.e., a mathematical model). It was intended to be a spatial model of a structure, in the same way that a model ship or airplane represents an actual ship or plane. Of course, the chemist does not think that molecules are made up of little colored balls held together by wires; there are many features of his model that are not to be taken literally. But, in general spatial configuration, it is regarded as a correct picture of the spatial configuration of atoms of the actual module. As has been shown, there are good reasons sometimes for taking such a model literally—a model of a solar system for example, or a crystal or molecule. Even when there are no grounds for such an interpretation, visual models can be extremely useful. The mind works intuitively, and it is often helpful for a scientist to think with the aid of visual pictures. At the same time, there must always be and awareness of the model's limitations. The building of a neat visual model is no guarantee of a theory's soundness, nor is the lack of a visual model and adequate reason to reject a theory. (Carnap, 1966, p. 176)

Carnap's exposition clearly indicates the value of a map in fostering pictures by which to visually conceptualize an intuitive idea. He also cautions that maps are not substitutes for reality, but pictures and as such they cannot be interpreted literally.

3 Using Maps

There are three uses of maps:

To DIRECT… where are we planning to go,
To LOCATE… where we are, along the way, and
To RECORD… where we have been.

These three uses indicate that a map is the beginning and end of test construction (Stone, 1995). In the beginning stages, a map defines our intentions. At the end, it is a realization of progress to date. In between are markers along the way. Maps of variables are never finished because they invite constant correction and improvement. When maps embody abstractions derived from experience they connect the world of the mind to the world of experience. Abstraction is validated by correspondence to experience and experience is understood by abstraction.

Mapping illustrates the dialogue that must take place between these two worlds in order to communicate constructively. A map is a visual, operational definition of a variable. While maps are necessarily only models, their pictorial representation invites continual correction, ever increasing their accuracy.

4 Graphs as Maps

Graphs of functions are maps showing the relationship between two variables. Graphs make it easy to see by looking whether a useful function is emerging.

Graphs make functions recognizable and familiar. We recognize linearity in a straight line, and special curves describe a parabola or a quadratic relation. Other well-known functions like the undulating curve for the sine are easily recognized by their shape. The graphs of functions are maps as familiar to their users as roadmaps are to motorists. They aid understanding by simplifying the process and allowing us to "see" a complex representation.

5 A Map Is an Analogy

Measurement is made by analogy. Our most efficient and utilitarian measures rely upon visual representation. The ruler, the watch face, the mercury column, and the dial are common analogies used to record length, time, temperature, and weight. The utilitarian success of analogy in these measuring tools is demonstrable by their ubiquity.

- The "intended map" of the variable is the idea, plan, and best formulation of our intentions.
- The "realized map" of the variable which is made from item calibrations and person measures implements the plan.
- Continuous dialogue between intention (idea) and realization (data) produces and maintains the validity of the variable.
- A "Map of a Variable" is. the scope and sequence of instruction because it shows how to sequence instruction and how to relate it to assessment.
- Progress from instruction and resultant learning can be located on the variable. Growth can be seen and measured.
- There are shortcomings to maps, especially evident if their use is "pushed" to extremes.

What writers like Carnap (1966) and Kaplan (1964) present in their discussions about the shortcomings of models also applies to maps. We must be careful not to expect too much of a map and ascribe more substance to what is produced than can be justified. Constant monitoring of map building is necessary. Monmonier's (1996) book "How to Lie With Maps" presents in a useful and amusing way the fallacies that can result from viewing a map as a "finished product" rather than as a "fiction," an approximation of the outcome and one that is in process and never completed.

Braithwaite (1956) has also cautioned, "The price of the employment of models [maps] is eternal vigilance."

Psychometric maps serve as the plan for instrument development and revision. The map of a variable is a blueprint for a test. When a map is logical and well constructed, its implementation can be straightforward in the form of ordered items.

Figure 1 is a flowchart of the steps in bringing a variable into existence. Its development is guided by a map of intention.

When the map is empirically verified, it documents a successful realization of an idea. The map of the variable pictures both the idea and its realization in the form of calibrated items and measured persons (Wright & Stone, 1996, Chap. 14). It embodies the construct validity of the instrument.

Figure 2 is the item/person map for the Knox Cube Test (Stone & Wright, 1980) generated by BIGSTEPS (Wright & Linacre, 1991 to date) This map as well as

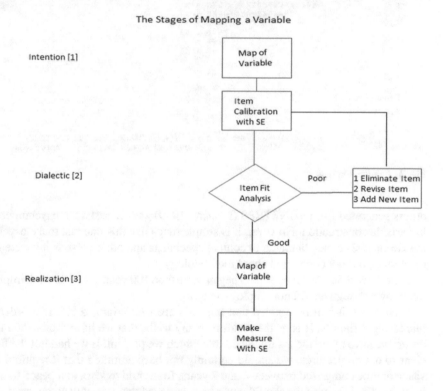

1. The "intended map" of the variable is the idea and plan of our intention.
2. There is continuous dialogue between the intention (idea) and realization (data} to maintain validity and quality control. The degree of correspondence between the maps of intention and realization indicates the degree of success achieved.
3. The "realized map" of the variable conveys in the item calibrations and person measures the best outcome to date.

Fig. 1 Flowchart for variable construction

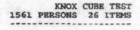

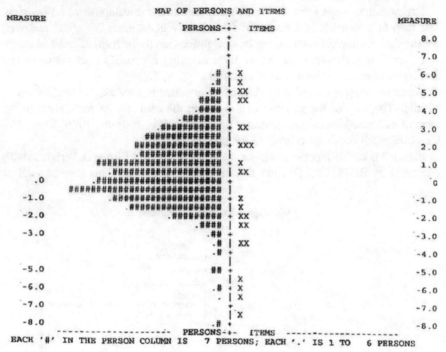

Fig. 2 Item/person map

others generated from WINSTEPS (Linacre, 1999) greatly assist the psychometrist in variable construction. However, it is simple maps like this one that make psychometric analysis understandable to content specialists and other persons interested in the results, but not concerned about methodology.

Binet's work in test development began more than 100 years ago. His work implies mapping although he did not employ the term.

First of all, it will be noticed that our tests are well arranged in a real order of increasing difficulty. It is as the result of many trials, that we have established this order; we have by no means imagined that which we present. If we had left the field clear to our conjectures, we should certainly not have admitted that it required the space of time comprised between 4 and 7 years, for a child to learn to repeat 5 figures in place of 3. Likewise we should never have believed that it is only at ten years that the majority of children are able to repeat the names of the months in correct order without forgetting any; or that it is only at 10 years that a child recognizes all the pieces of our money. (Binet, 1916, p. 329).

Binet clearly indicates how data from experience was used to establish a hierarchy of item difficulty. He makes special note of the requirement for "well-arranged" items expressing a "real order". Binet also relied on "numerous" replications of ordered items in order to produce the level of accuracy he desired.

One might almost say, 'It matters very little what the tests are so long as they are numerous' (Binet, 1916, p. 329).

Binet clearly stressed (1) item arrangement by difficulty order, (2) numerous items, sufficient for precision. How else can one be successful? There is no other way except to do as Binet did: begin with an idea for a variable, illustrate the variable by items, arrange them by their intended difficulty, and measure persons by their locations among the items. The hallmark of Binet's efforts is his early effort at benchmarking items and persons on a variable. He must have had a mental map of what he intended, although there is no indicate of one in his writings.

An early example of a psychometric map is Thurstone's "Scale of Seriousness of Offense" (1927, 1959) shown in Fig. 3. His map marks out the severity of offenses from "vagrancy" located at the bottom end to "rape" at the top.

The scale is further subdivided into three offense categories: (1) sex offenses, located at the top of the scale, (2) injury to the person, located from the top to the middle, and (3) property offenses, located from the middle of the scale and down. Thurstone's map provides insight into a hierarchy of criminal acts and a practical "ruler" for determining, not only the location of offenses, but the "distance" between them.

Figure 4 is a map of an achievement variable: WRAT3 (Wilkinson, 1993). This test of achievement measures (1) word naming, (2) arithmetic computation, and (3) spelling from dictation. Items are arranged according to difficulty These maps progress from left to right indicating increases in item difficult and person ability. The arrangement of items indicates the expected arrangement of persons according to their abilities. Less able persons will be located to the left of more able persons. Able persons will find the items on the left easier than those items further along to the right. These three variables follow developmental lines of learning, correspond to instructional goals and make test administration efficient and informative. The map of each WRAT3 variable is enhanced by sample items illustrating progressive difficulty and below the items is an equal-interval scale indicating the measures.

These maps have immediate application. Like the marks made on a door jamb to show the increasing height of a child, this map shows student progress on three achievement variables. The maps show order to the items and measures. Progress of pupils along this educational ruler is enhanced by criterion and normative locations. The grade and age norms show growth. The map provides useful information to students, teachers and parents.

Figure 5 is a reduction of the "map" of the Lexile Scale of Reading© copyright Metametrics (1995).

The master map is larger and more comprehensive and requires a chart greater than 2' by 3' in order to picture only some of the large amount of available information. Lexile calibration values have been computed for a substantial number of trade books, texts, and tests. The title column indicates the content validity of the scaling. The educational levels column shows the increase in difficulty corresponding to reading more difficult materials. Construct validity can be demonstrated by these map

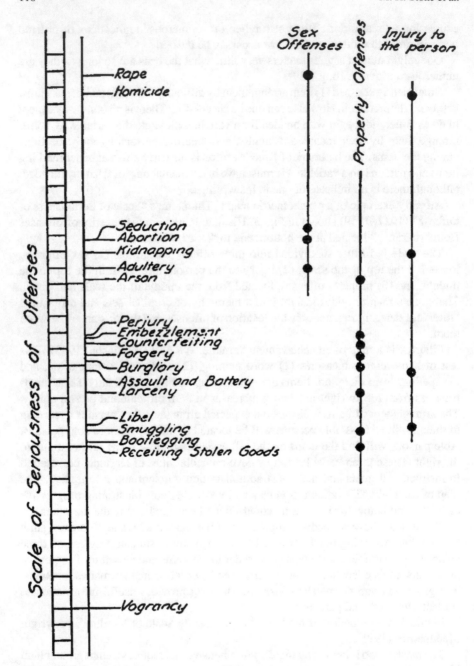

Fig. 3 Scale of seriousness of offense. From Thurstone (1959, p. 75)

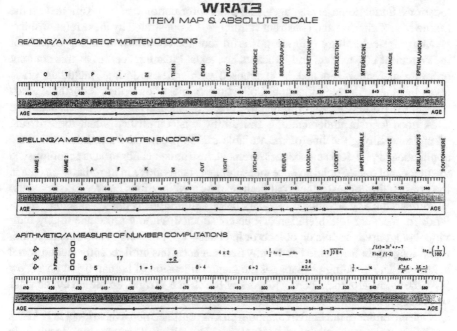

Fig. 4 Wide range achievement test: WRAT3. From Wilkinson (1993). The Wide Range Achievement Test (1993 Edition). Wilmington, DE: Wide Range

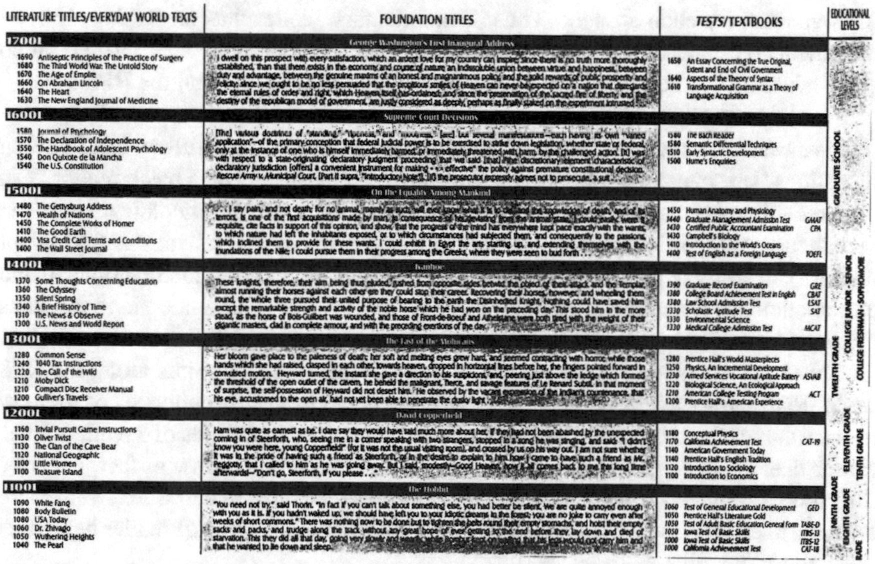

Fig. 5 The Lexile framework. From MetaMetrics (1995) [For a more legible version, see https://www.isbe.net/Documents/lexile.pdf]

locations. Educational levels, ages, and other information can be positioned on the Lexile Map. Criterion and construct validity are demonstrated by these relationships.

Mapping technology offers a powerful tool for conjointly ordering objects of measurement i.e., readers and indicants i.e., texts. Meaning accrues as this conjoint ordering of reader and text is juxtaposed with other orderings including grade level, income or job classification. Collections of these "orderings" constitute a rich interpretive framework for bringing meaning to the measurement of human behavior.

A good leap in understanding and utility is accomplished when the ordering of indicants along the line of the variable can be predicted from theory. In every application of physical science measurement, instrument calibration is accomplished via theory not data. Social science measurement stands alone in its reliance on data in the construction of instrument calibration and co-calibrations between instruments.

Perhaps the key advantage of theory based calibrations is that an absolute framework for measure interpretation can be constructed without reference to any individual or group measures on objects or indicants. The prospect of absolute measurement, long taken for granted in the physical sciences, has until recently eluded social scientists. The building of maps for the major dimension of human behavior is now possible because of the theoretical work of Rasch and Wright, amplified by the work of colleagues.

One pretender to the kind of mapping process outlined above is evident in NAEP's use of the Reading Proficiency Scale (RPS). The RPS is a transformed Rasch scale with an operating range of 0 to 500. NAEP describes performance at grades 4, 8, and 12 as rudimentary, basic, intermediate, adept or advance depending upon the RPS attained by each student. Thus, a rudimentary reader, has an RPS = 150, an intermediate reader at RPS = 250 and an advanced reader at RPS = 350. So far so good, since all we have done is "name" certain "anchor" points on the RPS scale.

Problems develop when reader performance on the RPS scale is described using relative language such as stating that a rudimentary reader "can follow brief written directions" or "can carry out simple, discrete reading tasks" or a basic reader "can understand specific or sequentially related information." An intermediate reader "can search for specific information, interrelate ideas, and make generalizations." An adept reader "can analyze and integrate less familiar material and provide reactions to and explanations of the text as a whole." An advanced reader "can understand the links between ideas even when those links are not explicitly stated."

These statements are not appropriate descriptions of scale points along the RPS scale. Rather they are good descriptions of the behavioral consequences of more or less accurately matching the demands of a text with the capabilities of a reader. Thus, rather than describing absolute scale positions, these annotations, in fact, describe differences between a reader measure and a text measure. When a text's measure exceeds a reader's measure, comprehension is low and the kinds of reader behaviors used above describe a "basic" result.

When a reader's measure exceeds a test's measure the kinds of reader behaviors used to describe adept and advanced readers are evident. The key point is that each of these behaviors can be elicited in the same reader simply by altering the level of text that is presented to the reader. Thus we can make a 400L (second grade level) reader adept by presenting a 100L text or a 400L reader rudimentary by presenting 800L text. Comprehension rate is always relative to the match between reader and text and it is this rate, rather than the reader's measure, that is appropriately described in behavioral and proficiency terms. Much confusion has resulted from a failure to recognize this distinction.

6 Summary

Successful item calibration and person measurement produces a map of the variable. The resulting map is no less a ruler than the ones constructed to measure length.

The map indicates the extent of content, criterion, and construct validity for the variable. The empirical calibration of items and the measures of persons should correspond to the original intent of item and person placement, but changes must be made when correspondence is not achieved. There should be continuous dialogue between the plan, person measures, and item calibrations. Variables are never created once and for all. Continuous monitoring of the variable is required in order to keep the map coherent and up-to-date. Support for reliability and validity does not rest in coefficients, but in substantiating demonstrations of relevant and stable indices for items and measures. Such indications must be continuously monitored in order to maintain the variable map and assure its relevancy.

References

Binet, A. (1916). *The development of intelligence in children*. Wilkins and Wilkins.

Braithwaite, R. (1956). *Scientific explanation*. Cambridge University Press.

Brown, L. (1949). *The story of maps*. Little, Brown and Company.

Carnap, R. (1966). *Philosophical foundations of physics: An introduction to the philosophy of science*. Basic Books.

Kaplan, A. (1964). *The conduct of inquiry*. Thomas Crowell.

Linacre, J. (1999). *WINSTEPS*. MESA Press.

Metametrics. (1995). *The Lexile framework*. Author.

Monmonier, M. (1996). *How to lie with maps* (2nd ed.). University of Chicago Press.

Stone, M., & Wright, B. (1980). *Knox's cube test*. Stoelting.

Stone, M. (1995). *Mapping variables*. Paper given at the Midwest Objective Measurement Seminar. The University of Chicago.

Thurstone, L. (1959). *The measurement of values*. The University of Chicago Press.

Tufte, E. (1983). *Visual explanations: Images and quantities, evidence and narrative*. Graphics Press.

Tufte, E. (1997). *The visual display of quantitative information*. Graphics Press.

Wilford, J. (1981). *The mapmakers*. Alfred Knopf, Inc.

Wilkinson, G. (1993). *Administration manual: WRAT-R*. Wide Range Inc.

Wright, B., & Linacre, J. (1991). *BIGSTEPS*. MESA Press.

Wright, B., & Stone, M. (1979). *Best test design*. MESA Press.

Wright, B., & Stone, M. (1996). *Measurement essentials*. Wide Range Inc.

Theory Referenced Measurement: Combining Substantive Theory and the Rasch Model

A. Jackson Stenner

Abstract A construct theory is the story we tell about what it means to move up and down the scale for a variable of interest (e.g., temperature, reading ability, short term memory). Why is it, for example, that items are ordered as they are on the item map? The story evolves as knowledge regarding the construct increases. We call both the process and the product of this evolutionary unfolding "construct definition" (Stenner et al., Journal of Educational Measurement 20:305–316, 1983). Advanced stages of construct definition are characterized by calibration equations (or specification equations) that operationalize and formalize a construct theory. These equations, make point predictions about item behavior or item ensemble distributions. The more closely theoretical calibrations coincide with empirical item difficulties, the more useful the construct theory and the more interesting the story. Twenty-five years of experience in developing the Lexile Framework for Reading enable us to distinguish five stages of thinking. Each subsequent stage can be characterized by an increasingly sophisticated use of substantive theory. Evidence that a construct theory and its associated technologies have reached a given stage or level can be found in the artifacts, instruments, and social networks that are realized at each level.

1 Level 1

At this stage there is no explicit theory as to why items are ordered as they are on the item map. Data are used to estimate both person measures and item difficulties. Just as with other actuarial sciences, empirically determined probabilities are of paramount importance. When data are found to fit a Rasch Model, relative differences among persons are independent of which items or occasions of measurement are used to make the measures. Location indeterminacy abounds: Each instrument/scale pairing for a specified construct has a uniquely determined "zero.". At Level 1, instruments

---

Paper Presented at the Pacific Coast Research Conference. 2002.

A. J. Stenner (✉)
MetaMetrics, Inc., Durham, NC, USA

W. P. Fisher and P. J. Massengill (eds.), *Explanatory Models, Unit Standards, and Personalized Learning in Educational Measurement*,
https://doi.org/10.1007/978-981-19-3747-7_9

don't share a common "zero" i.e., location parameter. A familiar artifact of this stage is the scale annotated with empirical item difficulties (Artifact 1). Most educational and psychological instruments in use today are Level 1 technologies.

2   Level 2

A construct theory can be formalized in a specification equation used to explain variation in item difficulties. If what causes variation in item difficulties can be reduced to an equation, then a vital piece of construct validity evidence has been secured. We argue elsewhere that the single most compelling piece of evidence for an instrument's construct validity is a specification equation that can account for a high proportion of observed variance in item difficulties (Stenner et al., 1983). Without such evidence only very weak correlational evidence can be marshaled for claims that "we know what we are measuring" and "we know how to build an indefinitely large number of theoretically parallel instruments that measure the same construct with the same precision of measurement".

Note that the causal status of a specification is tested by experimentally manipulating the variables in the specification equation and checking to see if the expected changes in item difficulty are, in fact, observed. Stone (2002) performed just such an experimental confirmation of the specification equation for the Knox Cube Test— Revised (KCT_R), when he designed new items to fill in holes in the item map and found the theoretical predictions coincided closely with observed difficulties. Can we imagine a more convincing demonstration that we know what we are measuring than when the construct theory and its associated specification equation accord well with experiments (Stenner & Smith, 1982; Stenner & Stone, 2003)?

Similar demonstrations have now been realized for hearing vocabulary (Stenner et al., 1983), reading (Stenner & Wright, 2002), quantitative reasoning (Enright and Sheehan, 2002), and abstract reasoning (Embretson, 2002). Artifacts that signal Level 2 use of theory are specification equations, RMSE's from regressions of observed item difficulties on theory, and evidence for causal status based on experimental manipulation of item design features (Artifact 2).

3   Level 3

The next stage in the evolving use of theory involves application of the specification equation to enrich scale annotations. We move beyond using empirical item difficulties as annotations. One example of this use of the specification equation is in the measurement of text readability in the Lexile Framework for Reading. In this application, a book or magazine article is conceptualized as a test made up of as many "imagined" items as there are paragraphs in the book. The specification equation is then used to generate theoretical calibrations for each paragraph which then stand in for empirical item difficulties (Stone et al., 1999).

For instance, the text measure for a book is the Lexile reader measure needed to produce a sum of the modeled probabilities of correct answers over paragraphs, qua items, equal to a relative raw score of 75%. We can imagine a thought experiment in which every paragraph (say 900) in a Harry Potter novel is turned into a reading test item. Each of the 900 items is then administered to 1000 targeted readers and empirical item difficulties are computed from a hugely complex connected data collection effort. The text measure for the Harry Potter novel (880L) is the amount of reading ability needed to get a raw score of 675/900 items correct, or a relative raw score of 75% (Artifact 3).

The specification equation is used in place of the tremendously complicated and expensive realization of the thought experiment for every book we want to measure. The machinery described above can also be applied to text collections (book bags or briefcases) to enable scale annotation with real world text demands (college, workplace, etc.).

Artifacts of a Level 3 use of theory include construct maps (Artifact 3) that annotate the reading scale with texts that, thanks to theory, can be imagined to be tests with theoretically derived item calibrations.

4   Level 4

In biochemistry, when a substance is successfully synthesized using amino acids and other building blocks, the structure of the purified entity is then commonly considered to be understood. That is, when the action of a natural substance can be matched by that of a synthetic counterpart, we argue that we understand the structure of the natural substance. Analogously, we argue that when a clone for an instrument can be built and the clone produces measures indistinguishable from those produced by the original instrument, then we can claim we understand the construct under study. What is unambiguously cumulative in the history of science is not data text or theory but is rather the gradual refinement of instrumentation (Ackerman, 1985).

In a Level 4 use of construct theory there is enough confidence in the theory and associated specification equation that a theoretical calibration takes the place of an empirical item difficulty for every item in the instrument or item bank. There are now numerous reading tests (e.g., Scholastic Reading Inventory- Interactive and the Pearson PA Series Reading Test) that use only theoretical calibrations. Evidence abounds that the reader measures produced by these theoretically calibrated instruments are indistinguishable from measures made using the more familiar empirically scaled instruments (Artifact 4). At Level 4, instruments developed by different laboratories and corporations share a common scale. The number of unique metrics for measuring the same construct (e.g., reading ability) diminishes.

5   Level 5

Level 5 use of theory builds on Level 4 to handle the case in which theory provides not individual item calibrations but rather a distribution of "potential" item calibrations. Again, the Lexile Framework has been used to build reading tests incorporating this more advanced use of theory. Imagine a Time magazine article that is 1500 words in length. Imagine a software program that can generate a large number of "cloze" items (see Artifact 5) for this article. A sample from this collection is served up to the reader when she chooses to read this article. As she reads, she chooses words to fill in the blanks (missing words) distributed throughout the article. How can counts correct from such and experience produce Lexile reader measures, when it is impossible to effect a one-to-one correspondence between a reader response and an item and at theoretical calibration, specific to that particular item? The answer is that the theory provides a distribution of possible item calibrations (specifically, a mean and standard deviation), and a particular count correct is converted into a Lexile reader measure by integrating over the theoretical distribution (Artifact 6).

6   In Conclusion

"There is nothing so practical as a good theory." Kurt Lewin.

References

Ackerman, R. J. (1985). *Data, Instruments, and Theory*. Princeton.

Embretson, S. E. (1998). A cognitive design system approach to generating valid tests: Application to abstract reasoning. *Psychological Methods, 3*, 380–396.

Enright, M. K., & Sheehan, K. M. (2002). Modeling the difficulty of quantitative reasoning items: implications for item generation. In Irvine S. H., & Kyllonen P. C. (Eds) *Item Generation for Test Development*. Hillsdale, NJ. Lawrence Erlbaum Associates.

Stenner, A. J., Burdick, H., Sanford, E., & Burdick, D. How accurate are Lexile text measures? Manuscript accepted. *Journal of Applied Measurement*.

Stenner, A. J., & Smith, M. (1982). Testing construct theories. *Perceptual and Motor Skills, 55*, 415–426.

Stenner, A. J., Smith, M., & Burdick, D. (1983). Toward a theory of construct definition. *Journal of Educational Measurement, 20*(4), 305–316.

Stenner, A. J., & Stone, M. H. (2003). Item specifications vs. item banking. transactions of the rasch SIG; *17*(3) 929–930.

Stenner, A. J.,& Wright, B. D. (2002) Readability, reading ability, and comprehension. Paper presented at the association of test publishers hall of fame induction for Benjamin Wright D., San Diego. In Wright, B. D., & Stone, M. H. (2004). Making Measures. (Chicago: Phaneron Press.

Stone, M. H. (2002) Knox Cube Test - revised. Itasca, IL. Stoelting.

Stone, M. H., Wright, B. D., & Stenner, A. J. (1999). Mapping variables. *Journal of Outcome Measurement, 3*(4), 308–322.

Matching Students to Text: The Targeted Reader

A. Jackson Stenner

Abstract Teachers make use of these two premises to match readers to text. Knowing a lot about text is helpful because "text matters" (Hiebert, 1998). But ordering or leveling text is only half the equation. We must also assess the level of the readers. These two activities are necessary so that the right books can be matched to the right reader at the right time. When teachers achieve this match intuitively, they are rewarded with students choosing to read more.

Teachers share two intuitions:

- Texts can be ordered according to the difficulty each presents for a reader.
- Readers can be assessed according to the success each will have with any particular text.

Teachers make use of these two premises to match readers to text. Knowing a lot about text is helpful because "text matters" (Hiebert, 1998). But ordering or leveling text is only half the equation. We must also assess the level of the readers. These two activities are necessary so that the right books can be matched to the right reader at the right time. When teachers achieve this match intuitively, they are rewarded with students choosing to read more.

When texts are selected that align with all the facets of the reading process, the reader is truly "targeted." The targeted reader benefits from a caring adult (teacher, library-media specialist, parent) who takes the time to understand the reader not just in the terms of reading level, but also in the terms of interests, motivation, developmental maturity, prior knowledge, purpose for reading, and available scaffolding support. An ideal of "targeted" context for reading practice and deepening comprehension can be created when this caring adult helps a student to select appropriate reading material (Five, 1986). The objective reality is that targeted readers comprehend a

---

Scholastic Red. 2002.

A. J. Stenner (✉)
MetaMetrics, Inc., Durham, NC, USA

© The Author(s) 2023   127
W. P. Fisher and P. J. Massengill (eds.), *Explanatory Models, Unit Standards, and Personalized Learning in Educational Measurement*,
https://doi.org/10.1007/978-981-19-3747-7_10

high percentage of the passages they read. The subjective reality is that they report confidence, capability, and control when reading. Finally, targeted readers choose to read, and thus read more and read better. Targeted reading is self-reinforcing, pleasurable, and productive. Poorly targeted reading can be discouraging, or worse— it can produce frustrated students who do not choose to read or like to read.

The best of my own teachers were gifted diagnosticians who seemed to have a second sight about the next chapter or book I should read. They built upon my strengths with just the right mix of success and failure, soaring and stumbling, clarity and confusion. As a learner, I felt centered and on target. However, it can take decades for teachers to polish intuition, to learn a 200-book classroom text collection from the lowest-level book to the highest, and to refine field-based techniques for leveling readers. And because the product of these thousands of hours of professional practice is a private, non-exchangeable metric for simultaneously ordering books and assessing readers, the profession at large does not advance.

1   The Lexile Framework® for Reading

The Lexile Framework for Reading is a system for measuring texts and readers by the same metric (a Lexile®). When a reader's Lexile measure and a book's Lexile measure are both known, a forecast can be made about the success that the reader will have with that book (Stenner, 1997). Over 40,000 books and 40 million articles now have Lexile measures, and tests such as the Scholastic Reading Inventory (SRI, print and electronic versions); the Harcourt SAT-9, SAT-10, MAT-8 and SDRT-4; the CTB/McGraw-Hill TerraNova Assessment Series (CAT/6, CTBS/5); the Riverside Publishing Educational Assessments (The Iowa Tests, GMRT-4); NWEA Achievement Level Tests, and other well-known reading achievement tests have been linked to the Lexile Framework. In addition, the Lexile Framework is used in states and districts throughout the nation. Such links make it possible for the users of these tests to request equivalent Lexile measures for any specific score. Today, more than 14 million students get at least one Lexile measure each year from a standardized test report. Educators, librarians, and parents can use a student's Lexile measure to search the Lexile Website and build a customized, targeted reading list for the reader. Former Assistant Superintendent of Schools in North Carolina, Dr. Suzanne Triplett, states:

> The Lexile Framework manifests what good teachers try to do anyway, which is to judge where a student is and find material that will challenge him adequately without being so difficult that he loses his motivation. The problem is that as children get into the latter stages of elementary school, the variance in texts and among students increases dramatically. The choice of material expands and the range of reading skills widens, so it becomes much harder for teachers to make accurate judgements about where children are and what materials are good choices for them. By using the Lexile Framework, schools can take the guesswork out of this equation and operationalize the selection of developmentally appropriate material for their students.

"Empowerment" has become a hackneyed word, but that's the key advantage of the Lexile Framework—it gives students, parents, teachers, and administrators accurate information that empowers them. With a Lexile measure, you know precisely where a student stands in terms of an absolute scale of reading comprehension, and you know exactly what steps that student needs to take to achieve higher levels of reading performance.

The Lexile Framework is a tool that can be combined with other tools, techniques, and strategies to optimize instruction. The Lexile Framework offers an open standard and a public, exchangeable metric for measuring text and readers.

2 The Lexile Map

The Lexile Map (Fig. 1) is a visual display of the reading continuum, ranging from early first grade texts (100L) to advanced graduate school texts (1200L). The Lexile Map combines nouns (books) with numbers (Lexile measure). Every book ever written in English has a theoretical location on this Map. Once measured, a book takes a unique and invariant position in relation to every other book. In this sense, a Lexile measure is absolute in that it is independent of other books that might be measured or reader performances that might be observed. Readers can be visualized as "in motion," moving up the Lexile Map, each on an individual growth trajectory as he or she encounters various new and enriching texts. If we were to plot a "poor" reader's growth trajectory, we would find that a high proportion of the assigned reading registers above the growth trajectory—sometimes far above (250L+). In contrast, for a "good" reader, we find a high proportion of assigned reading falling below the growth trajectory—often far below. The consequence is that the "poor" readers get reinforced in the belief that they can't read for meaning, and the "good" readers receive reinforcement that they can. There are, in absolute sense, no "good" or "poor" readers. Comprehension is relative; it is a simple function of the match between reader and text. We can control the text level and thereby gain control over the motivational consequences of reading "on" and "off" target.

3 The Shoe Store Story

Some time ago, I went into a shoe store and asked for a fifth-grade shoe. The clerk looked at me suspiciously, and asked if I knew how much shoe sizes varied among 11-year-olds. Furthermore, he pointed out that shoe size was not nearly as important as purpose, style, color, etc., but if I would specify the features I wanted and size, he could walk to the back and magically reappear with several options to my liking. He further noted—somewhat condescendingly—that the store used the same metric to measure feet and shoes, and that when there was a match between foot and shoe, the shoes got worn, there was no pain, and the customer was happy and became a repeat

Lexile Level: 1200L-1700L

Benchmark Literature: First Inaugural Address by George Washington   1700
The Good Earth by Pearl S. Buck   1530
The Life and Times of Frederick Douglass by Frederick Douglass   1400
Silent Spring by Rachel Carson   1340
Great Expectations by Charles Dickens   1200

Lexile Level: 1100L

The War of the Worlds by H.G. Wells   1170
Animal Farm by George Orwell   1170
Ethan Frome by Edith Wharton   1160
A Separate Peace by John Knowles   1110
Pride and Prejudice by Jane Austen   1100

Lexile Level: 1000L

Anne Frank: The Diary of a Young Girl by Anne Frank   1080
One More River to Cross by Jim Haskins   1070
20,000 Leagues Under the Sea by Jules Verne   1030
The Pearl by John Steinbeck   1010
Freak the Mighty by Rodman Philbrick   1000

Lexile Level: 900L

Exploring the Titanic by Robert Ballard   980
Beauty: A Retelling of the Story of Beauty and the Beast   970
by Robin McKinley
The Abracadabra Kid: A Writer's Life by Sid Fleischman   940
Dogsong by Gary Paulsen   930
Roll of Thunder, Hear My Cry by Mildred Taylor   920

Lexile Level: 800L

Anthony Burns: The Defeat and Triumph of a Fugitive Slave   860
by Virginia Hamilton
Julie of the Wolves by Jean Craighead George   860
Johnny Tremaine by Esther Forbes   840
Call It Courage by Armstrong Sperry   830
The Dark is Rising by Susan Cooper   820

Lexile Level: 700L

And Now Miguel by Joseph Krumgold   780
Red Scarf Girl: A Memoir of the Cultural Revolution by Ji-Li Jiang   780
Harriet the Spy by Louise Fitzhugh   760
Pacific Crossing by Gary Soto   750
From the Mixed Up Files of Mrs. Basil E. Frankweiler by E.L. Konigsburg   700

Fig. 1 Lexile leveled reading framework

Lexile Level 600L

Charlotte's Web by E.B. White	680
Henry Huggins by Beverly Cleary	670
Sadako and the Thousand Paper Cranes by Eleanor Coerr	630
Flossie and the Fox by Patricia McKissack	610
...If You Sailed on the Mayflower in 1620 by Ann McGovern	600

Lexile Level 500L

Buffalo Woman by Paul Gable	590
The True Story of the Three Little Pigs by A. Wolf by Jon Scieszka	570
Encyclopedia Brown, Boy Detective by Donald J. Sobol	560
Red Riding Hood by James Marshall	520
The Magic School Bus Inside the Earth by Joanna Cole	500

Lexile Level 400L

Madeline by Ludwig Bemelmans	480
Dinosaur Bones by Aliki	460
How My Parents Learned to Eat by Ina R. Friedman	450
Henry and Mudge and the Forever Sea by Cynthia Rylant	420
Frog and Toad Are Friends by Arnold Lobel	400

Lexile Level 300L

Babushka's Doll by Patricia Polacco	360
The Best Way to Play by Bill Cosby	360
Arthur's Nose by Marc Brown	350
Noisy Nora by Rosemary Wells	320
Pet Show! by Ezra Jack Keats	300

Lexile Level 200L

Mr. Rabbit and the Lovely Present by Charlotte Zolotow	280
The Cat in the Hat by Dr. Seuss	260
Play Ball, Amelia Bedelia by Peggy Parish	220
Clifford, the Big Red Dog by Norman Birdwell	220
Danny and the Dinosaur by Syd Hoff	200

Fig. 1 (continued)

customer. I called home and got my son's shoe size, and then asked the clerk for an 8½, red, hightop basketball shoe.

After a brief credit-card transaction, I had my shoes. Then I walked next door to my favorite bookstore and asked for a fifth-grade fantasy novel. Without hesitation, the clerk and I walked to a shelf where she gave me three choices. I chose "The Hobbit," a 1000L classic that I had read three times, and went home. My son, I later learned, reads at 850L. As I write this, my son is passionately practicing free throws in the driveway.

Today, we can apply the Lexile Framework to avoid this kind of mismatch. It is available to bring the art of good teaching and the science of technology to a classroom, library, or living room near you.

Matching students to texts at appropriate levels helps to increase their confidence, competence, and control over the reading process. The Lexile Framework is a reliable and tested tool designed to bridge two critical aspects of student reading achievement—leveling text difficulty and assessing the reading skills of each student.

References

Five, C. L. (1986). Fifth graders respond to a changed reading program. *Harvard Educational Review, 56*, 395–405.
Hiebert, E. H. (1998). *Text matters in learning to read* (CIERA Report 1-001).
Stenner, A. J. (1997). *The Lexile framework: A map to higher levels of achievement*. MetaMetrics.

Does the Reader Comprehend the Text Because the Reader Is Able or Because the Text Is Easy?

A. Jackson Stenner and Mark H. Stone

Abstract Does the reader comprehend the text because the reader is able or because the text is easy? Localizing the cause of comprehension in either the reader or the text is fraught with contradictions. A proposed solution uses a Rasch equation to models comprehension as the difference between a reader measure and text measure. Computing such a difference requires that reader and text are measured on a common scale. Thus, the puzzle is solved by positing a single continuum along which texts and readers can be conjointly ordered. A reader's comprehension of a text is a function of the difference between reader ability and text readability. This solution forces recognition that generalizations about reader performance can be text independent (reader ability) or text dependent (comprehension). The article explores how reader ability and text readability can be measured on a single continuum, and the implications that this formulation holds for reading theory, the teaching of reading, and the testing of reading.

1 Introduction

A *koan* is a Zen riddle designed to provoke meditation and open the mind to new ways of thinking and understanding. We begin by offering a koan: Does the reader comprehend the text because the reader is able or because the text is easy? To seek an answer to this riddle, we must question how we think about the relative contributions of reader and text to comprehension.

One possibility, which was popular for most of the last century up to about 25 years ago, is the "behaviorist" argument that the text is preeminent in dictating the meaning that a reader takes from a reading. To a behaviorist, readers are exchangeable, and

Paper Presented at the 13th International Objective Measurement Workshop, 2006.

A. J. Stenner (✉)
MetaMetrics, Inc., Durham, NC, USA

M. H. Stone
Adler School of Professional Psychology, Chicago, IL, USA

© The Author(s) 2023
W. P. Fisher and P. J. Massengill (eds.), *Explanatory Models, Unit Standards, and Personalized Learning in Educational Measurement*,
https://doi.org/10.1007/978-981-19-3747-7_11

133

what matters most is the schedule of challenges and rewards that surrounds the exercise of reading. A second possibility, which has been the dominant approach in reading research in the last few decades, is the "constructivist" argument that the reader's prior knowledge and expectations shape the meaning that is constructed from a reading, so that comprehension lies in the mind of the reader. A third possibility is that the riddle may be a trap, in the sense that reader and text measurement are incommensurable and so their contributions to reading comprehension cannot be usefully compared.

It is characteristic of riddles that they only rarely can be solved by adhering to the rules. The reading riddle asks for an either-or choice; that is, does comprehension result from reader ability *or* text readability? Our answer to the riddle is to break out of the either-or choice, and to answer that *both* reader and text share in accounting for comprehension. We begin this paper by arguing that much of contemporary research on reading, in which reading comprehension is placed in the mind of the reader, is based on a fundamental conceptual flaw that makes the results open to multiple interpretations. We will then argue that reader ability and text readability can be measured on a single numerical scale, called Lexiles. We introduce and discuss the Lexile Framework and both practical and conceptual aspects of how it measures text readability, reader ability, and comprehension. Using this framework, reader comprehension is determined by the gap between reader ability and text readability. Using both reader ability and text readability to explain reading comprehension leads to new perspectives for research on reading, the teaching of reading, and reading assessment.

2 The Problem of Placing Reading Comprehension in the Mind of the Reader

For the past quarter century, much reading research has ignored text when accounting for what happens when a reader engages a text. Instead, much contemporary reading research and theory attempts to solve the "reading riddle" by placing reading comprehension in the mind of the reader. From this "constructivist" perspective, meaning is constructed in the reader's head, and because readers vary greatly in their background knowledge, a single text evokes such varied meanings that any universal representation of text (which, say, a measurement of the readability of the text necessarily would be) is a chimera. Constructivists argue that text is needed to trigger a reader reaction, but what the reader brings to an encounter with text is the dominant explanatory mechanism in accounting for what the reader comprehends.

In this view, it becomes unnecessary even to contemplate measuring the readability of text. This perspective explains why approximately 3,000 pages of reading theory and research spread across three volumes and two decades in the three *Handbooks of Reading Research* (Barr et al., 1984, 1996; Kamil et al., 2000) can manage with less than a handful of references to text readability.

Similarly, the Rand Corporation published a study about framing future research on reading comprehension that includes one disparaging reference to readability (Snow, 2002): although "text" is frequently mentioned, its measurement is not. In the modern vernacular, "text is not where it's at" (Hiebert, 2004).

A weaker version of this ontology holds that although text measurement can be helpful in some circumstances, this admission should not mask the reality that the reader is the dominant actor in reading comprehension. In this version, the role of text is not openly dismissed, but rather quietly neglected. This strategy of quiet neglect has been rhetorically quite successful. After all, an open assertion that a well-researched construct with a century-long tradition, like the readability of text, should simply be dismissed from modem reading theory and practice would have required assuming a heavy intellectual burden of explanation and justification. Simple neglect, on the other hand, leaves the unwary with the incorrect notion that, in human science research, constructs come and go, and the time for looking at readability of text has passed.

However, choosing reader over text, as many contemporary reading theorists have done, leads to certain confusions (Stenner & Wright, 2002). To understand the implications of this problem for reading research, suppose an investigator is interested in the relationship between reading and summarizing. Data on 100 fourth graders' reading ability is collected by administering a grade-appropriate reading achievement test. After examinees read and answer written questions, they are asked to summarize orally the content of the passage.

These oral summaries are evaluated for quality by five raters. A correlation is computed between the reading scale scores and the summary measure, and the researchers discover a relatively high correlation of $r = 0.70$. The investigators conclude, "Good readers possess summarization skills that are much better developed than those of poor readers." Thousands of studies using this general design have allowed researchers to estimate statistical relationships between reading performance and dozens of so-called reading process correlates (Pressley, 2000).

But this research design suffers from a fatal flaw. There are two conceptually distinct reader behaviors: "reading ability", which does not depend on the text being read, and "reading comprehension", which does depend on the text being read. In any study where all examinees respond to the same text, those examinees with high measured reading ability also experience high comprehension rates, and those examinees with low reading ability experience correspondingly low comprehension rates. Thus, those who are identified as "good" readers in this experimental design are both readers with high reading ability and readers who enjoy (with this text) high comprehension rates. So called "poor" readers in this experimental design have both low reading ability, relative to the other fourth graders in the study, and experience low comprehension rates.

Even after finding a strong positive correlation between the written reading test and the oral summary, this research design cannot conclude whether the oral summarization performance is correlated with reading ability or comprehension rate. Perhaps persons with high reading ability summarize well regardless of what they read, or perhaps when readers engage text that they comprehend well—regardless of their reading ability—they summarize well.

Conversely, perhaps those with low reading ability have difficulty summarizing, or perhaps when anyone engages text that they comprehend poorly, they will summarize poorly, whatever their reading ability. In this latter interpretation, summarization performance results from reading comprehension and only appears to be correlated with reading ability (in the hypothetical study) because the research design confounds text-dependent performance and text-independent performance.

Thousands of studies have sought to contrast "good" or "mature" readers on the one hand and "poor" or "immature" readers on the other. Usually, the research report makes a claim about what good readers can do that poor readers cannot, but it is often not clear to what kind of reading behavior the epithets *good* and *poor* refer. Two extended quotations below summarize the reader characteristics that differentiate *good, skilled,* and *mature* readers on the one hand and *poor, less skilled,* and *immature* readers on the other. The first quotation is from Daneman (1996); the second is from (Pressley, 2000). Each quotation includes at least a dozen generalizations about good vs. poor readers. Consider which of these generalizations are text-independent claims about reading ability and which generalizations are text-dependent claims about reading comprehension—or if it is even possible to tell which argument is being made.

> Relatively few studies have examined exactly which of the high-level processes shared by reading and listening are responsible for the individual differences. However, from the little evidence we do have, a consistent picture seems to emerge. Poor readers are at a particular disadvantage when they have to execute a process that requires them to integrate newly encountered information with information encountered earlier in the text or retrieved from semantic memory. So, for example, poor readers have problems interrelating successive topics (Lorch et al., 1987) and integrating information to derive the overall gist or main theme of a passage (Daneman & Carpenter, 1980; Oakhill, 1982; Palincsar & Brown, 1984; Smiley et al., 1977). They have more difficulty making inferences (Masson & Miller, 1983; Oakhill et al., 1986) and tend to make fewer of them during text comprehension (Oakhill, 1982). Poor readers also have more difficulty computing the referent for a pronoun (Daneman & Carpenter, 1980; Oakhill et al., 1986). Other researchers have found that poor readers do not demand informational coherence and consistency in a text, and often fail to detect, let alone repair, semantic inconsistencies (Garner, 1980).

> Good readers are extremely active as they read, as is apparent whenever excellent adult readers are asked to think aloud as they go through text (Pressley & Afflerbach, 1995). Good readers are aware of why they are reading a text, gain an overview of the text before reading, make predictions about the upcoming text, read selectively based on their overview, associate ideas in text to what they already know, note whether their predictions and expectations about text content are being met, revise their prior knowledge when compelling new ideas conflicting with prior knowledge are encountered, figure out the meanings of unfamiliar vocabulary based on context clues, underline and reread and make notes and paraphrase to remember important points, interpret the text, evaluate its quality, review important points as they conclude reading, and think about how ideas encountered in the text might be used in the future. Young and less skilled readers, in contrast, exhibit a lack of such activity (e.g., Cordón & Day, 1996).

Neither quotation contains any reference to text measurement, so the presumption appears to be that claims about good and bad readers are independent of the text. But even a reader with a fairly high reading ability might do a poor job of comprehending, say, the relatively difficult text of a Supreme Court decision. Conversely, even a reader of fairly low ability may provide a rich summary of a children's story like *Frog and Toad Are Friends* (Lobel, 1979).

To the unwary, a research design in which all participants have the same text may seem like a perfect way to ensure that text plays no role in drawing inferences about good and bad readers, but this approach is deeply misguided. In the attempt to focus on only the reader, rather than embracing the duality of reading ability and text readability, comprehension rate and reading ability are hopelessly confounded. We are left with the current predicament: thousands of studies that calculate correlations between a reading test performance and some process that may be connected to reading, and no way at all to determine which of these correlations support text-dependent generalizations about reading comprehension and which support text-independent generalizations about reading ability.

The Lexile Framework, discussed in the next section, will separate reading ability, text readability, and reading comprehension. Thus, it requires that claims about "good" and "poor" readers must be qualified as to whether the discussion is about reading ability or reading comprehension, and further requires that claims about "easy" or "hard" texts must be qualified as to whether they refer to the text in isolation from readers or whether they refer to a particular sample of readers having difficulty comprehending the text (Stenner & Stone, 2003; Stenner et al., 1988; Stone et al., 1999; Wright & Stone, 2004).

3   A Lexile Framework Primer

When a reader confronts a text, is required to carry out some task in response to that text, and the reader's response is subsequently rated by some mechanism, the situation may appear so amorphous and complex that it is easy to despair that any simple model can account for and measure what happens. Yet all science progresses by inventing workable simplifications of complex reality. The Lexile Framework for Reading purports to measure in a common unit, called Lexiles, the traits of reader ability and text readability. Based on these measures, reading comprehension can be calculated based on the gap between reader ability and text readability. When reader ability far exceeds text readability, then comprehension should approach unity. Conversely, when text measure far exceeds reader measure, then the probability of little or no comprehension should approach unity.

The way of formulating the connection from reader ability and text readability to reader comprehension raises a number of concerns, which will be discussed in this section: (1) the readability of text must be measured; (2) reading ability must be measured, which given the readability of the text will lead to a prediction about reading comprehension; (3) the conceptual framework and assumptions that allow a researcher to subtract text readability from reader ability and calculate a comprehension rate must be spelled out; (4) the approach should be flexible enough to include other aspects of the reading environment like subjective assessments of reading and different tasks for measuring reading. This section will discuss how the Lexile framework addresses these concerns.

3.1   Measuring the Readability of Text

All systems of communication through symbols share two features: a semantic component and a syntactic component. In language, the semantic units are words, which are organized according to rules of syntax into thought units and sentences (Carver, 1974). Thus, the readability of text is governed largely by the familiarity of the words and by the complexity of the syntactic structures used in constructing the message.

Regarding the semantic component, most methods of measuring difficulty use proxies for the probability that a person will encounter a word in context and infer its meaning (Bormuth, 1966). "Exposure theory," which is the main explanation for the development of receptive or hearing vocabulary, is based on how often people are exposed to words (Miller & Gildea, 1987; Stenner et al., 1983). Knowing the frequency of words as they are used in written and oral communication provides the best means of inferring the likelihood that a word will be encountered and become a part of an individual's receptive vocabulary. Klare (1963) built the case for the semantic component varying along a continuum of familiarity to rarity, a concept that was further developed by Carroll et al. (1971), whose word-frequency study examined the reoccurrences of words in a five-million-word corpus of running text.

Sentence length is a powerful proxy for the syntactic complexity of a passage. However, sentence length is not the underlying causal influence (Chall, 1988). Davidson and Kantor (1982), for example, illustrated that reducing sentence length can increase difficulty and vice versa. The underlying factor in the difficulty of syntax probably involves the load placed on short-term memory, and a body of evidence suggests that sentence length is a good proxy for the demands of structural complexity on verbal short term memory (Crain & Shankweiler, 1988; Klare, 1963; Liberman et al., 1982; Shankweiler & Crain, 1986).

The Lexile equation combines into an algebraic equation the measurements of word frequency and sentence length drawn from a certain text, and produces a measure in Lexiles. A simple children's book would have a text measure of about 200L; a complex and specialized work might have a text measure of 1700L or more. The body of text used for measuring the likelihood of encountering particular words makes use of a 550-million-word corpus, comprising the full text for more than 30,000 books. The Lexile Analyzer—a software program for measuring text readability—is freely available for noncommercial use at lexile.com. Also, the Lexile website contains a search program that allows you to insert book titles and see their Lexile measures, or to insert a range of Lexile scores and see a list of book titles in that range.

3.2 Measuring Reader Ability and Comprehension

Reader ability is defined in reference to the text readability—although not in reference to any particular tests. For example, *Harry Potter* has a text readability of 880L. Imagine that the book was turned into a test with 1000 questions, each question appearing in a standard Lexile form called the "native form," illustrated in Fig. 1. We then say that a reader with the ability to answer correctly 750 of the 1000 items correctly has a Lexile reading ability of 880L.

An important conceptual insight of the Lexile framework is that text itself could be viewed as a virtual test made up of standard-sized "native form" 125-word passages. (If the 125th word is not the end of a sentence, the passage is extended until it reaches the sentence-ending punctuation.) The degree of difficulty of each passage can be calibrated using the Lexile equation that takes vocabulary and syntax into account. Thus, this approach produces a measure of readability expressed in the same metric used to express reader ability.

Choosing one type of reading task, like the "native item" format, does not cause a loss of generality. Any reading behavior that systematically varies as a function of the reader/text difference can in principle be used to measure reading ability; for example, the accuracy with which words are read aloud from text can be used as a measure of reading. As we will discuss in a later section, the Lexile approach can allow making comparisons across reading tasks, so that the results from different tasks can still be measured in Lexiles.

A passage of text can be understood as entailing a large number of propositions, each of which can form the basis for making a "native item" reading test question. The set of all such allowable items for a passage is the "ensemble." Each individual test item in the ensemble can be imagined to have an observed difficulty, and the average over this distribution of difficulties is the ensemble mean. The Lexile Theory claims that these ensemble means are predictable from knowledge of the semantic and syntactic features of text passages (Stenner et al., 2004). Thus, the Lexile Theory replaces statements about individual reading item difficulties with statements about ensembles (Kac, 1959). Because text readability is a property of the entire text passage and not a characteristic of any particular test item, the ensemble

Wilbur likes Charlotte better and better each day. Her campaign againstinsects seemed sensible and useful. Hardly anybody around the farm had a good word to say for a fly. Flies spent their time pestering others.The cows hated them. The horses hated them. The sheep loathed them.Mr. And Mrs. Zuckerman were always complaining about them, and putting up screens. **Everyone _____ about them.**

A. agreed
B. gathered
C. laughed
D. learned

From *Charlotte's Web* by E. B. White, 1952, New York: Harper

Fig. 1 An example Lexile test item

interpretation produces the right level of aggregation for the empirical variation that the Lexile Theory attempts to explain.

The choice of a 75% comprehension rate on the virtual test items is arbitrary, but highly useful. At a conceptual level, this thought experiment of breaking a text into virtual test questions enables combining a substantive theory of text difficulty, based on word familiarity and syntax, with a psychometric model for reader ability. At the practical level, 75% comprehension represents a balance between reading skill and difficulty of text that allows the reading experience to be both successful and challenging.

When given a measure of text readability, it is often useful to imagine what reader characteristics match this text readability; for example, a text at 880L like Harry Potter would match the 880L reading ability of a fiftieth percentile sixth grader. Similarly, when given a reader measure (1200L), imagine the texts that this reader can read with 75% comprehension; someone with reading ability of 1200L, for example, could read *Cold Mountain* (1210L) or *The Trumpeter of Krakow* (1200L) with 75% comprehension.

The gap between reader ability and text difficulty will determine the extent of comprehension, as measured by the proportion of correct answers on the reading test. For example, a 1200L reader can read text such as in *USA Today* (1200L) with 75% comprehension and would have 92% comprehension of *Harry Potter* (880L) but would only have 60% comprehension of the typical College Board SAT text (1330L). Figure 2 shows the expected level of comprehension for a reader with ability of 750L with texts at different levels of readability.

Finally, this framework implies that differences in text readability can be traded off for differences in reader ability to hold comprehension constant, which as we shall see is a protean concept with important implications for reading theory and practice.

Reader Ability	Text Readability	Title	Expected Comprehension Rate
750L	250L	*Play Ball Amelia Bedelia*	95%
750L	500L	*Harold and The Purple Crayon*	90%
750L	750L	*The Adventures of Pinocchio*	75%
750L	1000L	*Island of The Blue Dolphins*	50%
750L	1250L	*The Midwife's Apprentice*	25%

Fig. 2 Expected comprehension rates for a 750L reader reading books with varying text measures

3.3 The Conceptual Framework

Perhaps the preeminent idea underlying the Lexile Framework is the measurement of reading ability and text readability on a common scale, followed closely by the realization that text and task used to evaluate reading must be conceptually separated if the measurement of reading is to be unified (Stenner et al., 1983, 1988; Wright & Stone, 1979). This movement from measures of reading ability and text readability to a measure of reading comprehension rests on a solid and well-developed conceptual framework.

This conceptual framework is based on four key assumptions. The first assumption is *sufficiency*, which means that the number of correct answers on a reading test contains all of the information in the response record that is informative about a reader's ability (Andersen, 1977; Fisher, 1922). In particular, sufficiency means that which specific items the reader got correct or incorrect contributes no information about the reader's ability beyond what is given by the total count correct. In addition, the sufficiency assumption means that the measure of text readability summarizes everything about the text that is important in accounting for comprehension.

The second assumption is *separability*, which means reader ability and text readability have separate effects that can be isolated from one another (Rasch, 1960 [1980]). The third assumption, *specific objectivity*, means that reader measures can be estimated independently of the particular text used, the particular form of the response (for example, whether the response is an essay or some other response), and the method used to rate the essay or constructed response (Rasch, 1960 [1980]).

A final assumption of *latent additivity* means that the measures of reader ability and text readability, when related to reading comprehension, are connected to one another by addition or subtraction (Luce & Tuckey, 1964). This represents a test of the quantitative hypothesis that the construct being measured adds up in the same way that the numbers representing it do (Michell, 1999). An important implication of latent additivity is that reader ability and text readability are measured on a common scale and differences between them can be traded off to hold success rate constant. Lest the assumption of an additive representation seem overly restrictive, we note that an additive representation can be transformed into a multiplicative representation and vice versa quite easily (by using logarithms).

Thus, the assumption of an additive form actually includes the possibility of a multiplicative relationship, suitably transformed. Because it is often easier to conceptualize addition (or subtraction) than multiplication (or division), from this point on we will stipulate that reader ability and text readability are related through subtraction and have equal potential in affecting the outcome of an encounter between reader and text. Thus:

reader ability − text readability = probability of comprehending the text.

Only one model meets the four assumptions set forth above, and thus allows us to subtract text readability from reader ability to determine the probability of comprehension. It is one of a family of models named after the Danish mathematician, Rasch (1960 [1980]). The Rasch model is a mathematical function that relates the probability of a correct response of an item/text to the difference between one reader parameter (in this case, reader ability) and one text parameter (in this case, text readability). This probability may be interpreted as a comprehension rate for the item/text. Reader comprehension for a multi-item passage is a function of the sum of these modeled probabilities of a correct answer. A high reader ability relative to the text readability produces a high probability of a correct response to a test item—that is, a high comprehension rate. Conversely, a low reader ability relative to a high text readability results in a low-model probability of a correct answer. In this latter case when the probabilities associated with a number of these reader/paragraph and counters are summed, the result is a low forecasted comprehension rate. This probabilistic perspective thus admits reasonable uncertainty regarding what may happen with a particular reader, reading a particular text on a particular occasion, while focusing attention on average performances of readers of a particular ability, on typical occasions, reading texts of a particular difficulty.

The Rasch model states a set of requirements for the way observations and theory combine in a probability model to make meaningful measures. The Rasch Model combines the three components in any definition of measurement—observation, theory, and measure—into a simple, elegant, and, in some important respects, unique representation (Wright & Stone, 2004).

This algebraic framework will also allow for testing the assumptions behind the idea that reader comprehension can be modeled as the gap between reader ability and text readability. For example, it will test whether differences in text readability can be traded off against differences in reader ability to keep comprehension constant. These kinds of trade-offs or invariance's are only possible if additivity obtains.

Text readabilities computed using up-to-date technology like the Lexile method are highly reliable (Stenner et al., 2004). Indeed, uncertainty over text readability can be effectively ignored in many applications of the Lexile Framework. Uncertainty in reader abilities, as opposed to text readabilities, is by far the major source of error in forecasted comprehension.

The combination of the Lexile Framework and the Rasch model is thus not a matter of collecting a body of data and then trying "to model the data" with strategies like fitting various functional forms, inserting different variables, and trying out different interaction terms between variables. Instead, the Lexile Framework in cooperation with the Rasch model set forth a set of requirements that data must meet if the data are to be useful in constructing meaningful reader measures. If the data do not meet these requirements, then the appropriate response is to question the way that the observations were made and ask what contaminants to the observation process might have introduced unintended dependencies in the data.

The overall goal of this conceptual framework is to create measures of reading ability and text readability that transcend the initial conditions of measurement. This standard holds true for most measurement procedures in use in the so-called "hard" sciences—for example, thermometers for measuring temperature. The way is now clear to achieve this standard in the measurement of reading. Measures of reading should provide the same result, even if they are carried out using different methods, just like temperature measurements using Fahrenheit or Celsius scales, or using different types of physical thermometers, provide a common result. By using the Lexile framework, we can leave behind the particulars of the method and moment of measurement as we move forward to use the measure in description and prediction of reader behavior.

3.4   Different Tasks and Subjective Raters

Along with measuring text readability and reader ability, a full version of the Lexile Framework will also address two other important aspects. A full model of reading would include not only reading ability and text difficulty, but also whether the reading task involved something other than the "native form," and whether a computer based or human rating system was employed to judge a readers performance. For example, observations on readers might include the read-aloud accuracy rate, words read correctly per minute, number of correct answers to multiple choice questions, quality rating of summary, and retelling. These tasks are all different from the 125-word "native form" at the basis of the Lexile framework. However, these different task types can be shown to measure the same reading ability as is measured by traditional reading tests and therefore, can be rescaled in Lexiles (Stenner & Wright, 2002).

In its simplest manifestation, which has been the focus up to this point, the Lexile approach uses a single multiple-choice task type, thus eliminating the need for a rater/observer parameter. Moreover, the task type can be restricted to a basic common format, the so-called "native item" format illustrated earlier. Thus, in its simple form, the Lexile Framework reduces to a measure of reader ability and a measure of text readability. But the Lexile framework can be used to adjust for the severity of subjective raters or observers and to adjust for the difficulty of the tasks demanded across different tests.

An expanded version of the Rasch model enables the measurement of four facets of reading—readers, texts, tasks, and raters—on a common scale (Linacre, 1987). If data fit the multi-faceted Rasch model, so that the four key assumptions discussed in the previous section are met, then (1) each facet can be estimated independently of the other facets; (2) the facets can be added to each other; (3) differences in measurements on one facet can be traded off for equal differences on other facet(s) to hold constant the probability of reading comprehension. In this case, the many-facet Rasch model enables the estimation of measures of reader ability, text readability, the reading task, and the subjectivity of the rater all expressed in equal interval Lexile measures.

For an intuitive sense of how such conversions might work, note that it is straightforward to describe how performance on other tasks will correspond to a particular comprehension rate on a "native form" Lexile test. For example, perhaps a 75% comprehension rate on a "native form" Lexile test represents performance equal to 98% word call accuracy in a read-aloud study. There are probably dozens of task types that can be ordered on a "task continuum" for reading, all measuring the same reading ability construct but doing it with added easiness or hardness relative to the native item format. Once the demands of a certain task type have been located on the task continuum, then test results using this type of task can be converted into Lexiles (Wright & Masters, 1982; Wright & Stone, 1979).

There are numerous philosophical, mathematical, and practical implications of the way that substantive reading theory and the many-facet Rasch Model are combined in the Lexile Framework for Reading. The reader interested in more extensive treatments might begin with Linacre (1989) and Boomsma et al. (2001).

4   Implications for Reading Research

The last 50 years of reading research has amassed a staggering array of studies, which seek to correlate some measure of reading performance with some "process variable," which is a putative cause or effect of reading performance. These relationships are sometimes presented as correlations between variables and sometimes presented as differences in the averages between groups of "good" and "poor" readers. However, results presented in either form can be re-expressed in the other form, and so in this section, we will for convenience refer only to correlations.

About the process variables that are more or less correlated with reading performance, Stanovich (1986) asked an embarrassingly simple question: "Is there evidence that the correlate is in fact a cause (worthy of instructional focus), or is it in fact an effect or consequence of reading performance?" After all, if the process variable causes better reading performance, then it might be worth spending precious instructional time and money on trying to use that variable to improve reading performance, but if the variable is a result of better reading performance, then while it may serve as a marker of pedagogical success, it cannot be used to improve reading performance.

The previous discussion has argued that there are two kinds of reading performance: "reading ability," which is a reader trait that is independent of text, and "comprehension," which is modeled as a function of the difference between reader ability and text readability. These two kinds of reading performance can be crossed with four kinds of relationships: (1) The process variable causes one of the two reading performances; (2) The process variable is caused by (is a consequence of or is an effect of) one of the two reading performances; (3) There is a reciprocal causal relationship between the process variable and one of the two reading performances; or (4) An observed correlation between the process variable and reading performance is spurious (perhaps due to a third variable that is causing both of them).

Figure 3 juxtaposes the two kinds of reading performance—reading ability and comprehension—with three kinds of relationship—cause, effect, and reciprocal causation. (Spurious relationships are not considered further in this discussion.) The candidate variables listed in this table should be taken as provisional. We have not conducted a proper review of the literature to solidify their placement in the classification. Thus, we offer these candidate variables to provoke thought and, perhaps, to inflict some gentle bruises upon prevailing intuitions.

	Causes	Effects	Reciprocal Causation
	(Type 1)	(Type 2)	(Type 3)
Trait:	Accessibility of text	Last Grade completed	Vocabulary
Reading Ability (text independent)	Targeting text on reader ability (long-term)	Many process variables Life earnings	Amount of text read Syntactic facility
State:	(Type 4)	(Type 5)	(Type 6)
Reading Comprehension (text dependent)	Reader ability minus text readability Rereading	Eye movement Retelling Inferencing Many process Variables: accuracy, fluency, etc.	Engagement Motivation Distractibility

Fig. 3 A framework for thinking about reading performance, its correlates, and causations

Type 1 relationships are candidate causes of reading ability. The first mechanism suggests that readers cannot improve their reading if they cannot access text (Krashen, 2002). A second candidate cause of reading ability is that exposure to text that is well matched to the reader's ability promotes faster development than repeated exposure to text that is too hard or too easy (Carver, 2000).

Type 2 relationships include effects of reading ability. For example, continuing in college is a difficult prospect for an 800L reader, because college texts have measures in the 1200L–1400L range, which results in an expected comprehension rate approaching 25% (Williamson, 2004). In this scenario, a decision to forego college is an effect of low reading ability. Many correlates of reading ability that have been proposed as causes are probably effects of comprehension (Stanovich, 1986).

Type 3 relationships share a reciprocal causation with reading ability. Vocabulary grows, primarily, as a result of listening up to about 600L–700L, and then reading becomes a major avenue for vocabulary growth. However, because vocabulary (semantic facility) is one of the two key variables in the Lexile equations, it is viewed as causal for reading ability. A virtuous circle characterizes high ability readers: they read more so their vocabulary grows and because their vocabulary grows, they can comprehend increasingly difficult texts.

Type 4 relationships include those variables that, when experimentally manipulated, change the reader's comprehension. The Lexile theory asserts that the "match" between reader ability and text readability is the major cause of comprehension.

Type 5 relationships are probably vast in number. Many constructs have been proposed as causes of reading performance that are, in fact, effects of reader comprehension. As one example, Stanovich (1986, p. 365) reviewed the evolving view in the research on eye movements as a cause of reading performance:

The relationship of certain eye movement patterns to reading fluency has repeatedly, and erroneously, been interpreted as indicating that reading ability was determined by the efficiency of the eye movements themselves. For example, researchers have repeatedly found that less skilled readers make more regressive eye movements, make more fixations per line of text, and have longer fixation durations than skilled readers (Rayner, 1985a, 1985b). The assumption that these particular eye movement characteristics were a cause of reading disability led to the now thoroughly discredited "eye movement training" programs that repeatedly have been advanced as "cures" for reading disabilities. Of course, we now recognize that eye movement patterns represent a perfect example of a causal connection running in the opposite direction. Poor readers do show the inefficient characteristics listed above; but they also comprehend text more poorly. In fact, we now know that eye movements rather closely reflect the efficiency of ongoing reading—with the number of regressions and fixations per line increasing as the material becomes more difficult, and decreasing as reading efficiency increases (Aman & Singh, 1983; Just & Carpenter, 1980; Olson et al., 1983; Rayner, 1978, 1985a, 1985b; Stanley et al., 1983; Tinker, 1958)—and this is true for all readers, regardless of their skill level.

In short, "eye movement" migrated from what we are calling a Type 1 relationship to a Type 5 relationship—that is, from a cause of reading ability to an effect of reading comprehension. We wonder how many of the currently popular "reading process variables" or "comprehension processes" are in fact consequences of reader comprehension rates.

We should avoid, however, being too quick to dismiss all process variables that are correlated with reading comprehension as unimportant if they are not causal. In some cases, effects of reading comprehension may be important outcomes in their own right. For example, perhaps manipulations of reading fluency are not a useful way to cause greater comprehension, but gains in comprehension can help to encourage greater reading speed—and reading speed (or fluency) may be important in its own right. Given two prospective employees both reading at 1300L, the employer may choose the one who reads at 250 wpm over one who reads at 175 wpm.

Finally, Type 6 process variables exhibit reciprocal causation with reading comprehension. A reader who is motivated to read about basketball will comprehend more and, because of the comprehension, will be further motivated to read. Again, a virtuous circle on a shorter time scale develops.

The classification scheme in Fig. 3 has a number of useful applications. It emphasizes that reading research should always specify whether reading ability or comprehension is under study, and should attempt to specify the causal status of each process variable. This matrix of causes and consequences of reading ability and comprehension can be used to consider instructional implications of research findings. Also, the classification scheme may prove useful in carrying out meta-analyses that attempt to summarize the results of many studies in the reading literature, by helping such analyses avoid the "apples-and-oranges criticism" that they are jumbling together causes and effects for different reading performances.

Reading research has more than academic implications. Its results play out in the development of reading programs and technologies and influence how teachers view the reading process, both of which influence how reading is taught. Indeed, today's top-selling reading technologies each include one or more process variables that were originally developed in academic research before finding their way from the researcher's lab to the publisher's boardroom.

But there are also more subtle examples of how research on reading ability, text readability, and comprehension might directly affect pedagogy. For example, the dominant instructional model used in U.S. middle school through college courses, with the exception of some classes that use laboratories or case studies, relies on core textbooks. All students use the same textbook, regardless of their reading ability.

However, in a typical classroom, student reading abilities vary over an 800L–900L range which implies the stunning conclusion that comprehension rates vary within a class from less than 25% to more than 95%. Thus, many students cannot access the majority of content in the textbook because their comprehension rates are too low.

If students were better targeted to textbook material according to their reading ability, would their content knowledge increase? Krashen (2002, p. 30) argues the skeptical view that "there is no evidence at all for the 'targeting hypothesis' as a cause of non-reading," and states, "In fact, what little data there is on this issue suggests that matching for reading level is not the problem." Carver (2000), on the other hand, devoted a chapter in his book, *The Causes of High and Low Reading Achievement*, to reviewing the research that the "match"—that is, that reader ability matches text readability—is causally implicated in a wide range of important behaviors, not the least of which is the improvement of reading ability itself.

If the "match" is, in fact, causally implicated in promoting growth in reading ability and also a controlling influence in how much science, social studies, health, and so on a student learns in different classes, then the one-textbook-fits-all instructional philosophy may need to be revisited.

More broadly, the prospect of linking reading test scores to particular books (or more generically, connecting readers to text) has been the single most compelling application that has sustained the unification of reading. Teachers and parents want reading test scores to be more actionable and refrigerator-friendly. It is empowering for teachers, students, and parents to be able to forecast the success that a reader will have with a text.

5 Implications for Reading Assessment

Reading assessments may be the most common tests in education, so a change in perspective could have substantial consequences for test theory and practice.

The evidence seems overwhelming that we can usefully treat reading ability, text readability, and comprehension as if they are one-dimensional constructs. The strongest support for such a treatment comes from the fact that when using the Lexile measures, differences between two reader ability measures can be traded off for an equivalent difference in two text readability measures, while holding comprehension constant.

Consequently, there is no justification for reporting separate components of reading ability, other than the semantic and syntactic facility measures that are implicit within the Lexile measures.

If reading ability is "one thing," then it makes sense to unify the measurement and reporting of this quantity around a single metric. At present, hundreds of reading tests report in proprietary and non-exchangeable metrics. The life of the professional educator can be simplified tremendously by unifying the reading construct. In the 1700s, countries unified the measurement of temperature (Celsius and Fahrenheit). In the 1800s, nations unified the measurement of time (Greenwich mean time). All major norm-referenced reading tests have now been linked to the Lexile scale so as to enable test publishers to report in both their own proprietary metrics and a common supplemental metric. In particular, the linking technology already exists to translate reading scores into Lexiles from the SAT-10, Iowa Tests, Terra Nova, NWEA-MAPS, Gates MacGinitie, Stanford Diagnostic Reading Test, and Scholastic Reading Inventory, among others. Moreover, dozens of text publishers and text aggregators have adopted the Lexile scale as the standard for representing the readability of text and 19 million K-12 students now get a Lexile measure at least once a year.

Once unification of reading assessments has been realized, it will prove useful to link progress assessments to the common metric, so that reading assessment isn't done only once a year, but monthly and even weekly. Periodic classroom-based progress assessments that report in the same metric as the high-stakes instruments will shift the reporting focus away from status ("how well is the student reading on this May morning?") to growth ("on what growth trajectory is this reader, and what do we forecast reading ability to be at high school graduation?"). Indeed, a common metric spanning the full developmental continuum makes it possible to build individual growth trajectories for each student.

When the causes of test item difficulty are known, it even becomes possible to engineer reading test items on demand, either by having human item writers follow a protocol or by teaching a computer to develop reading items with associated theoretical calibrations. The implications of this possibility are far-reaching and a little unsettling. Suppose, for example, that a future edition of a high stakes reading test did not involve a test nor bank of questions, but instead comprised a set of rules for generating appropriate test items, along with an adaptive algorithm for choosing the next best text/question for an examinee depending on what reading ability is revealed, and a richly annotated scale that links reported reader ability to appropriate texts (Stenner & Stone, 2003).

Reproducible measures of reader ability can thus be made from counts correct on items that have never been administered before and may never be used again. In this scenario, test items are disposable commodities. Test security is guaranteed because no one sees the items until the computer builds them. In this case, releasing "the test" would mean either releasing sample items from the test or releasing the item-generating algorithm itself.

Perhaps the main difficulty with this approach is that among the randomly generated individual test questions, some will inevitably work better than others to measure ability. Because each item is used only once with one examinee, there would be no means of comparing whether some questions create a wider or narrower spread of answers than other questions, or whether certain groups of students perform better on a certain question. With each test item used only once, the problem of comparing text questions that are better or worse would be less accessible to study. But the hope would be that in a text of reasonable length, these random differences would largely balance out.

This ultimate unification in the measurement of reading ability need not diminish the role of normative metrics for reading ability such as percentiles, stanines (in which students are ranked in nine groups, with the bottom three being below-average, the middle three being average, and the upper three being above average), and "normal curve equivalent" scores (in which students are measured along a scale from 1 to 99 based on how their score would rank them in a normal-curve distribution). However, measuring reading performance in terms of grade-equivalents should be abandoned in favor of text based descriptions of reader performance, such as "This reader can read typical third grade text (500L–600L) with 75% comprehension." With this formulation, it becomes possible to give a coherent answer to a common refrain of parents and school board members, "What does it mean to be on grade level?" For example, minimally acceptable end of third grade performance can be described as the level needed to read fourth grade textbooks with 60% or better comprehension, which emphasizes that the reading goal is not being set relative to other students, but is in place to ensure that students can handle the text readability of the grade to which they are being promoted.

6 Conclusion

We claim that the measurement of reading is philosophically, mathematically, and practically analogous to the measurement of temperature. Temperature theory is adequately developed so that thermometers can be constructed without reference to any data. We know enough about liquid expansion coefficients, the gas laws, glass conductivity, and fluid viscosity to construct a remarkably precise temperature measurement device with recourse only to theory. Routine manufacture of thermometers occurs without even checking the calibrations against data with known values prior to shipping the instruments to customers. Furthermore, a second instrument developer can follow the same specification and produce another thermometer that measures temperature on the same scale and with a precision comparable to the first instrument. If two people are measured with two different thermometers, our confidence in temperature measurement is such that the readings can be usefully compared.

In recent decades, reading measurement has proceeded in a radically different manner: instruments (collections of items) are field-tested on thousands of readers, and empirical difficulties are estimated for each item. The potential is now upon us to move reading assessment from a data-driven to a theory-driven enterprise. The consequences of this shift for research and practice are hard to over-estimate (Stone, 2002). The Lexile approach is built on a theory of what makes text readable (word choice and syntax), how a measure of reader ability can be connected to this measure of text readability, and how to combine the readability of text and the ability of readers into a useful measure of comprehension. The ultimate result of building measures of reading on a strong theoretical base may be that if two people are tested for reading ability with two different tests, their scores will be comparable.

Moreover, if anyone wants to generate a new reading test, then as long as they follow the same underlying specification, anyone using that test will have comparable results as well.

Text matters! Paradoxically, the last 25 years of reading research has celebrated the role of text but, for the most part, avoided measuring it. In much of the current literature, "reading performance" conflates reader ability, which does not depend on text, and reader comprehension, which does. However, reader ability and reading comprehension are conceptually and operationally separable. We have described how reader ability and text readability, measured in the common scale of Lexiles, can be used to model reading comprehension.

References

Aman, M. G., & Singh, N. N. (1983). Specific reading disorders: Concepts of etiology reconsidered. *Advances in Learning & Behavioral Disabilities*.

Andersen, E. B. (1977). Sufficient statistics and latent trait models. *Psychometrika, 42*(1), 69–81.

Barr, R., Kimil, M. L., & Mosenthal, P. B. (Eds.). (1984). *Handbook of reading research* (Vol. I). Longman.

Barr, R., Kimil, M. L., Mosenthal, P. B., & Pearson, P. D. (1996). *Handbook of reading research* (Vol. II). Lawrence Erlbaum.

Boomsma, A., van Duijn, M. A. J., & Snigders, T. A. B. (Eds.). (2001). *Essays on item response theory*. Springer.

Bormuth, J. R. (1966). Readability: New approach. *Reading Research Quarterly, 7*, 79–132.

Carroll, J. B., Davies, P., & Richman, B. (1971). *The word frequency book*. Houghton Mifflin.

Carver, R. P. (1974). Measuring the primary effect of reading: Reading storage technique, understanding judgments and cloze. *Journal of Reading Behavior, 6*, 249–274.

Carver, R. P. (2000). *The causes of high and low reading achievement*. Lawrence Erlbaum.

Chall, J. S. (1988). The beginning years. In B. L. Zakaluk & S. J. Samuels (Eds.), *Readability: Its past, present, and future*. International Reading Association.

Cordón, L. A., & Day, J. D. (1996). Strategy use on standardized reading comprehension tests. *Journal of Educational Psychology, 88*(2), 288.

Crain, S. & Shankweiler, D. (1988). Syntactic complexity and reading acquisition. In A. Davidson & G. M. Green (Eds.), *Linguistic complexity and text comprehension: Readability issues reconsidered*. Erlbaum Associates.

Daneman, M. (1996). Individual differences in reading skills, IV. In M. L. Kamil, P. B. Mosenthal, P. D. Pearson & R. Barr (Eds.), *Handbook of reading research* (Vol. III, pp. 512–538). Erlbaum Associates.

Daneman, M., & Carpenter, P. A. (1980). Individual differences in working memory and reading. *Journal of verbal learning and verbal behavior, 19*(4), 450–466.

Davidson, A., & Kantor, R. N. (1982). On the failure of readability formulas to define readable text: A case study from adaptations. *Reading Research Quarterly, 17,* 187–209.

Fisher, R. A. (1922). On the mathematical foundations of theoretical statistics. *Philosophical Transactions of the Royal Society of London, A, 222,* 309–368.

Garner, R. (1980). Monitoring of understanding: An investigation of good and poor readers' awareness of induced miscomprehension of text. *Journal of Reading Behavior, 12*(1), 55–63.

Hiebert, E. H. (2004). Standards, assessment and text difficulty. In A. E. Fastrup & S. J. Samuels (Eds.), *What research has to say about reading instruction* (3rd ed.). Instructional Reading Association.

Just, M. A., & Carpenter, P. A. (1980). A theory of reading: from eye fixations to comprehension. *Psychological Review, 87*(4), 329.

Kamil, M. L. (1959). *Statistical independence in probability, analysis, and number theory.* Wiley.

Kamil, M. L., Mosenthal, P. B., Pearson, P. D., & Barr, R. (2000). *Handbook of reading research* (Vol. III). Lawrence Erlbaum.

Klare, G. R. (1963). *The measurement of readability.* Iowa State University Press.

Krashen, S. (2002). The Lexile framework: The controversy continues. *California School Library Journal, 25*(2), 29–31.

Liberman, I. Y., Mann, V. A., Shankweiler, D., & Werfelman, M. (1982). Children's memory for recurring linguistic and non-linguistic material in relation to reading ability. *Cortex, 18,* 367–375.

Linacre, J. M. (1987). *An extension of the Rasch model to multi-facet situations.* University of Chicago Department of Education.

Linacre, J. M. (1989). *Many-faceted Rasch measurement.* MESA Press.

Lobel, A. (1979). *Frog and toad are friends.* HarperCollins.

Lorch Jr, R. F., Lorch, E. P., & Mogan, A. M. (1987). Task effects and individual differences in on-line processing of the topic structure of a text. *Discourse Processes, 10*(1), 63–80.

Luce, R. D., & Tukey, J. W. (1964). Simultaneous conjoint measurement: A new type of fundamental measurement. *Journal of Mathematical Psychology, 1,* 1–27.

Masson, M. E., & Miller, J. A. (1983). Working memory and individual differences in comprehension and memory of text. *Journal of Educational Psychology, 75*(2), 314.

Michell, J. (1999). *Measurement in psychology: A critical history of a methodological concept.* University Press.

Miller, G. A., & Gildea, P. M. (1987). How children learn words. *Scientific American, 257,* 94–99.

Oakhill, J. (1982). Constructive processes in skilled and less skilled comprehenders' memory for sentences british. *Journal of Psychology, 73*(1), 13–20.

Oakhill, J., Yuill, N., & Parkin, A. (1986). On the nature of the difference between skilled and less-skilled comprehenders. *Journal of Research in Reading, 9*(2), 80–91.

Olson, R. K., Kliegl, R., & Davidson, B. J. (1983). Dyslexic and normal readers' eye movements. *Journal of Experimental Psychology: Human Perception and Performance, 9*(5), 816.

Palincsar, A. S., & Brown, A. L. (1984). Reciprocal teaching of comprehension-fostering and comprehension-monitoring activities. *Cognition and Instruction, 1,* 117–175.

Pressley, M. (2000). What should comprehension instruction be the instruction of? In M. L. Kamil, P. B. Mosenthal, P. D. Pearson, & R. Barr (Eds.), *Handbook of reading research* (Vol. III, pp. 545–561). Erlbaum.

Pressley, M., & Afflerbach, P. (1995). *Verbal protocols of reading: The nature of constructively responsive reading.* Hillsdale, NJ: Lawrence Erlbaum Associates.

Rasch, G. (1960 [1980]). *Probabilistic models for some intelligence and attainment tests.* University of Chicago Press.

Rayner, K. (1978). Eye movements in reading and information processing. *Psychological Bulletin, 85*(3), 618–660.

Rayner, K. (1985a). Do faulty eye movements cause dyslexia?. *Developmental Neuropsychology, 1*(1), 3–15.

Rayner, K. (1985b). The role of eye movements in learning to read and reading disability. *Remedial and Special Education, 6*(6), 53–60.

Shankwiler, D., & Crain, S. (1986). Language mechanisms and reading disorder: A modular approach. *Cognition, 14*, 139–168.

Smiley, S. S., Oakley, D. D., Worthen, D., Campione, J. C., & Brown, A. L. (1977). Recall of thematically relevant material by adolescent good and poor readers as a function of written versus oral presentation. *Journal of Educational Psychology, 69*(4), 381.

Snow, C. E. (2002). *Reading for understanding: Toward a R&D program in reading comprehension.* Rand Corporation.

Stanley, G., Smith, G. A., & Howell, E. A. (1983). Eye-movements and sequential tracking in dyslexic and control children. *British Journal of Psychology, 74*(2), 181–187.

Stanovich, K. E. (1986). Matthew effects in reading: Some consequences of individual differences in the acquisition of literacy. *Reading Research Quarterly, 21*(4), 360–406.

Stenner, A. J., Smith, M., & Burdick, D. S. (1983). Toward a theory of construct definition. *Journal of Educational Measurement, 20*(4), 305–315.

Stenner, A. J., & Stone, M. H. (2003, in press). Item specifications vs. item banking. *Rasch Measurement Transactions, 17*(3), 929–930.

Stenner, A. J., & Wright, B. D. (2002). Readability, reading ability, and comprehension. In B. D. Wright & M. H. Stone. *Making measures.* Phaneron Press.

Stenner, A. J., Horabin, I., Smith, D. R., & Smith, M. (1988). Most comprehension tests do measure reading comprehension: A response to McLean and Goldstein. *Phi Delta Kappan, 765–769.*

Stenner, A. J., Burdick, H., Sanford, E., & Burdick, D. S. (2004). *How accurate are Lexile text measures?* In press, *Journal of Applied Measurement.*

Stone, M. H., Wright, B. D., & Stenner, A. J. (1999). Mapping variables. *Journal of Outcome Measurement, 3*(4), 308–322.

Stone, M. H. (2002). *The Knox cube test: A manual for clinical and experimental uses.* Stoelting Company.

Tinker, M. A. (1958). Recent studies of eye movements in reading. *Psychological Bulletin, 55*(4), 215.

Williamson, G. L. (2004). Student readiness for postsecondary options. [Online].

Wright, B. D., & Masters, G. N. (1982). *Rating scale analysis: Rasch measurement.* MESA Press.

Wright, B. D., & Stone, M. H. (1979). *Best test design.* MESA Press.

Wright, B. D., & Stone, M. H. (2004). *Making measures.* Phareon Press.

From Model to Measurement with Dichotomous Items

Don Burdick, A. Jackson Stenner, and Andrew Kyngdon

Abstract Psychometric models typically represent encounters between persons and dichotomous items as a random variable with two possible outcomes, one of which can be labeled success. For a given item, the stipulation that each person has a probability of success defines a construct on persons. This model specification defines the construct, but measurement is not yet achieved. The path to measurement must involve replication; unlike coin-tossing, this cannot be attained by repeating the encounter between the same person and the same item. Such replication can only be achieved with more items whose features are included in the model specifications. That is, the model must incorporate multiple items. This chapter examines multi-item model specifications that support the goal of measurement. The objective is to select the model that best facilitates the development of reliable measuring instruments. From this perspective, the Rasch model has important features compared to other models.

1 The Atomic Model

This chapter examines the role of the Rasch (1960) model for dichotomous data from the perspective of first principles concerning the measurement of psychometric constructs. The chapter therefore begins with an atomic model for a single, dichotomous item. The descriptive term *atomic* implies that single-item models appear as basic units in more elaborate so-called molecular models that incorporate multiple items. Typically, 1 of the 2 outcomes for the dichotomous item is the favored or

Journal of Applied Measurement, 11(2). 2010.

D. Burdick · A. Kyngdon
Metametrics, Inc., Durham, NC, USA

D. Burdick
Department of Mathematics, Duke University, Durham, NC, USA

A. J. Stenner (✉)
MetaMetrics, Inc., Durham, NC, USA

153

successful response, and this is denoted as 1. The atomic model defines a construct on persons by specifying that each person has a probability of responding 1 to the item, and this probability varies from person to person. This success probability can, if we choose, be taken as the person parameter, that is, the true measure of a person on the construct defined by the atomic model. The variability from person to person is a sine qua non for the construct-defining property of the atomic model. A model for the tossing of a coin, in which each person has the same probability of obtaining the favored outcome (i.e., heads), does not define a construct on persons (Wood, 1978).

Two important observations can be made about the atomic model and its associated construct: (1) the quantity that represents the construct is a latent variable (i.e., it is not directly observable as data); and (2) the essential character of this latent variable is ordinal. We consider the implications of both of these observations in some detail, beginning with the latter.

The ordinal character referred to in the second observation above does not imply that the atomic model has only order relationships among the objects of measurement (i.e., persons). The success probabilities are a valid numerical representation of the construct. *Ordinal character* instead implies that any monotonic transformation of the success probabilities would be as valid a representation of the construct as the success probabilities themselves. Thus, the probabilities could be converted to logits or probits, perhaps followed by multiplicative rescaling. Transforming the probabilities from an atomic model into logits yields a Rasch model with the difficulty for the single item set to zero. Setting the difficulty parameter to something other than zero in the Rasch model for a single item produces an alternative monotonic transformation. If that transformation is followed by a multiplicative rescaling,[1] the result is the two-parameter model for a single item with specified location and discrimination parameters (Birnbaum, 1968). Because of the ordinal character of the construct obtained from the atomic model, there is no inherent reason to prefer any particular monotonic transformation of the probabilities compared to any other. Within the confines of the atomic model, the choice of a numerical structure to represent the order relationships is a matter of personal preference.

The first of the two important observations has critical implications for measurement. When the objective is measurement, it is not enough to define the construct in terms of a latent variable. The sine qua non for measurement is the ability to use observable data to discriminate reliably between objects that differ on the construct by an amount large enough to matter. In almost all situations, the one bit of information produced by the observable response to a single dichotomous item does not yield the reliable discrimination required for measurement. Reliable discrimination requires replication, and that in turn requires we broaden the atomic model to accommodate more items.

---

[1] Multiplicative rescaling will change the numerical distance between person measures without changing the substantive difference. And analogy with temperature helps to clarify this point. An object that is warmer than another by 5 °C is warmer by 9 °F; 9 is larger than 5, but the substantive difference in temperature is unaffected by the multiplicative rescaling.

Some of the previously introduced concepts (e.g., a construct difference that is considered large enough to matter) as well as concepts to be introduced subsequently are context-dependent. It is helpful to have a prototype example to which to refer when discussing these context-dependent concepts. As such an example, consider an item from a reading test in which a passage is presented followed by a task, scored as correct or incorrect, to assess the reader's understanding of the passage. The construct is reader ability. We assume that better readers have higher probabilities of succeeding at the assessment task.

The inadequacy of the discrimination obtainable from a single dichotomous item is easy to see in this example. The difference between a reader with a 40% chance of success and a reader with an 80% chance of success is almost certainly large enough to matter. There is a less than even chance of successful detection of this difference with the single item. Hence, replication is needed.

The purpose of replication is to generate new information that can be used to estimate the value of the latent variable with less uncertainty. In many other dichotomous experiments (e.g., coin-tossing), the experiment can be repeated and the second outcome assumed to be independent of the first, but the assumption of independence for a repeat encounter is not appropriate in the current context. If a reader is presented again with the same passage and the same assessment task, the same outcome is virtually assured to occur. When the second outcome is independent of the first, it provides new information that can be used to reduce uncertainty about the latent variable; when the second outcome is perfectly predictable from the first, no new information is produced. This dependence problem can be overcome by introducing a new passage and task as the replication, but in this situation, the single-item atomic model is no longer adequate. There are now at least two atoms to consider. Continuing the metaphor, a model that represents two or more dichotomous items will be called *molecular*.

2 Molecular Models

With two dichotomous items, call them Item A and Item B, each could be represented by its own atomic model. Thus, each person has two probabilities, p_A and p_B, that represent the probability of a correct response on Item A and Item B, respectively. These two latent variables could potentially represent two distinct constructs. A third possible construct, represented by the latent variable equal to the sum of the two probabilities, comes readily to mind.

The replication needed to achieve measurement requires multiple items whose atomic models represent the same construct. What conditions must be imposed on the molecular model to ensure that its constituent atomic models define the same construct? The answer lies in the essential ordinal character of the construct obtained from the atomic model.

In the two-atom molecular model, the set of success probabilities for Item A generates a rank ordering of persons that allows the possibility of ties. A similar statement applies to Item B. The two atomic models define the same construct if and only if the success probabilities for the two items generate the same rank ordering of persons, including ties. A set of two or more items whose atomic models define the same construct is said to satisfy the unidimensionality condition. It is worth noting that unidimensionality may depend on the population of persons. A set of items that satisfies unidimensionality for a given population may not be unidimensional when the population is extended.

When the unidimensionality condition is satisfied for a two-atom molecular model, the constructs derived from the two atomic models and the construct obtained from the sum of the two success probabilities are all identical. The latent variables that arise from the two atomic models are related by a strictly monotonic transformation. The proof of this assertion is straightforward. Define a function h as follows:

If a person has success probabilities p_A and p_B, then $h(p_A) = p_B$.

The preservation of ties implies that h is unambiguously defined. The preservation of rank order implies that h is strictly monotonic. The ordinal character of the construct implies that any strictly monotonic transformation of a valid numerical representation is another valid numerical representation of the same construct.

More than two items of replication are generally necessary to achieve the measurement objective. A multi-item instrument can provide adequate replication if the items satisfy the unidimensionality condition (Stout, 1990). The question is whether unidimensionality is a reasonable assumption in a given context. For an answer in the context of the prototype example, consider a test of reader ability that consists of 40 dichotomous items. The assumption of unidimensionality asserts that examinees can be ordered by reading ability such that for every item on the test, more able readers have higher success probabilities than less able readers.

3 Parameterizations

As mentioned above, any monotonic transformation of the success probabilities in an atomic model can be used as a numerical representation of the person parameter for the construct defined by that model. Let $\theta = \theta(p)$ be the result of applying a monotonic transformation to a success probability p. The change from p to θ can be regarded as a reparameterization of the atomic model. The inverse function $p = p(\theta)$, which is also monotonic, is called the *item characteristic function*. Note that the form of the item characteristic curve depends on the reparameterization, and the selection of the reparameterizing monotonic transformation is arbitrary. If success probabilities are converted to logits, the item characteristic curve will be the logistic ogive. If success probabilities are converted to probits, the item characteristic curve will be the normal ogive.

In a molecular model that incorporates multiple items, each person has multiple success probabilities, one for each item in the model. If the model satisfies the unidimensionality condition, it unambiguously determines a single construct of ordinal character. When a parameterization of that construct is selected, each person can be mapped to the construct with a parametric value that determines all of that person's success probabilities.

For example, the construct from a unidimensional molecular model could be parameterized in terms of the success probabilities for a single canonical item, say Item A. For any other item in the model, say Item B, there is a monotonic function, h_B that satisfies $p_B = h_B(p_A)$ for all persons. For this parameterization, the item characteristic curve for Item A will be a straight line from the origin to the point (1, 1). For Item B, the item characteristic curve need not be a straight line but will start at the origin and increase monotonically until it reaches (1, 1). If instead values for a latent variable θ are obtained by transforming the success probabilities for Item A into logits or probits, the item characteristic curve $p_A(\theta)$ for A will be the logistic or normal ogive, and Item B's characteristic curve will be $h_B(p_A(\theta))$.

4 Unidimensionality, Replication, and Measurement

At this point it is appropriate to return to the prototype example to examine the progress toward the goal of achieving measurement. Two questions should be raised when a new item is included in the assessment instrument: (1) Is the new item on the construct? and (2) Does its inclusion provide new information about the person's location on the construct (i.e., the person parameter)? To achieve our goal of measurement, we need affirmative answers to both questions. The first question can be answered yes when the unidimensionality condition holds, and the second can be answered yes if the new item has the property of local independence with the other items in the instrument. Local independence occurs when the responses to two items by the same person are statistically independent. In a model in which the person parameter is not represented as a random variable, local independence is identical to statistical independence. If the person parameter is a random variable, as in some Bayesian models, local independence is a conditional independence, given the person parameter. In either case, the addition of new items with the local independence property produces replication that leads to a reduction in the standard error of measurement.

Achieving the measurement objective requires replication, but how much replication is required to achieve reliable discrimination between objects that differ by an amount large enough to matter? The answer depends on the context. The finer the differences to be discriminated and the higher the desired level of confidence in the discrimination, the more replication is needed. In the prototype example, if a valid reading test does not have sufficient reliability for the purpose at hand, it may be possible to find more items with affirmative answers to Questions #1 and #2 that could be appended to the test to increase the reliability to the desired level.

At this stage, the unidimensional molecular model can be enhanced with a replication capability by assuming an inexhaustible supply of new on-construct items. Without making further assumptions, what are the measurement implications of this replication-enhanced, unidimensional model? The answer is that reliable discrimination can be achieved between two objects whose construct values differ but only to the extent of determining the order relationship between the two objects. In the prototype example, suppose John and Alice have different reading abilities. If replication is available via an inexhaustible supply of reading items, but all that can be said of these items is that they have local independence and are on construct (the unidimensionality assumption), then a test that consists of a sufficient number of these items can determine who is the better reader but not by how much. Just being on construct (i.e., ordering persons identically) is not an adequately stringent assumption about items to measure any more detailed information about persons than their order relationships. Measurement on an interval scale requires more specificity in the assumptions about the atomic models for the items.

5   The Stringency Construct for Model Specifications

It is possible and often useful to contemplate a stringency construct defined for statistical models. Ordinal position on this construct is determined by the stringency of the assumptions incorporated into the statistical model. These assumptions are called the model's specification, and a so-called *tight* specification has more stringent assumptions than a loose one. The molecular model that asserts only that there is some encounter-specific probability of success associated with each encounter between a person and an item is a loose specification—the loosest under consideration. Incorporating local independence and unidimensionality tightens the specification but only enough to allow ordinal measurement even though the model is based on a numerical latent variable. The Rasch model is clearly an even tighter specification because it assumes more about the items' atomic models.

   If we denote by ULI the model that assumes only unidimensionality and local independence, there is a stringency difference between ULI and the Rasch model. Two questions naturally arise: (1) What further tightening assumptions must be added to the ULI model to yield the Rasch model specification? and (2) Are there any important model specifications located between ULI and the Rasch model on the stringency construct?

   The answer to the second question is yes. When the roles of persons and items are reversed in the molecular model, the success probabilities associated with a person can constitute an atomic model for a construct on items. There are as many such atomic models as there are persons. Application of the assumption of unidimensionality to these atomic models tightens the specification. Models can satisfy person unidimensionality without satisfying item unidimensionality. Models that satisfy both unidimensionality conditions are called *doubly monotonic models*. As the discussion in the next section demonstrates, the doubly monotonic model specification is tighter than person unidimensionality but looser than the Rasch model specification.

6 Doubly Monotonic Models

In the prototype example, it may happen that for each reader, the success probability is lower for Item B than for Item A. As a possible explanation for this feature, perhaps the passage in Item B presents more of a challenge to comprehension than does the passage for Item A (e.g., more complex syntax and more difficult vocabulary). This suggests the possibility that a text readability construct for items might be obtainable from a molecular model that incorporates multiple persons and items. So far, persons have been the objects of measurement and items have been regarded as instruments for measuring the person construct. These roles can be reversed. Consider a molecular model with multiple persons and items and a success probability for each encounter between a person and an item. There is an atomic model associated with each person, and each of these atomic models defines a construct on items. The unidimensionality condition will be satisfied if the rank order of items is consistent for all persons' atomic models. In the prototype example, this unidimensionality of the persons' atomic models implies that the text readability rank of two passages will not depend on who happens to be reading them.

A molecular model with multiple items and persons has two sets of atomic models: one set provides rankings of persons for each item and the other provides rankings of items for each person. If unidimensionality is satisfied for both sets, it is called a *doubly monotonic model*. If a doubly monotonic model is capable of replication, then ordinal measurement is enabled for persons if enough items are included and for items if enough persons are included.

When a molecular model has double monotonicity, the item characteristic curves do not cross regardless of the parameterization. Conversely, if a unidimensional model does not have double monotonicity, the item characteristic curves will be monotonic, but at least one pair of curves will cross. In a doubly monotonic model, the person characteristic curves also do not cross. If the atomic models for persons define a unidimensional construct on items, the person characteristic curves will be monotonic. If there is any crossing of these person characteristic curves, the molecular model will not define a unidimensional construct on persons.

7 Numerical Conjoint Measurement Models

The Rasch model is an example of a numerical conjoint measurement model. With the Rasch model, when the success probabilities are transformed to logits, the result is an array of numbers that can be expressed as differences between a person parameter that does not change from item to item and an item parameter that does not change from person to person. In other words, the interaction in the array of success probabilities can be removed by transforming them to logits. This feature, existence of a monotonic transformation to remove the interaction from the array of success probabilities, characterizes a numerical conjoint measurement model. A numerical conjoint measurement model is necessarily doubly monotonic.

The absence of interaction means that the transformed success probabilities can be expressed as differences between a latent variable for persons and a latent variable for items. Consequently, the constructs for persons and items (e.g., reader ability and text readability in the prototype example), can be expressed on a common interval scale. The absence of interaction also implies that item and person characteristic curves are horizontally parallel, that is, any item characteristic curve can be transformed to any other by moving it either left or right. A similar statement applies to person characteristic curves. The parallelism implies the trade-off of a difference between two reader measures for an identical difference between two text measures to hold constant the success probability (i.e., comprehension rate).

The Rasch model specification is not the only one with the numerical conjoint measurement property. For every numerical conjoint measurement model, there is a monotonic transformation that removes the interaction from the array of success probabilities. An important feature of the logistic transformation, and thus a feature unique to the Rasch model, is the property that the raw scores for persons and the item p values are sufficient statistics for estimating the respective latent variables.

This is not to say that the Rasch model is an instantiation of the *theory* of conjoint measurement (Luce & Tukey, 1964; Krantz, Luce, Suppes, & Tversky, 1971). The Rasch model is not concerned with the ordinal and equivalence relations necessary and sufficient for additive representation (i.e., those entailed by the hierarchy of cancellation axioms (Scott, 1964)).[2]

8   The Score Sufficiency Condition and Its Implications

Sufficiency is an important technical term in the language of statistical inference. It is especially important in the current context because of its implications with respect to the Rasch model, which is unique in having the property that the raw score—the total number of correct responses—is a sufficient statistic for estimation of the person parameter. Raw score sufficiency means that, once we know the total number of correct responses, we can learn nothing more about the person parameter from the response pattern.

The precise statement of this result is as follows. Assume that the multi-item, multi-person molecular model is unidimensional for persons and that local independence holds. If, in addition, the raw score is a sufficient statistic for the person parameter, then the molecular model is necessarily a Rasch model. In other words, in the context of a unidimensional molecular model with local independence, the reparameterization obtained by transforming the success probabilities to logits produces horizontally parallel item characteristic curves.

---

[2] For a debate on this issue, see Borsboom and Zand Scholten (2008), Kyngdon (2008a, 2008b), and Michell (2008).

The proof is rather straightforward. Let item A and item B represent two arbitrarily selected items, and let $p_A(\theta)$ and $p_B(\theta)$ denote the success probabilities for these items expressed as an arbitrary monotonic function of a person parameter θ. Raw score sufficiency implies that the conditional probability distribution of response patterns with the same raw score does not depend on the person parameter. Consider a response pattern in which item A is answered correctly but item B is not in comparison to a pattern in which the only change is to reverse these two responses (i.e., item B correct, item A not). These two patterns have the same raw score. The ratio of the probabilities of these patterns in the conditional distribution is the same as their ratio in the unconditional distribution.

If this ratio is denoted by R, it is defined by the equation:

$$p_A(1 - p_B) = R(1 - p_A)p_B, \tag{1}$$

where the common factors for the responses to other items have been canceled out. Dividing this equaion by $(1-p_A)\,(1-p_B)$ and taking logarithms yields:

$$\ln\frac{p_A}{(1 - p_A)} = \ln\frac{p_B}{(1 - p_B)} + \ln R.$$

Because the ratio R applies to the conditional as well as to the unconditional distribution and the unconditional distribution is independent of the person parameter, R cannot depend on the person parameter θ. This implies that the item characteristic curves are horizontally parallel when the success probabilities are transformed to logits. That is the defining characteristic of the Rasch model.

9 Tightening via Theory

Although the Rasch model, in which person abilities and item difficulties are parameters to be estimated from data, is the tightest model yet considered, further tightening of the model specification is possible and desirable. As an alternative to a data-based empirical approach for the estimation of item difficulties, a theory might be used to predict an item's difficulty from characteristics of the item. In the prototype example, a readability formula for the text passage might be used as a predictor of item difficulty.

Tightening a model's specifications is, however, a two-edged sword. This process increases the number of ways in which a model can be wrong, which can hamper a model's usefulness. On the other hand, tightening can enhance a model's capability for measurement, which adds value to the model. What then is the enhanced capability afforded by a theory of item difficulty? For an answer to this question, consider the prototype example and the enhancement that occurs at various stages of tightening the model specification.

In the prototype example, the assumption of unidimensionality under replication allows the comparison between John and Alice as to who is the better reader. That specification is too loose to allow a data-based answer to the question of how much better. The Rasch model specification with person and item parameters to be estimated is tighter still and with the assumption of adequate replication for both items and persons, this model can answer the question of how much better Alice is than John. The answer to this question does not change when other items that satisfy the specification are used instead of the original items. The capability of providing a measure of the difference in reading ability between John and Alice independent of the items (qua instrument) used to effect the measurement is called *specific objectivity* (Rasch, 1977).

How well does Alice read? This is a question about Alice's reading ability apart from any comparison with the reading ability of John or anyone else. Suppose the only data available from which to infer an answer to this question are Alice's responses to a set of dichotomous reading items. The Rasch model specification with undetermined item difficulties does not have the capability to provide an answer. When the Rasch model is tightened by using theory to determine item difficulties, the question can be answered using only the data from Alice's responses. The enhancement provided by the theory-based determination of item difficulties is substantial. Alice now has an absolute measure on the reading ability construct, as distinct from her measure relative to another person's, which is independent of the items (qua instrument) used to make the measure.

The use of substantive theory in the form of a construct specification equation (Stenner et al., 1983) adds stringency to the model specification. Other uses of the specification equation include: (1) Explaining the variation detected by an instrument. The specification equation includes just those features of the measurement context that cause variation in success probabilities. In the prototype example, the construct theory states that as we move up the scale, we will encounter text that places higher syntactic and semantic demands on the reader. The specification equation includes proxies for these two text features. As these text features are manipulated, the theory predicts changes in the observed item difficulties. Some argue that there is no more compelling validity evidence than causal control over the variation an instrument detects; (2) Bringing nontest behavior into the measurement frame of reference. In the Lexile Framework for Reading, books are imagined to be tests with theoretical calibrations provided by the specification equation. The Rasch model is solved for the reader ability given an arbitrary but useful relative raw score of 75% and the theoretical item calibrations. The resulting reader measure required to answer correctly 75% of the virtual test items is the text readability measure assigned to the book. Thus, an important nontest behavior, comprehension of a particular book by a particular reader, can be forecasted; and (3) Generating item calibrations for reading items that have been built by a software program. This application enables one-off instruments to be used with each examinee and then disposed of as with disposable (single-use) thermometers. The specification equation and item engineering rules maintain the unit from instrument to instrument (Stenner et al., 2006).

10 Applying the Framework

There are important issues to consider when formulating and evaluating models for the purpose of effecting measurement from dichotomous data. Dimensionality, differential item functioning, and the number of parameters needed to represent item characteristics are examples of such issues. The developments presented in this chapter can provide a framework for addressing these issues when we formulate and evaluate models.

As an example of the framework in action, suppose the items on a test of verbal ability have been organized into subtests labeled *comprehension* and *vocabulary*. Do these subtests measure two distinct person constructs or is the full test unidimensional? The framework provides a basis for the conduct of data analysis to answer this question. At issue is the question of whether the ordering of persons is the same for the two subtests. Unidimensionality within subtest and the ability to replicate are all the assumptions needed to make it possible for measurement to provide the answer. In practice, however, there could be a reason to tighten the specification. Observed between-subtest differences in the rankings of persons could occur despite overall unidimensionality as a result of measurement error. An assessment of the statistical significance of the departure from overall unidimensionality may require more stringent assumptions than within-subtest unidimensionality about the interrelationships between the atomic models of the items.

This is just one example. Other important questions can be similarly addressed from the perspective of the framework.

11 Summary and Conclusion

A typical psychometric model represents the encounter between a person and a dichotomous item as a probability. For a given item, the probabilities associated with persons define a construct that is ordinal in character, that is, the probabilities can be arbitrarily subjected to a monotonic transformation without changing the character of the construct. Although a construct may be defined by the model for a single item, the single item model is insufficient for measurement, which requires replication, and replication requires multiple items that all define the same construct. Because single item models are the basic building blocks for multiple item models, it is natural to use the terms *atomic* and *molecular* for single item and multiple item models, respectively.

Molecular models involve assumptions about the relationships between the atomic models of the items, and these assumptions vary as to their stringency. This variation enables us to locate model specifications on an ordinal stringency construct where less stringent specifications are described as *loose* and more stringent ones as *tight*. Tighter specifications have less data-fitting flexibility but compensate with features that enhance their usefulness for measurement. The Rasch model is a tight specification that enables conjoint measurement. Although other, equally tight specifications enable conjoint measurement, the Rasch model is the only one for which the raw scores and item *p*-values provide all of the information that is relevant to the measurement of persons and items on the same scale. When the Rasch model is further tightened with item difficulties specified by theory, each person can be measured on an absolute scale with a measurement derived from single-use items.

For a given measurement application, the choice of model is likely to depend on particulars of the context. The concepts of atomic and molecular models and the location of a model specification on the stringency construct provide a framework that can help guide the choice of a model. The framework can also help to guide analyses of issues such as multidimensionality and differential item functioning that can threaten the validity of model-based inferences.

References

Birnbaum, A. (1968). Some latent trait models and their use in inferring an examinee's ability. In F. M. Lord & M. R. Novick (Eds.), *Statistical theories of mental test scores* (pp. 395–479). Addison-Wesley.

Borsboom, D., & Zand Scholten, A. (2008). The Rasch model and conjoint measurement theory from the perspective of psychometrics. *Theory and Psychology, 18*, 111–117.

Krantz, D. H., Luce, R. D, Suppes, P., & Tversky, A. (1971). *Foundations of measurement, Vol. I: Additive and polynomial representations*. New York: Academic Press.

Kyngdon, A. (2008a). The Rasch model from the perspective of the representational theory of measurement. *Theory and Psychology, 18*, 89–109.

Kyngdon, A. (2008b). Conjoint measurement, error and the Rasch model: A reply to Michell and Borsboom and Zand Scholten. *Theory and Psychology, 18*, 125–131.

Luce, R. D., & Tukey, J. W. (1964). Simultaneous conjoint measurement: A new scale type of fundamental measurement. *Journal of Mathematical Psychology, 1*, 1–27.

Michell, J. (2008). Conjoint measurement and the Rasch paradox: A response to Kyngdon. *Theory and Psychology, I, 8*, 119–124.

Rasch, G. (1960). *Probabilistic models for some intelligence and attainment tests*. Copenhagen, Denmark: Danish Institute for Educational Research (Expanded edition). Chicago: University of Chicago Press.

Rasch, G. (1977). On specific objectivity: An attempt at formalising the request for generality and validity of scientific statements. *Danish Yearbook of Philosophy, 14*, 58–94.

Scott, D. (1964). Measurement models and linear inequalities. *Journal of Mathematical Psychology, 1*, 233–247.

Stenner, A. J., Burdick, H., Sanford, E. E., & Burdick, D. S. (2006). How accurate are Lexile text measures? *Journal of Applied Measurement, 7*, 307–322.

Stenner, A. J., Smith, M., & Burdick, D. S. (1983). Toward a theory of construct definition. *Journal of Education Measurement, 20*, 305–315.

Stout, W. (1990). A new response theory modeling approach with applications to unidimensionality assessment and ability estimates. *Psychometrika, 55*, 293–325.

Wood, R. (1978). Fitting the Rasch model: A heady tale. *British Journal of Mathematical and Statistical Psychology, 31*, 27–32.

Generally Objective Measurement of Human Temperature and Reading Ability: Some Corollaries

A. Jackson Stenner and Mark Stone

Abstract We argue that a goal of measurement is general objectivity: point estimates of a person's measure (height, temperature, and reader ability) should be independent of the instrument and independent of the sample in which the person happens to find herself. In contrast, Rasch's concept of specific objectivity requires only differences (i.e., comparisons) between person measures to be independent of the instrument. We present a canonical case in which there is no overlap between instruments and persons: each person is measured by a unique instrument. We then show what is required to estimate measures in this degenerate case. The canonical case encourages a simplification and reconceptualization of validity and reliability. Not surprisingly, this reconceptualization looks a lot like the way physicists and chemometricians think about validity and measurement error. We animate this presentation with a technology that blurs the distinction between instruction, assessment, and generally objective measurement of reader ability. We encourage adaptation of this model to health outcomes measurement.

In this paper we look closely at two latent variables (temperature and reading) and two instruments used in their measurement. At first glance the conceptual foundations for these constructs seem to represent quite different entities. A thermometer measures human temperature widely assumed to be a physical attribute, while a test for measuring reader ability clearly represents a mental attribute. Common sense tends to assert fundamentally dissimilar ontologies. We intend to show that underneath this surface dissimilarity there are striking parallels that can be exploited to illuminate the "oughts" of human science measurement. Moreover, our expectation is this comparison will offer important insights to health outcome researchers by showing

Journal of Applied Measurement, 11(3). 2010.

A. J. Stenner (✉)
MetaMetrics, Inc., Durham, NC, USA

M. Stone
Adler School of Professional Psychology, Chicago, IL, USA

the efficiency of linear measurement for establishing closed knowledge systems. Closed knowledge systems offer important benefits to effective rehabilitation by isolating key constructs affecting patient functional status. Their parameterization in a closed system diminishes treatment uncertainty and increases measurement efficiency. Both thermodynamics and Lexile Theory emphasize latent constructs that are operationally defined not only by units but comprehensive "deep structure" substantive theories, which guide general system formulation. Both thermodynamics and Lexile Theory are systems that manipulate only a few key constructs, which are applicable across broad classes of physical and mental activity. This approach to latent traits suggest health outcome measurement is also probably defined by several deep structure constructs that permeate outcome measurement. Closed systems conforming to substantive theories that follow thermodynamics and Lexile models can likely be developed to govern these constructs.

Before comparing and contrasting temperature and reader ability we assert the physical science construct *temperature* and the human science construct *reader ability* share common philosophical foundations. Both latent constructs signify real entities whose causal action can be manipulated and their effects observed on interval scales. Neither construct should be conceived of as "just a useful fiction." Thus, we reject a constructivist interpretation for either construct. Both constructs are attributes of human beings, and we are entity realists. The relationship between the latent variable and the measurement outcome is causal at both the intra-individual and interindividual levels, stated more formally, the conditional probability distribution of the measurement outcome given the latent variable is to be given a *stochastic subject interpretation* (Holland, 1990). We reject a repeated sampling interpretation of the conditional probability distribution. We assert for both temperature and reading ability the measurement model takes the same form within and between persons, what Ellis and Van den Wollenberg (1993) called the *local homogeneneity assumption*. We assert this assumption is testable in the theory referenced measurement context (Stenner et al., 1983).

Figure 1 presents an aspect chart for two latent variables: temperature and reader ability. For our purposes we assume that the object of measurement in both cases is a person. In each case the instrument (thermometer or reading test) is brought into contact with the person. A measurement outcome (number of cavities (0–45) that fail to reflect green light or count correct on 45 theoretically calibrated reading items) is recorded by a professional (nurse or teacher). Note that both instruments have been targeted on the appropriate range for each person. The thermometer measures from 96° to 104.8 °F and let's assume the reading test has been targeted for a typical fourth grader (500–900 L). The substantive theory provides the link between the measurement outcome and the measure denominated in a conventional unit (degrees Fahrenheit (°F) or Lexiles (L)). Note that substantive theory is used to build the respective instrument and to convert cavity counts and counts correct on the reading test into a measure. Cavity counts and counts correct are sufficient statistics for their respective parameters, i.e., measures. Sufficient statistics exhaust the information in the data that is relevant to estimate the parameter/measure. In each case a point estimate of the measure is produced for the person without recourse to information

Anatomy of Two Measurement Procedures		
Aspect/Construct	Temperature	Reader Ability
Object of measurement	Person	Person
Instrument	Thermometer	Reading test
Measurement outcome	Number of theory calibrated cavities (0-45) that fail to reflect green light	Count correct on a collection of theory calibrated test items
Substantive theory	Thermodynamic Theory	Lexile Theory
Unit of measurement	Degree Fahrenheit (°F)	Lexile (L)
Correspondence table/ Calibration equation	Exploits a chemical reaction and light absorption to table temperature as a function of a sufficient statistic	Exploits semantic and syntactic features of test items to table reading ability as a function (Rasch model) of a sufficient statistic
Measure/Quantity	Measurement outcome converted into aquantity via substantive theory	Measurement outcome converted into a quantity via substantive theory
Readable technology	NexTemp Thermometer™	MyReadingWeb™
General objectivity	Point estimates of temperature are independent of the thermometer	Point estimates of reader ability areindependent of the reading test

Fig. 1 Temperature versus reader ability measurements

about other persons' measures. Instrument calibrations come from theory and individual instruments may be disposable—have never been used before and will never be used again. We turn now to a more detailed look at the Nextemp® Thermometer and My Reading Web™ to further draw out the parallel structures underlying the measurement of these two latent variables.

1 Human Body Temperature

Temperature is a physical property of systems including the human system that imperfectly corresponds to the human sensation of hot and cold. Temperature as a latent variable is a principal parameter in thermodynamic theory. Like other latent variables temperature cannot be directly observed but its effects can be and thermometers are instruments that detect these effects. A thermometer has two key components: The sensor (e.g., cavities of fluid that differentially reflect light) which detects changes in temperature and a correspondence table that converts the measurement outcome (cavity count) into a scale value (degrees Celsius) via theory (thermodynamic). A wide range of so-called primary thermometers have been built each relying on radically different physical effects (electrical resistance, expansion coefficients of two metals, velocity of sound in a monatomic gas, and gamma ray emission in a magnetic field). For primary thermometers the relationship between a measurement outcome and its measure is so well understood that temperature readings can be computed directly from the measurement outcomes. So-called "secondary" thermometers (e.g., mercury thermometers) produce measurement outcomes whose relationship to temperature is not yet so well understood that temperature readings can be directly computed. In these by far more common cases secondary thermometer readings are often calibrated against a primary thermometer and a correspondence table is generated that links the measurement outcome to temperature expressed in, say, degrees Celsius. Celsius measures can then be converted directly into Kelvin, Fahrenheit, Rankine, Delisle, Newton or other metrics that find use in, for example, special engineering applications or high energy physics.

In 1861 Carl Wunderlich reported in a study of one million persons that the average human body temperature was 37.0 °C. This value converts precisely to 98.6 °F and is believed to be the source for the putative "normal" temperature. Mackowiak et al. (1992) measured 148 healthy men and women multiple times each day for three consecutive days with electronic oral thermometers. The authors found that the average temperature was 36.8 °C which converts to 98.2 °F. They speculate that Wunderlich rounded his "average" up to 37 °C and then the rounded measure was converted to 98.6 F and subsequently popularized. Mackowiak, Wasserman and Levine recommend that the popular value of 98.6 F for oral temperature be abandoned in favor of the new value of 98.2 and that "normal" ranges be similarly revised.

Figure 2 provides a black line sketch of a NexTemp® disposable thermometer for measuring human temperature. The NexTemp thermometer is a thin, flexible, paddle-shaped plastic strip containing multiple cavities. In the Fahrenheit version, the 45 cavities are arranged in a double matrix at the functioning end of the unit. The columns are spaced at 0.2 °F intervals covering the range of 96.0–104.8 °F. Each cavity contains a chemical composition comprised of three cholesteric liquid crystal compounds and a varying concentration of a soluble additive. These chemical compositions have discrete and repeatable change-of-state properties, the temperatures of which are determined by the concentrations of the additive. Additive concentrations are varied in accordance with an empirically established formula to produce a series of change-of-state temperatures consistent with the indicated temperature points on the device. The chemicals are fully encapsulated by a clear polymeric film, which allows observation of the physical change but prevents any user contact with the chemicals. When the thermometer is placed in an environment within its measure range, such as 98.2 °F (37.0 °C), the chemicals in all of the cavities up to and including 98.2 °F (37.0 °C) change from a liquid crystal to an isotropic clear liquid state. This change of state is accompanied by an optical change that is easily viewed by a user. The green component of white light is reflected from the liquid crystal state but is transmitted through the isotropic liquid state and absorbed by the black background. As a result, those cavities containing compositions with threshold temperatures up to and including 98.2 °F (37.0 °C) appear black, whereas those with transition temperatures of 98.2° (37.0 °C) and higher continue to appear green (Medical Indicators, pp. 1–2).

In-vitro accuracy of the NexTemp liquid crystal thermometer equals or exceeds glassmercury and electronic thermometers. More than one hundred million production units have shown agreement with calibrated water baths to within 0.2 °F in

Fig. 2 NexTemp thermometer

the range of 98.0–102.0 °F and within 0.4 °F elsewhere (0.1 °C in the range of 37.0–39.00 C and within 0.2 °C elsewhere).

In-vivo tests in the U.S., Japan and Italy resulted in excellent agreement with measurements using specially calibrated glass-mercury thermometers. The mean difference between the NexTemp thermometers and the calibrated glass-mercury equilibrium device was only 0.12 °F (0.07 °C). The NexTemp thermometer also achieves equilibrium very rapidly, due to its small "drawdown" (cooling effect on tissue upon introduction of a room-temperature device) and the small amount of energy required to make the physical phase transition.

A competitive marketing analysis reported the following advantages of this new technology over the market dominant older mercury in a tube technology. These "technical advantages" will prove useful in our comparison and contrast with reading test technology:

- *Cost*—The NexTemp temperature measurement technology provides lower costs when compared to other temperature devices.
- *Safety*—The safety advantages of NexTemp technology are substantial. There is no danger, as with a conventional thermometer, of glass ingestion or mercury poisoning if a child bites the active part of the unit. NexTemp® and its packaging are latex-free.
- *Speed and ease-of-use*—The NexTemp thermometer is quick, portable, non-breakable and easy to use (e.g., no shakedown or resetting).
- *Reduced chance of cross-contamination of patients*—The NexTemp disposable product comes individually wrapped and is intended to be used and then discarded. The reusable version of the product is for single patient use over time with cleaning between uses (Medical Indicators, p. 4).

2   Reader Ability

Of approximately 6000 spoken languages in the world only about 200 are written and many fewer than that have an extensive test base. *Reader ability* is the capacity of the individual to make meaning from text. Reader ability like other latent variables cannot be directly observed. Rather its existence must be inferred from its effects on measurement outcomes (count correct). A reading test is an instrument designed to detect variation in reading ability. A reading test has three key components:

(1) Text (e.g., a newspaper article on global warming), (2) a response requirement (e.g., answering multiple choice questions embedded in the passage), and (3) a correspondence table that converts the measurement outcome (counts correct) to

a scale value (Lexiles) via a theory (Lexile Framework for Reading). Although the ubiquitous multiple choice item type dominates as the response requirement in the measurement of reading ability other task types have found use for specific ranges of the reading ability scale, including: retelling, written summaries, short answers, oral reading rate and cloze. Because the notion of scale unification is still foreign in the human sciences common practice associates a unique measurement scale with every published instrument. Many dissertations are written that involve development of a new instrument and a new scale purportedly measuring the intended construct in an equal interval metric unique to that instrument. At this writing the authors are unaware of any dissertation that reports on the unification of multiple measures of the same latent variable (e.g., depression, anxiety, spatial reasoning, mathematical ability) denominated in a common unit of measure. Failure to separate the instrument from the scale has had pernicious effects on human science. Because different reading instruments often employ different task types different test names (test of reading ability, test of reading comprehension, test of reading achievement) and the aforementioned different scales it should not surprise us that many test users and reading researchers perceive that these various reading tests measure different latent variables. The unwary are fooled by the fact that looks can be deceiving.

WEB based reading technology (MyReadingWeb™) has been developed to accommodate Lexile measurement. Accompanying text and associated machine generated cloze items is a "key" to score (mark correct or incorrect) reader responses and a correspondence table relating count correct to Lexiles. Any continuous prose can be loaded into MyReadingWeb™ and the software will instantly turn text into a reading measurement instrument. The first step involves measuring the text be it an article, chapter, or book. The Lexile Analyzer computes various statistics on word frequency and sentence length. These statistics are then combined in an equation that returns a Lexile measure for the text. MyReadingWeb™ uses this measure to "cloze" vocabulary words that are at a comfortable range for a reader's vocabulary who has a reading ability equal to the text readability of the article. A-part-of-speech parser then chooses three foils (incorrect answers) that are the same part of speech and have a similar Lexile level as the closed word but that are not synonyms or antonyms of the closed word. Finally, the four choices are randomized before presentation. On average a response requirement is imposed on the reader every 50–70 words of running text.

Counts correct on a text are evaluated by a dichotomous Rasch model in which the location and dispersion parameters of an item difficulty distribution are treated as known. The Lexile measure that maximizes the likelihood of the data is the reader measure reported for each particular encounter between reader and text. A Baysean growth model is used to combine individual article measures within and across days over the complete MyReadingWeb™ history for a reader.

Table 1 presents results from a large study designed to test how well Lexile Theory could predict percent correct on machine generated items like those described above. First grade through twelfth grade students (N = 1,743) read a total of 289,345 articles comprised of 194,968,617 words.

Table 1 Lexile theory of prediction of machine generated items

	Student	Mean reader		Encounters	Words	Time		WPM		Observed	Expected	Observed	Expected
	Count	Measure (L)	sd	Encounters	Words	Spent	WPM	sd	Items	Correct	Correct	Performance (%)	Performance (%)
Overall	1,743	1071	344	289,345	194,968,617	2 y 157 d 23 h 16 m	150	82	3,051,341	2,245,741	2,291,787	73.90	75.11
Grade 1	4	739	105	47	16,203	2 h 36 m	103	44	477	306	308	60.54	64.53
Grade 2	217	586	295	14,449	3,127,596	29 d 21 h 17 m	72	43	119,037	81,523	85,242	69.02	71.61
Grade 3	174	810	284	22,286	5,644,421	47 d 18 m	89	51	193,840	130,819	135,400	68.29	69.85
Grade 4	186	946	295	32,936	10,932,254	74 d 19 h 24 m	108	58	302,708	202,905	210,746	68.26	69.62
Grade 5	164	1074	221	34,864	15,026,873	92 d 10 h 12 m	121	59	340,605	243,056	251,237	72.35	73.76
Grade 6	175	1130	180	33,650	16,335,039	80 d 20 h 48 m	144	66	338,226	245,232	252,667	73.68	74.70
Grade 7	171	1171	229	19,485	9,352,170	43 d 1 h 36 m	153	70	173,593	127,235	131,027	74.36	75.48

(continued)

Table 1 (continued)

	Student Count	Mean reader Measure (L)	Mean reader sd	Encounters	Words	Time Spent	WPM	WPM sd	Items	Observed Correct	Expected Correct	Observed Performance (%)	Expected Performance (%)
Grade 8	164	1281	252	17,083	8,725,553	39 d 19 h 37 m	158	75	150,612	112,762	114,806	76.09	76.23
Grade 9	149	1285	254	22,815	19,490,193	81 d 19 h 21 m	167	80	264,169	203,037	206,715	77.26	78.25
Grade 10	130	1268	229	23,225	21,477,331	89 d 11 h 49 m	182	80	264,252	199,222	201,473	75.44	76.24
Grade 11	102	1324	151	23,906	26,811,304	104 d 19 h 16 m	191	82	312,976	240,951	246,715	77.38	78.83
Grade 12	107	1353	157	28,394	35,844,222	128 d 3 h 13 m	206	84	384,070	298,797	306,723	78.56	79.86
Graduated				16,205	22,185,458	75 d 17 h 49 m	219	86	206,776	159,896	163,092	78.18	78.87

Note Oasis-Reading Data by Cohort-Corinth (**MS**) (Data From 2007-06-01 to 2010-04-26)

They spent 2 years 157 days 23 h 16 min in the program and averaged 150 words read per minute (WPM). Of the 3,051,341 unique cloze items generated by the computer the participants answered 2,245,741 or 73.90% of the items correctly. The model forecasted 2,291,787 correct or 75.11%. Figure 3 presents a histogram of differences between theory and observation. One thousand and five (1,005) students had observed counts correct within ±3% of the model expectation for a subsample of 1,325 students.

The best explanation for the close agreement between theoretical comprehension rate and observed comprehension rate is that the Lexile theory and Rasch model are cooperating in providing (1) good text measures for the articles, (2) good reader measures for the students, and (3) well modeled comprehension rates. The cooperation between substantive theory and the Rasch model evidences cross sectional developmental consistency, i.e., the theory works throughout the reading range reflected in Table 1 (100–1500 L). Invoking the "no miracles argument" currently fashionable among philosophers of science of the realist persuasion: the congruence between theory and observation is explained by the fact that the theory is at least approximately right.

In summary, the measurement of human temperature and reading ability if conceptualized in a particular way can be seen to share a common deep structure. Both

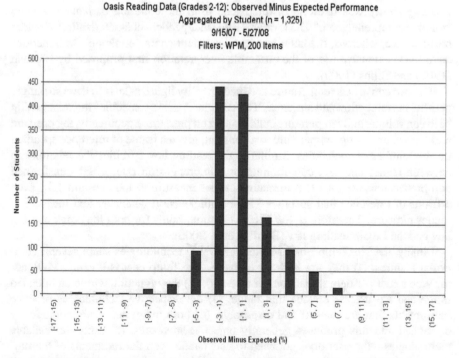

Fig. 3 Differences between Lexile theory and observations

constructs are latent variables which assign a causal role to an unobservable attribute of persons. Conditioning on the latent variable renders the measurement outcomes (cavity count and count correct) statistically independent. Temperature and reading ability are real entities that can be manipulated and the effects of these manipulations can be detected. Persons possess a true value on each construct that is approached by repeated measurement but is never precisely determined. The two attributes apply equally well to between person variation and within person variation. For human temperature measurement with the NexTemp® technology we can trade off a change in the amount of the soluble additive for a change in temperature to hold the number of cavities that "turn black" constant. Similarly, for reading we can trade-off a difference in reader ability for an equivalent difference in text readability to hold constant the count correct. In both cases enough is known about the measurement procedure and the relevant active processes that two persons with equal temperature or reading abilities can be made to produce different measurement outcomes (cavity count or count correct) by systematically manipulating the respective instruments. In short the two latent variables are under precise experimental control and that is why an indefinitely large number of parallel instruments can be manufactured for each construct.

Both thermodynamic theory and Lexile reading theory force a distinction not made before the theories were put forth. The sensation of hot and cold was formalized as temperature and was later distinguished from the common parlance synonym "heat." Reading ability was likewise distinguished from the common parlance synonym "reading comprehension." The former is a text independent characterization of reader performance, whereas, the latter is a text dependent characterization. Both theories have made extensive use of the ensemble interpretation first proposed by Einstein (1902) and Gibbs (1902).

The two constructs temperature and reader ability figure in laws that are strikingly parallel in conception and structure. The combined gas law specifies the relationship between volume and temperature conditioning on pressure. Specifically, log pressure and log volume—log temperature = a constant, given a frame of reference specified by the number of molecules. Similarly, the reading law specifies the relationship between reader and text conditioning on comprehension rate. Logit transformed comprehension rate plus text measure—reader measure = the constant 1.1, given a frame of reference that specifies 75% comprehension whenever text measure = reader measure. Therefore $a + b - c =$ constant, holds for both the combined gas law and the Lexile reading law (Burdick et al. 2006).

Finally, the NexTemp® and MyReadingWeb™ technologies share several additional features: (1) they are both inexpensive (NexTemp cost is 9 cents, MyReadingWeb's cost per item is fractions of a penny), (2) they function within an intended range and are useless outside that range, (3) the respective technologies produce instruments that are one-off and disposable, (4) both instruments are theoretically calibrated and this produces generally objective measures, (5) both are readable technologies—the user does not need to understand even the rudiments of thermodynamic theory or Lexile theory to produce valid and useful measures, and (6) both can measure growth or change within and between persons.

In conclusion, although closed knowledge systems have not been developed for health outcomes, corollaries between temperature and reading ability presented in this paper should offer valuable insights to their formulation. For example, functional assessment in rehabilitation currently defined by separate CAT measures could be consolidated into a closed system defined by deep structure common across functional measures. Separate functional items could be reduced to a single structure that includes not only outcome but intervention constructs. In this system, variation of intervention treatment values would be systematically related to outcome measures. Consequently, treatment effectiveness could be specified in advance of delivery and terminated at maximum effectiveness.

Acknowledgements Authors extend special thanks to Nikolaus Bezruczko for improving this paper.

References

Burdick, D. S., Stone, M. H., & Stenner, A. J. (2006). The combined gas law and a Rasch reading law. *Rasch Measurement Transactions, 20*(2), 1059–1060.

Einstein, A. (1902). Kinetische Theorie des Warmegleichgewichtes und des zweiten Hauptsatz der Thermodynamik. *Annalen der Physik, 9,* 417–433. English translation in BECK (1989), 30–47

Ellis, J. L., & van den Wollenberg, A. L. (1993). Local homogeneity in latent trait models: A characterization of the homogeneous monotone IRT model. *Psychometrika, 58,* 417–429.

Gibbs, J. W. (1902). *Elementary principles in statistical mechanics.* Yale University Press.

Holland, P. W. (1990). On the sampling theory foundations of item response theory models. *Pshychometrika, 55,* 577–601.

Mackowiak, P. A., Wasserman, S. S., & Levine, M. M. (1992). A critical appraisal of 98.6 degrees F, the upper limit of the normal body temperature, and other legacies of Carl Reinhold August Wunderlich. *JAMA, 268*(12), 1578–1580.

Medical Indicators. (2006). www.medicalindicators.com/pdf/NT-FC-Tech-bulletin.pdf

Stenner, A. J., Smith, M., & Burdick, D. S. (1983). Toward a theory of construct definition. *Journal of Educational Measurement, 20,* 305–315.

A Technology Roadmap for Intangible Assets Metrology

William P. Fisher Jr. and A. Jackson Stenner

Abstract Measurement plays a vital role in the creation of markets, one that hinges on efficiencies gained via universal availability of precise and accurate information on product quantity and quality. Fulfilling the potential of these ideals requires close attention to measurement and the role of technology in science and the economy. The practical value of a strong theory of instrument calibration and metrological traceability stems from the capacity to mediate relationships in ways that align, coordinate, and integrate different firms' expectations, investments, and capital budgeting decisions over the long term. Improvements in the measurement of reading ability exhibit patterns analogous to Moore's Law, which has guided expectations in the micro-processor industry for almost 50 years. The state of the art in reading measurement serves as a model for generalizing the mediating role of instruments in making markets for other forms of intangible assets. These remarks provide only a preliminary sketch of the kinds of information that are both available and needed for making more efficient markets for human, social, and natural capital. Nevertheless, these initial steps project new horizons in the arts and sciences of measuring and managing intangible assets.

1 Introduction

Standards ensure the performance, conformity, and safety of innovative new products and processes. Manufacturing and the provision of services require standards to coordinate the matching of services (as in telecommunications), the fitting of parts,

Joint International IMEKO TC1+TC7+TC13 Symposium, 2011.

W. P. Fisher Jr. (✉)
Living Capital Metrics LLC, Sausalito, CA, USA

Graduate School of Education, BEAR Center, University of California, Berkeley, CA, USA

A. J. Stenner
MetaMetrics, Inc., Durham, NC, USA

W. P. Fisher and P. J. Massengill (eds.), *Explanatory Models, Unit Standards, and Personalized Learning in Educational Measurement*,
https://doi.org/10.1007/978-981-19-3747-7_14

179

or the gauging of expectations (Allen & Sriram, 2000). Measurement, then, plays an essential economic role in the creation of markets centering on the efficiencies gained from the universal availability of precise, accurate, and uniformly interpretable information on product quantity and quality (Barzel, 1982; Benham & Benham, 2000; Callon, 2002; Miller & O'Leary, 2007). Clear, fully enforced property rights and transparent representations of ownership are other forms of standards that reduce the costs of transactions further by removing sources of unpredictable variation in social factors (Ashworth, 2004; Beges et al., 2011; Birch, 2008; Lengnick-Hall et al., 2004). When objective measurement is available in the context of enforceable property rights and proof of ownership, economic transactions can be contracted most efficiently in the marketplace (Baker et al., 2001; Jensen, 2003). The emergence of objective measures of individual abilities, motivations, and health, along with service outcomes, organizational performance and environmental quality, present a wide array of new potential applications of this principle.

Proven technical capacities for systematic and continuous improvements in the quality of objective measures enable the alignment, coordination, and integration of expectations, investments, and capital budgeting decisions over the long term. The relationship between standards and innovation is complex and dynamic, but a general framework conducive to innovation requires close attention to standards. The trajectory of ongoing improvements in instrumentation in the psychosocial and environmental sciences suggests a basis for a technology road map capable of supporting the creation of new efficiencies in human, social, and natural capital markets. New efficiencies are demanded by macroeconomic models that redefine labour and land as human and natural capital, respectively, and that add a fourth form of capital—social—to the usual three-capitals (land, labour, and manufactured) framework.

These models enhance sensitivity to the full complexity of intangible assets, enable the conservation and growth of their irreplaceable value, and frame economics in terms of genuine progress, real wealth, sustainability, and social responsibility not captured in accounting and market indexes restricted to the value of property and manufactured capital. Of special interest is the fact that the technical features of improvements in rigorously defined and realized quantification are likely to be able to support the coordination of capital budgeting decisions in ways analogous to those found in, for instance, the microprocessor industry relative to Moore's Law.

The state of reading measurement (Burdick et al., 2006; Stenner et al., 2006) is sufficiently advanced for it to serve as a model in extrapolating the principle to further developments in the creation of literacy capital markets, and for generalizing the mediating role of instruments in creating markets to other constructs and forms of capital in the psychosocial, health, and environmental sciences.

Instruments, metrological standards, and associated conceptual images play vitally important mediating roles in economic success. For instance, the technology roadmap for the microprocessor industry, based in Moore's Law and its projection of doubled microprocessor speeds every two years, has successfully guided semiconductor market expectations and coordinated research investment decisions for over

40 years (Miller & O'Leary, 2007). Moore's Law is more than a technical guideline—it has served as a business model for an entire industry for almost 50 years. This paper pro- poses the form similar laws and technology roadmaps will have to take to be capable of guiding innovation at both the technical level and at the broader level of human, social, and natural capital markets, comprehensively integrated economic models, accounting frameworks, and investment platforms.

The fulfilment of the potential presented by these intentions requires close attention to measurement and the role of technology in linking science and the economy (Callon, 2002; Miller & O'Leary, 2007). Of particular concern is the capacity of certain kinds of instruments to mediate relationships in ways that align, coordinate, and integrate different firms' expectations, investments, and capital budgeting decisions over the long term.

Instruments capable of mediating relationships in these ways are an object of study in the social studies, history, and philosophy of science and technology. In this work, the usual sense of technology as a product of science is reversed (Bud & Cozzens, 1992; Hankins & Silverman, 1999; Ihde, 1983; Ihde & Selinger, 2003; Latour, 2005; Price & Science, 1986; Rabkin, 1992). Instead of seeing science as rigidly tied to data and rule-following behaviours, the term technoscience refers to a multifaceted domain of activities in which theory, data, and instruments each in turn serves to mediate the relation of the other two (Ackermann, 1985; Ihde, 1991, 1998).

2 Measurement, Mediating Instruments, and Making Markets

In psychosocial research to date, there has been little recognition of the potential scientific and economic value of universally accessible, uniformly defined, and constant units. This article draws from the history of the microprocessor industry to project a model of how instruments measuring in such units can link science and the economy by coordinating capital budgeting decisions within and between firms. Links between the psychosocial sciences and industries such as education and health care are underdeveloped in large part because of insufficient attention to the mediating role some kinds of instruments are able to play in aligning investments across firms and agencies in an industry.

Instruments capable of mediating relationships do so by telling the story of a shared history and by envisioning future developments reliably enough to reduce the financial risks associated with the large investments required. In the microprocessor industry, for instance, Moore's Law describes a constant and predictable relation between increased functionality and reduced costs. From 1965 on, Moore's Law projected a detailed image of commercially viable applications and products that attracted investments across a wide swath of the economy. When it became clear in the early 1990s that the physical limits of existing technologies might disrupt or even end this improvement cycle, the Semiconductor Industry Association convened a special meeting aimed at creating a detailed common vision, a roadmap, for the next 15 years' developments in semiconductor technology (Miller & O'Leary, 2007).

This roadmap made it possible for the industry to navigate a paradigm shift in its basic technology with no associated economic upheaval and with the continuation of the historically established pattern of increased functionality and lower costs. Education, healthcare, government, and other industries requiring intensive human and social capital investments lack analogous ongoing improvements in their primary products' reliability, precision, and cost control. Where the microprocessor industry is able to reduce costs and improve quality while maintaining or improving profitability, education, healthcare, and social services seem only to always cost more, with little or no associated improvement in objective measures of quality.

To what extent might this be due to the fact that these industries have not yet produced mediating instruments like those available in other industries? If such instruments are necessary for articulating a shared history of past technical improvements and economies, and a shared vision of future ones, should not their development be a high priority? Within any economy, individual actors are able to contribute to the collective estimation of value only insofar as the information they have at hand is sufficient to the task. Ideally, with that information, those demanding higher quality can identify and pursue it, rewarding producers of the higher quality. Without that information, purchasers are unable to distinguish varying levels of quality consistently, so investments in improved products are not only unrewarded, they are discouraged. Philanthropic capital markets have lately been described in these terms (Goldberg, 2009).

Not yet having satisfactory mediating instruments in industries relying heavily on intangible assets is not proof of the impossibility of obtaining them. There are strong motivations for considering what appropriate mediating instruments would look like in human- and social-capital-intensive industries. Foremost among these motivations is a potential for correcting the significant capital misallocations caused when individual organizations make isolated investment decisions that cannot be coordinated across geographically distant groups' competing proprietary interests and temporally separated inputs and outputs.

The question is one of how to align investment decisions without compromising confidential budgeting processes or dictating choices. Simply sharing data on outcomes is a proven failure (Ho, 2008; Murray, 2006) and was never attractive to for-profit enterprises for which such information is of proprietary value. But instead of focusing on performance measured in locally idiosyncratic units incapable of supporting standard product definitions, might not a better alternative be found in defining a constant unit of increased learning, functionality, or health, and evaluating quality and cost relative to it? The key to creating coherent industry-wide communities and markets is measurement. Fryback (Fryback, 1993; Kindig, 1999) succinctly put the point, observing that the U.S. health care industry is a $900 + billion [over $2.5 trillion in 2009 (Data, 2011)] endeavor that does not know how to measure its main product: health. Without a good measure of output we cannot truly optimize efficiency across the many different demands on resources.

Quantification in health care is almost universally approached using methods inadequate to the task, resulting in ordinal and scale-dependent scores that cannot capitalize on the many advantages of invariant, individual-level measures (Andrich, 2004). Though data-based statistical studies informing policy have their place, virtually no effort or resources have been invested in developing individual-level instruments traceable to universally uniform metrics that define the outcome products of health care, education, and other industries heavily invested in human, social, and natural capital markets. It is well recognized that these metrics are key to efficiently harmonizing quality improvement, diagnostic, and purchasing decisions and behaviours (Berwick et al., 2003). Marshalling the resources needed to develop, implement them, and maintain them, however, seems oddly difficult to do until it is recognized that such a project must be conceived and brought to fruition on a collective level and against the grain of cultural presuppositions as to the objective measurability of intangible assets (Cooter, 2000; Fisher, 2009).

Probabilistic models used in scaling and equating different tests, surveys, and assessments to common additive metrics offer a body of unexamined resources relevant to the need for mediating instruments in the domains of human, social, and natural capital markets (Fisher, 2009). Miller and O'Leary (2007) complement the accounting literature's overly narrow perspective on capital budgeting processes with the fruitful lines of inquiry opened up in the history, philosophy, and social studies of science. In this work, mathematical models and instruments are valued for their embodiment of the local and specific material practices through which mediation is realized.

In these practices, instruments capable of serving as reliable and meaningful media must simultaneously represent a phenomenon faithfully and facilitate predictable control over it. Though the philosophy of science has long focused attention on the nature of objective representation, the history and social studies of science have, over the last 30 years or so, shifted attention to the role of technology in theory development and in determining the outcome of experiments. By definition, instruments capable of mediating must exhibit properties of structural invariance across the locally defined contexts of different organizations' particular investments, policies, workforces, and articulations of the relevant issues. It is only through the conjoint processes of representation and intervention that, for instance, the steam engine became the medium facilitating development of work in the sense of engineering mechanics and in the economic sense of a new source of labour (Wise, 1988). The medium is the message here, in the sense that mediating instruments like the steam engine both represent the lawful regularity of the scientific phenomenon and provide a predictable means of intervening in the production of it.

The unique importance and value of Rasch's models for measurement lie precisely here. Rasch-calibrated instruments have long been in use on a wide scale in applications that combine the representation of measured amounts for accountability purposes with instructional or therapeutic interventions that take advantage of the meaningful mapping of abilities relative to curricular or therapeutic challenges (Alonzo & Steedle, 2009; Chang & Chan, 1995; Kennedy & Wilson, 2007; Leclercq, 1980). These models are structured as analogies of scientific laws' three-variable multiplicative form (Burdick et al., 2006; Fisher, 2010a) and so enable experimental tests of possible causal relations (Bunderson & Newby, 2009; Stenner & Smith, 1982). When data fit such a model, demonstrably linear units of measurement may be calibrated and maintained across instrument configurations or brands, and across measured samples. Clear thinking about the measured construct is facilitated by the invariant constancy of the unit of measurement—one more unit always means one more unit of the same size. When instruments measuring the same thing are tuned to the same scale, mediation is achieved in the comparability of processes and outcomes within and across subsamples of measured cases.

Linear performance measures are recommended as essential to outcome-based budgeting (Jensen, 2003), and will require structurally invariant units capable of mediating comparisons in this way. Without instruments mediating meaningfully comparable relationships, it is impossible to effectively link science and the economy by coordinating capital budgeting decisions. The lessons so forcefully demonstrated over the course of the history of the microprocessor industry need to be learned and applied in many other industries.

The potential for a new class of mediating instruments resides here, where the autonomy of the actors and agencies forming a techno-economic network is respected and uncompromised. Rasch's parameter separation theorem is a scientific counterpart of Irving Fisher's economic separability theorem (Fisher, 2011). It is essential to realize that Rasch's equations model the stochastically invariant uniformity of behaviours, performances or decisions of individuals (people, communities, firms, etc.), and are not statistical models of group-level relations and associations between variables (Fisher, 2010b). Data fit a Rasch model and mediation is effected so far as the phenomenon measured (an ability, attitude, performance, etc.) retains its proper ties across samples and instrument brands or configurations. Given this fit, the unit of measurement becomes a common currency for the exchange of value within a market defined by the model parameters (Fisher, 2011). How could this implicit and virtual market be made explicit and actual? By devising mediating instruments linking separate actors and arenas in a way that conforms to the requirements of the techno-economic forecasts of a projection like Moore's Law or of a technology roadmap based in such a law. As Miller and O'Leary (2007) say,

> Markets are not spontaneously generated by the exchange activity of buyers and sellers. Rather, skilled actors produce institutional arrangements, the rules, roles and relationships that make market exchange possible. The institutions define the market, rather than the reverse.

What are the rules, roles and relationships that skilled actors need to arrange for their institutions to define efficient markets for human, social, and natural capital? What are the rules, the roles, and the relationships that make market exchange possible for these forms of intangible assets? How can standard product definitions for the outcomes of education, healthcare, and social services be agreed upon? Where are the lawful patterns of regularities that can be depended on to remain constant enough over time, space, firms, and individuals to support industry-wide standardizations of measures and products based on them? What trajectories can be mapped that would enable projections accurate enough for firms and agencies to rely on in planning products years in advance?

Answers to questions such as these provide an initial sketch of the kind of grounded, hands-on details of the information that must be obtained if the endless inflationary spirals of human- and social-capital- intensive industries are ever to be brought under control and transformed into profitable producers of authentic value and wealth.

3   The Rasch Reading Law and Stenner's Law

It is a basic fact of contemporary life that the technologies we employ every day are so complex that hardly anyone understands how they do what they do. Technological miracles are commonplace events, from transportation to entertainment, from health care to industry. And we usually suffer little in the way of adverse consequences from not knowing how automatic transmissions, thermometers, or digital video reproduction works. It is enough to know how to use the tool.

This passive acceptance of technical details beyond our ken extends as well into areas in which standards, methods, and products are much less well defined. And so managers, executives, researchers, teachers, clinicians, and others who need measurement but who are unaware of its technicalities tend to be passive consumers accepting the lowest common denominator of measurement quality.

And just as the mass market of measurement consumers is typically passive and uninformed, in complementary fashion the supply side is fragmented and contentious. There is little agreement among measurement experts as to which quantitative methods set the standard as the state of the art. Virtually any method can be justified in terms of some body of research and practice, so the confused consumer accepts whatever is easily available or is most likely to support a preconceived agenda.

It may be possible, however, to separate the measurement wheat from the chaff. For instance, measurement consumers may value a means of distinguishing among methods that emphasizes their interests in, and reasons for, measuring. Such a continuum of methods could be one that ranges from the least meaningful and generalizable to the most meaningful and generalizable, which is equivalent to ranging from the most to the least dependent on the local particulars of the specific questions asked, sample responding, judges rating, etc.

The aesthetics, simplicity, meaningfulness, rigor, and practical consequences of strong theoretical requirements for instrument calibration provide such criteria for choices as to models and methods (Andrich, 2002, 2004; Busemeyer & Wang, 2000; Myung, 2000; Myung & Pitt, 2004; Wright, 1997, 1999). These criteria could be used to develop and guide explicit considerations of data quality, construct theory, instrument calibration, quantitative comparisons, measurement standard metrics, etc. along a continuum from the most passive and least objective to the most actively involved and most objective.

The passive approach to measurement typically starts from and prioritizes content validity. The questions asked on tests, surveys, and assessments are considered relevant primarily on the basis of the words they use and the concepts they appear to address. Evidence that the questions actually cohere together and measure the same thing is typically deemed of secondary importance, if it is recognized at all. If there is any awareness of the existence of axiomatically prescribed measurement requirements, these are not considered to be essential. That is, if failures of invariance are observed, they usually provoke a turn to less stringent data treatments instead of a push to remove or prevent them. Little or no measurement or construct theory is implemented, meaning that all results remain dependent on local samples of items and people. Passively approaching measurement in this way is then encumbered by the need for repeated data gathering and analysis, and by the local dependency the results. Researchers working in this mode are akin to the woodcutters who say they are too busy cutting trees to sharpen their saws.

An alternative, active approach to measurement starts from and prioritizes construct validity and the satisfaction of the axiomatic measurement requirements. Failures of invariance provoke further questioning, and there is significant practical use of measurement and construct theory. Results are then independent of local samples, sometimes to the point that researchers and practical applications are not encumbered with usual test- or survey-based data gathering and analysis.

3.1 Six Developmental Stages

As is often the case, this black and white portrayal tells far from the whole story. There are multiple shades of grey in the contrast between passive and active approaches to measurement. The actual range of implementations is much more diverse than the simple binary contrast would suggest. Spelling out the variation that exists could be helpful for making deliberate, conscious choices and decisions in measurement practice.

It is inevitable that we would start from the materials we have at hand, and that we would then move through a hierarchy of increasing efficiency and predictive control as understanding of any given variable grows. Previous considerations of the problem have offered different categorizations for the transformations characterizing development on this continuum. Stenner and Horabin (Stenner & Horabin, 1992) distinguish between (1) impressionistic and qualitative, nominal gradations found in the earliest conceptualizations of temperature, (2) local, data-based quantitative measures of temperature, and (3) generalized, universally uniform, theory-based quantitative measures of temperature.

The latter is prized for the way that thermodynamic theory enables the calibration of individual thermometers with no need for testing each one in empiric studies of its performance. Theory makes it possible to know in advance what the results of such tests would be with enough precision to greatly reduce the burden and expenses of instrument calibration.

Reflecting on the history of psychosocial measurement in this context, it then becomes apparent that these three stages can be further broken down. The distinguishing features for each of six stages in the evolution of measurement systems are expanded from a previously described five stage conception (Stenner et al., 2006).

In Stage 1, conceptions of measurement are not critically developed, but stem from passively acquired examples. At this level, what you see is what you get, in the sense that item content defines measurement; advanced notions of additivity, invariance, etc. are not tested; the meanings of the scores and percentages that are treated as measures are locally dependent on the particular sample measured and items used; and there is no theory of the construct measured. Data must be gathered and analyzed to have results of any kind.

In Stage 2, measurement concepts are slightly less passively adopted. Additivity, invariance, etc. may be tested, but falsification of these hypotheses effectively derails the measurement effort in favour of statistical models with interaction effects, which are accepted as viable alternatives. Typically little or no attention is paid at this stage to the item hierarchy or the construct definition. An initial awareness of measurement theory is not complemented by any construct specification theory.

In Stage 3, measurement concepts are more actively and critically developed, but instruments still tend to be designed relative to content, not construct, specifications. Additivity and invariance principles are tested, and falsification of the additive hypothesis provokes questions as to why, where, and how those failures occurred. Models with interaction effects are not accepted as viable alternatives, and significant attention will be paid to the item hierarchy and construct definition, but item calibrations remain empirical. Though there is more significant use of measurement theory, construct theory is underdeveloped, so no predictive power is available.

In Stage 4, the conceptualization of measurement becomes more active than passive. Initial efforts to (re-)design an instrument relative to construct specifications occur at this level. Additivity, invariance, etc. are explicitly tested and are built into construct manifestation expectations. The falsification of the additive hypothesis provokes questions as to why and corrective action, models with interaction effects are not accepted as viable alternatives, significant attention is paid to the item hierarchy and construct definition relative to instrument design, but empirical calibrations remain the norm. Some construct theory gives rise to limited predictive power. Commercial applications that are not instrument-dependent (as in computer adaptive implementations) exist at this level.

In Stage 5, all of the Stage 4 features appear in the context of a significantly active approach to measurement. The item hierarchy is translated into a construct theory, and a construct specification equation predicts item difficulties and person measures apart from empirical data. These features are used routinely in commercial applications.

In Stage 6, the most purely active approach to measurement, all of the Stage 4 and 5 features are brought to bear relative to construct specification equations that predict the mean difficulties of ensembles of items each embodying a particular combination of components. Commercial applications of this kind have been in development for several years.

Various degrees of theoretical investment at each stage can be further specified, along with speculations as to the extent of application frequency in mainstream and commercial instrument development. Stage 1, with no effective measurement or construct theory, remains the mainstream, most popular approach in terms of its application frequency, which likely exceeds 90% of all efforts aimed at quantifying human, social, or natural capital. It is, however, commercially the least popular in application frequency (<10%?) in high stakes educational and psychological testing.

Stage 2, implementing very limited use of measurement theory and no construct theory is the next most popular mainstream psychosocial application frequency at perhaps eight percent, overall. It also has a somewhat higher commercial application frequency (10–20%).

Stage 3, with a strong use of measurement theory and little or no construct theory, may be used as much as one or two percent of the time in mainstream applications, and may be dominant methodologically in commercial applications (55–65%?).

Stage 4's strong use of measurement theory and use of some construct theory in informing instrument design have very limited psychosocial application frequency in mainstream applications (<0.5%?) but have made some significant starts in commercial applications (3–5%?). Stage 5's strong theoretical understanding of constructs is virtually unknown in mainstream psychosocial application, but has also begun to see some commercial developments. Stage 6's mature theoretical understanding of constructs is only just emerging in some well-supported commercial applications.

3.2   The Rasch Reading Law

Measurement theory sets the stage for thinking about constructs by focusing attention on the meaningfulness of the quantities produced, by facilitating the construction of supporting evidence, by testing construct hunches, and by supporting theory development. Construct theory then sets the stage for following through on measurement theory's fundamental principles by making it possible to more fully transcend local particulars of respondent and item samples. It does so by recognizing that failures of invariance are valuable as anomalous exceptions that "prove" (L. *probus*, test goodness of) the rule embodied in the measurement technology.

That is, data-model misfit is not considered to result from model failure, but from uninterpretable inconsistencies in the data stemming from under- eveloped theory and/or low quality data. Thus, failure to fit a model of fundamental measurement is not a sign of the end of the conversation or of the measurement effort. Rather, negative results of this kind provide needed checks on the strength of the object to withstand the rigors of propagation across media, which is the ultimate goal of having each different manufacturer's tool capable of functioning as a medium traceable to the same reference standard metric (Latour, 1987, 2005).

The predictability of a trajectory for the evolution of measurement allows the specification of a law capable of shaping fundamental expectations as to in- creases in the power and complexity of psychosocial measurement technology, and the timing of those increases. This practical law is applicable to business relationships in a manner analogous to the way the basic law describes scientific relationships. This is so even if the definition of work in engineering mechanics is of little immediate interest in gauging the economic value of labour. Despite the lack of immediate relevance, the practical utility of the widely used horsepower measure of engine pulling capacity depends on the scientific validity of the proportionate relations between mass, force, and acceleration in Newton's laws.

The same simultaneous instantiation of scientific and economic value must be possible for instruments to mediate relationships in ways that can effectively and efficiently coordinate capital budgeting decisions. Thus, the Rasch Reading Law describes invariantly proportionate ratios between reading comprehension, text complexity, and reader ability (Burdick et al., 2006; Stenner et al., 2006). As text complexity increases (the words used become less commonly encountered, and sentence length increases), reading comprehension rates decrease relative to a fixed reading ability measure. Conversely, given a fixed text complexity, reading comprehension rates increase as reading ability increases.

The practical value of this law is realized insofar as it then becomes possible to employ it productively in both (a) representing students' reading abilities in summative accountability measures and (b) intervening in ways likely to change those measures in formative instructional applications (Alonzo & Steedle, 2009; Chang & Chan, 1995; Kennedy & Wilson, 2007; Leclercq, 1980). Concerning the latter, it is well understood that learning is inherently a matter of leveraging what is already known (the alphabet, numbers, words, grammar, arithmetical operations, etc.) to frame and understand what is not yet known (new vocabulary, constructions, specific problems, etc.). It is therefore vitally important to target instruction at the sweet spot where enough is known to support comprehension, but where what is not known is still substantial enough to make the lesson challenging. This range along the measurement continuum just above the student's measure is known as the Zone of Proximal Development (Vygotsky, 1978) and is valued for indicating the range of curriculum content the student is developmentally ready to learn (Griffin, 2007). When measures are appropriately targeted, learning is maximized and measurement error is minimized. The same kind of strategy has proven useful in prescribing rehabilitation therapies (Chang & Chan, 1995) and likely has other as yet unexplored applications.

Targeting will be a key element in any future technology roadmap for education. Though there is no substitute for attention to other substantive aspects of the educational process, this indicator is of potentially central importance as a summary indicator of how accurately and precisely educational outcomes are represented, and how efficiently instructional interventions are implemented.

Rasch measurement isolates and focuses attention on empirical and theoretically tractable test item difficulty scale orders and positions. Then it estimates student abilities relative to that scale and describes them in terms of the probabilities of successful comprehension up and down the scale, *whether or not all of the items potentially available have actually been administered.* The goal of education, after all, is not to teach students only how to deal with the actual concrete problems encountered in instruction and assessment. The goal is rather to teach students how to manage any and all problems of a given type at a given level of difficulty.

Though a dialectic between part and whole is necessary, we cheat students and society when education becomes fixated on particular content and neglects the larger context in which skills are to be applied. The overall principle is effectively one of mass customization. Instruction and assessment, or any bidirectional method of simultaneous representation and intervention, benefits from forms of quantification coordinating substantive content with metrics that remain stable and constant no matter which particular test, survey, or assessment items are involved. The same principles apply in any other enterprise focused on intangible outcomes, such as health care, social services, or human resource management. We short change ourselves by failing to demand mediating instruments enabling a kind of virtual coordination of improvement, purchasing, hiring, and other investment decisions across different individuals, firms, agencies, and arenas in the economy. The architecture of probabilistic models open to the integration of new items and samples embodies the principles of invariance characteristic of the mediating instruments needed for aligning legally and geographically separated firms' decisions within a common inferential framework.

3.3 Stenner's Law

Of course, even though it has been almost 60 years since Rasch (1960) first did his foundational research (Andrich, 1988; Bond & Fox, 2007; Wright, 1985) on reading, integrating assessment and instruction on the basis of the Rasch Reading Law is not yet the norm in educational practice. Accordingly, most instruction is not integrated with assessment, and few examination results are reported so as to illustrate the alignment of a developmental continuum with the curriculum. Furthermore, and more specifically, most reading instruction is not appropriately targeted at individual students' Zones of Proximal Development. This is problematic, given that reading abilities within elementary school classrooms can easily range from two grade levels below to two grade levels above the reading difficulty of the textbook.

Figure 1, modelled on the first of two figures in Moore's original 1965 paper (Moore, 1965, 1975), shows a hypothetical but not unrealistic projection of the relation of average targeting accuracy with cost, by decade, from 1990 to 2030. Precision measurement is considered here to be realized when the targeted comprehension rate is realized to within 5%. Few reading tests were adaptively administered or well targeted before 1990; though a few were, computerization of test administration was difficult and expensive, as was (and remains) printed test production. Further, even fewer tests were administered for diagnostic or formative purposes before 1990, which is just as well as few would have been able to provide information useful for those applications.

It is plausible to suppose that, as the quality of testing has improved in the years after 1990, costs have been reduced and the targeting accuracy of assessment items

Fig. 1 Hypothetical projection of mean percentages of students comprehending text at a rate of 60% or less by average relative cost of producing a single precision reading measure by year

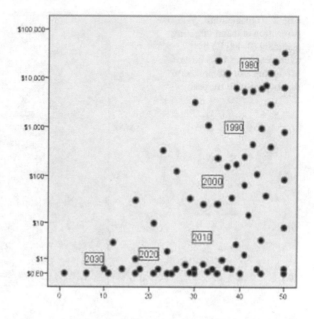

and instructional text has been enhanced, so the difference between the average item difficulty and the average measure of the targeted student approaches 0, to the left. The upper limit of targeting accuracy remains constant because the impact of new methods of test construction and administration are unevenly distributed. Costs are driven down as theory is able to inform the automatic production and administration of targeted text and test items in computerized contexts effectively integrating assessment and instruction. Costs may be dropping by an order of magnitude every decade, with the rate in reductions in mistargeting slowing as it nears 0. At some future date, accurate targeting may become universal, and the right, off-target end of the range may also drop to near 0.

Figure 2, is also patterned on the first figure in Moore (Moore, 1965), is a variation on the same information as that shown in Fig. 1. Mistargeted text and test items may bore able readers encountering material that is much too easy, but poor readers unable to make any headway with readings far too complex for them to comprehend are doomed to learn little or nothing. Figure 2 is thus intended to convert Fig. 1's targeting information into the implied percentage of students comprehending text at a rate of 60% or less.

Figure 3, patterned on the second figure in Moore (1965), describes what may be referred to as Stenner's Law: the expectation that the number of precision reading measures estimated will double every two years, with no associated increase in cost. The figure has historical validity in that the line begins not long after the 1960 introduction of Rasch's work in Chicago, is in the range of 350,000 in the 1970s, during the Anchor Test Study (Jaeger, 1973; Rentz & Bashaw, 1977), and is about 20–30 million in the period of 2005–2008, which is approximately how many measures

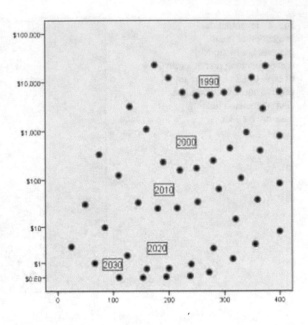

Fig. 2 Hypothetical projection of mean targeting accuracy (0–400 L) by average relative US$ cost of producing a single precision reading measure by year

were being produced annually at the time by users of the Lexile Framework for Reading (Stenner et al., 2006).

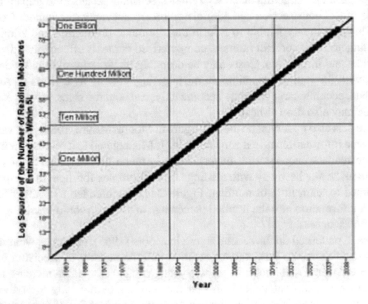

Fig. 3 Rate of increase in number of precision reading measures estimated

4   A Technology Roadmap for Intangible Assets

New and urgent demands challenged Moore's Law when it was realized in the 1990s that the physical limits of silicon could potentially disrupt the expectations that had allowed the microprocessor industry to coordinate its investment decisions so consistently for over 20 years (Schulz, 1999). The threat of a crisis led to the convening of an industry-wide meeting in 1992 by Gordon Moore, then chairman of the Semiconductor Industry Association's technology committee (Miller & O'Leary, 2007). This and subsequent meetings of the group resulted in the annual publication of an International Technology Roadmap for Semiconductors.

These charts provided a level of specificity and detail not present in the more bare-bones projections of Moore's Law. The established history of past successes, combined with new uncertainties compelled leaders in the field to seek out a basis on which new mediating instruments might be founded. Risks associated with evaluating several different methods of resolving the technical problem of continued reductions in microprocessor size and cost had to be mitigated so that no firms found themselves making large capital investments with no product or customers in sight (which, unfortunately, is the status quo in education, healthcare, and other industries making intensive investments in human, social, and natural capital).

Table 1 presents a reading measurement variation on the 2001 version of the semiconductor industry's roadmap (Miller & O'Leary, 2007). The basic structure of the table (the columns labelled "Year of first production" and "Technology node", and the subheadings focusing on "Expected shifts in product functionality and cost", Introductory volumes, and Innovations) is identical with the produced by the semiconductor industry. The remaining elements have been changed to focus on the kinds of functionality, cost, and innovations that have taken place historically in the domain of reading instruction and assessment.

Some of these suggested elements may prove less important in articulating a shared history and projecting an accessible vision of the future, and others may be needed. The point here is less one of specifying what exactly should be tracked and is more focused on conveying the general conceptual framework in.

which new possibilities for coordinating investment decisions in education might be explored. Plainly, it would be essential for the major stake holding actors and agencies involved in education, from academia to business to government, to themselves determine the actual contents of a roadmap such as this.

5   Conclusion

The mediation of individual and organizational levels of analysis, and of the organizational and inter-organizational levels, is facilitated by Rasch measurement. Miller and O'Leary (2007) document the use of Moore's Law in the microprocessor industry in the creation of technology roadmaps that lay out the structure, processes, and

Table 1 Sketch of a possible technology roadmap for literacy education (elements to be determined)

	Year of first production					
Technology node	1980	1990	2000	2010	2020	2030
Expected shifts in product functionality and cost—high volume targeted items						
Number of steps in item production and administration						
Multiple per 3-year technology cycle						
Affordable production cost per item at introduction						
Rate of cost reduction per cycle (%)						
Annualized cost reduction/item (%)						
Improvements in instructional targeting accuracy						
Reduction in students reading at <60% comprehension (%)						
Increase in number of precision reading measures (%)						
Available item volume at introduction						
Uncalibrated						
Calibrated bank						
From specification equation						
Innovations in item administration						
Paper and pencil adaptive formats						
Computer adaptive (bank)						
Theory adaptive (specified)						
Innovations integrating assessment & instruction						
Examinations aligned with curricula (%)						
Differentiated instruction aligned with adaptive tests (%)						
Writing assessment included (%)						

outcomes that have to be aligned at all three levels to coordinate an entire industry's economic success. Such roadmaps need to be created for each major form of human, social, and natural capital, with the associated alignments and coordinations put in play at all levels of every firm, industry, and government.

It has been suggested that economic recovery in the wake of the Great Recession could be driven by a new major technological breakthrough, one of the size and scope of the IT revolution of the 1990s. This would be a kind of Manhattan Project or international public works program, providing the unifying sense of a mission aimed at restoring and fulfilling the promises of democracy, justice, freedom, and prosperity. Industry-wide systems of metrological reference standards for human, social, and natural capital fit the bill. Such systems would be a new technological breakthrough on the scale of the initial IT revolution. They would also be a natural outgrowth of existing IT systems, an extension of existing global trade standards, and would require large investments from major corporations and governments. In addition, stepping beyond those suggestions that have appeared in the popular press, systematic and objective methods of measuring intangible assets would help meet the widely recognized need for socially responsible and sustainable business practices.

Better measurement will play a vital role in reducing transaction costs, making human, social, and natural capital markets more efficient by facilitating the coordination of autonomous budgeting decisions. It will also be essential to fostering new forms of innovation, as the shared standards and common product definitions made possible by advanced measurement systems enable people to think and act together collectively in common languages.

Striking advances have been made in measurement practice in recent years. Many still assume that assigning numbers to observations suffices as measurement, and that there have been no developments worthy of note in measurement theory or practice for decades. Nothing could be further from the truth.

Theory makes it possible to know in advance what the results of empirical calibration tests would be with enough precision to greatly reduce the burden and expenses associated with maintaining a unit of measurement. There likely would be no electrical industry at all if the properties of every centimetre of cable and every appliance had to be experimentally tested. This principle has been employed in measuring human, social, and natural capital for some time, but has not yet been adopted on a wide scale.

This might change with the introduction of Stenner's Law, and in conjunction with technology roadmaps for literacy capital and for other forms of human, social, and natural capital that project rates of increase in psychosocial measurement functionality and frame an investment appraisal process ensuring the ongoing creation of markets for advanced calibration services for the next 10–20 years or more.

A progression of increasing complexity, meaning, efficiency, and utility can be used as a basis for a technology roadmap that will enable the coordination and alignment of various services and products in the domain of intangible assets. A map to the theory and practice of calibrating instruments for the measurement of intangible forms of capital is needed to provide guidance in quantifying constructs such as literacy, health, and environmental quality. We manage what we measure, so when we begin measuring well what we want to manage well, we'll all be better off.

Acknowledgements Thanks to Michael Everett for his support of this work. Initial drafts of this paper were composed while William Fisher was an employee of MetaMetrics, Inc.

References

Ackermann, J. R. (1985). *Data, instruments, and theory: A dialectical approach to understanding science.* Princeton.

Allen, R. H., & Sriram, R. D. (2000). The role of standards in innovation. *Technological Forecasting and Social Change, 64*, 171–181.

Alonzo, A. C., & Steedle, J. T. (2009). Developing and assessing a force and motion learning progression. *Science Education, 93*, 389–421.

Andrich, D. (1988). *Rasch models for measurement.* (vols. series no. 07–068). Sage University Paper Series on Quantitative Applications in the Social Sciences. Sage Publications.

Andrich, D. (2002). Understanding resistance to the data-model relationship in Rasch's paradigm: A reflection for the next generation. *Journal of Applied Measurement, 3*(3), 325–359.

Andrich, D. (2004). Controversy and the Rasch model: A characteristic of incompatible paradigms? *Medical Care, 42*(1), I-7–I-16.

Ashworth, W. J. (2004). Metrology and the state: Science, revenue, and commerce. *Science, 306*(5700), 1314–1317.

Baker, G. P., Gibbons, R., & Murphy, K. J. (2001). Bringing the market inside the firm? *American Economic Review, 91*(2), 212–218.

Barzel, Y. (1982). Measurement costs and the organization of markets. *Journal of Law and Economics, 25*, 27–48.

Beges, G., Drnovsek, J., Pendrill, L.R. (2011). Optimising calibration and measurement capabilities in terms of economics s in conformity assessment. *Accreditation and Quality Assurance: Journal for Quality, Comparability and Reliability in Chemical Measurement, 15*(3), 147–154.

Benham, A., & Benham, L. (2000). Measuring the costs of exchange. In C. Ménard (Ed.), *Institutions, contracts and organizations: Perspectives from new institutional economics* (pp. 367–375). Edward Elgar.

Berwick, D. M., James, B., & Coye, M. J. (2003). Connections between quality measurement and improvement. *Medical Care, 41*(1), 130–138.

Birch, J. A. (2003). Benefit of legal metrology for the economy and society (International Committee of Legal Metrology). International Organization of Legal Metrology. Accessed 28 Oct 2008. http://www.oiml.org/publications/E/birch/E002-e03.pdf

Bond, T., & Fox, C. (2007). *Applying the Rasch model: Fundamental measurement in the human sciences* (2d ed.). Lawrence Erlbaum Associates.

Bud, R., & Cozzens, S. E. (Eds.). (1992). *Invisible connections: Instruments, institutions, and science.* Washington, USA, SPIE Optical Engineering Press.

Bunderson, C. V., & Newby, V. A. (2009). The relationships among design experiments, invariant measurement scales, and domain theories. *Journal of Applied Measurement, 10*(2), 117–137.

Burdick, D. S., Stone, M. H., & Stenner, A. J. (2006). The combined gas law and a Rasch reading law. *Rasch Measurement Transactions, 20*(2), 1059–1060.

Busemeyer, J. R., & Wang, Y. M. (2000). Model comparisons and model selections based on generalization criterion methodology. *Journal of Mathematical Psychology, 44*(1), 171–189.

Callon, M. (2002). From science as an economic activity to socioeconomics of scientific research: The dynamics of emergent and consolidated techno-economic networks. In P. Mirowski & E.-M. Sent (Eds.), *Science bought and sold: Essays in the economics of science.* University of Chicago Press.

Chang, W. C., & Chan, C. (1995). Rasch analysis for outcomes measures: Some methodological considerations. *Archives of Physical Medicine and Rehabilitation, 76*(10), 934–939.

Cooter, R. D. (2000). Law from order: Economic development and the jurisprudence of social norms. In M. Olson & S. Kahkonen (Eds.), *A not-so- dismal science: A broader view of economies and societies* (pp. 228–244). Oxford University Press.

Fisher, W. P., Jr. (2009). Invariance and traceability for measures of human, social, and natural capital: Theory and application. *Measurement, 42*(9), 1278–1287.

Fisher, W. P., Jr. (2010b). Statistics and measurement: Clarifying the differences. *Rasch Measurement Transactions, 23*(4), 1229–1230.

Fisher, W. P., Jr. (2011). Bringing human, social, and natural capital to life: Practical consequences and opportunities. *Journal of Applied Measurement, 12*(1), 49–66.

Fisher, W. P., Jr. (2010a). The standard model in the history of the natural sciences, econometrics, and the social sciences. *Journal of Physics, Conference Series, 238*(1). http://iopscience.iop.org/1742-6596/238/1/012016/pdf/1742-6596_238_1_012016.pdf

Fryback, D. (1993). QALYs, HYEs, and the loss of innocence. *Medical Decision Making, 13*(4), 271–272.

Goldberg, S. H. (2009). *Billions of drops in millions of buckets: Why philanthropy doesn't advance social progress.* Wiley.

Griffin, P. (2007). The comfort of competence and the uncertainty of assessment. *Studies in Educational Evaluation, 33,* 87–99.

Hankins, T. L., & Silverman, R. J. (1999). *Instruments and the imagination.* Princeton.

Ho, A. D. (2008). The problem with 'proficiency': Limitations of statistics and policy under no child left behind. *Educational Researcher, 37*(6), 351–360.

Ihde, D. (1983). The historical and ontological priority of technology over science. In D. Ihde (Ed.), *Existential technics* (pp. 25–46). State University of New York Press.

Ihde, D. (1991). *Instrumental realism: The interface between philosophy of science and philosophy of technology.* Indiana, Indiana University Press.

Ihde, D. (1998). *Expanding hermeneutics: Visualism in science.* Illinois, Northwestern University Press.

Ihde, D., & Selinger, E. (Eds.) (2003). *Chasing technoscience: Matrix for materiality.* (Indiana Series in Philosophy of Technology). Indiana University Press.

Jaeger, R. M. (1973). The national test equating study in reading (The Anchor Test Study). *Measurement in Education, 4,* 1–8.

Jensen, M. C. (2003). Paying people to lie: The truth about the budgeting process. *European Financial Management, 9*(3), 379–406.

Kennedy, C. A., Wilson, M. (2007). Using progress variables to interpret student achievement and progress. Berkeley, Evaluation & Assessment Research Center. Paper Series, No. 2006–12–01. University of California, Berkeley BEAR Center, 51.

Kindig, D. A. (1999). Purchasing population health: Aligning financial incentives to improve health outcomes. *Nursing Outlook, 47*(1), 15–22.

Latour, B. (1987). *Science in action: How to follow scientists and engineers through society.* Cambridge University Press.

Latour, B. (2005). *Reassembling the social: An introduction to actor-network-theory.* Oxford University Press.

Leclercq, D. (1980). Computerised tailored testing: Structured and calibrated item banks for summative and formative evaluation. *European Journal of Education, 15*(3), 251–260.

Lengnick-Hall, C. A., Lengnick Hall, M. L., & Abdinnour-Helm, S. (2004). The role of social and intellectual capital in achieving competitive advantage through enterprise resource planning (ERP) systems. *Journal of Engineering Technology Management, 21,* 307–330.

Miller, P., & O'Leary, T. (2007). Mediating instruments and making markets: Capital budgeting, science and the economy. *Accounting, Organizations, and Society, 32*(7–8), 701–734.

Moore, G. (1965). Cramming more components onto integrated circuits. *Electronics, 38*(8), 114–117.

Moore, G. (1975). Progress in digital integrated electronics. In *International electronics devices meeting (IEEE) Technical Digest* (Vol. 21, pp. 11–13).

Murray, C. (2006). By the numbers: Acid tests: No child left behind is beyond uninformative. It is deceptive. *Wall Street Journal.*

Myung, I. J. (2000). Importance of complexity in model selection. *Journal of Mathematical Psychology, 44*(1), 190–204.

Myung, I. J., & Pitt, M. (2004). Model comparison methods. In L. Brand & M. L. Johnson (Eds.), *Methods in enzymology, evaluation, testing and selection. Numerical computer methods, part D* (Vol. 383, pp. 351–365). Academic Press.

National Health Expenditure Data: NHE Fact Sheet. (2011). https://www.cms.gov/NationalHealthExpendData/25_NHE_Fact_Sheet.asp. Accessed 30 June 2011.

Price, D. J. D. S. (1986). Of sealing wax and string. In L. Science (Ed.), *Big science–and beyond* (pp. 237–253). Columbia University Press.

Rabkin, Y. M. (1992). Rediscovering the instrument: Research, industry, and education. In R. Bud & S. E. Cozzens (Eds.), *Invisible connections: Instruments, institutions, and science* (pp. 57–82). SPIE Optical Engineering Press.

Rasch, G. (1960). *Probabilistic models for some intelligence and attainment tests* (Reprint, with Foreword and Afterword by B.D. Wright, University of Chicago Press, 1980). Danmarks Paedogogiske Institut.

Rentz, R. R., & Bashaw, E. L. (1977). The national reference scale for reading: An application of the Rasch model. *Journal of Educational Measurement, 14*(2), 161–179.

Schulz, M. (1999). The end of the road for silicon? *Nature, 399*, 729–730.

Stenner, A. J., & Horabin, I. (1992). Three stages of construct definition. *Rasch Measurement Transactions, 6*(3), 229.

Stenner, A. J., & Smith, M., III. (1982). Testing construct theories. *Perceptual and Motor Skills, 55*, 415–426.

Stenner, A. J., Burdick, H., Sanford, E. E., & Burdick, D. S. (2006). "How accurate are Lexile text measures? *Journal of Applied Measurement, 7*(3), 307–322.

Vygotsky, L. S. (1978). *Mind and society: The development of higher mental processes*. Harvard University Press.

Wise, M. N. (1988). Mediating machines. *Science in Context, 2*(1), 77–113.

Wright, B. D. (1985). Additivity in psychological measurement. In Roskam (Ed.), *Measurement and personality assessment* (pp. 101–112). Elsevier Science Ltd.

Wright, B. D. (1997). A history of social science measurement. *Educational Measurement: Issues and Practice, 16*(4), 33–45, 52.

Wright, B. D. (1999). Fundamental measurement for psychology. In S. E. Embretson & S. L. Hershberger (Eds), *The new rules of measurement: What every educator and psychologist should know* (pp. 65–104). Lawrence Erlbaum Associates.

How to Model and Test for the Mechanisms That Make Measurement Systems Tick

A. Jackson Stenner, Mark Stone, and Donald Burdick

Abstract One must provide information about the conditions under which [the measurement outcome] would change or be different. It follows that the generalizations that figure in explanations [of measurement outcomes] must be change-relating... Both explainers [e.g., person parameters and item parameters] and what is explained [measurement outcomes] must be capable of change, and such changes must be connected in the right way (Woodward, 2003). Rasch's unidimensional models for measurement tell us how to connect object measures, instrument calibrations, and measurement outcomes. Substantive theory tells us what interventions or changes to the instrument must offset a change to the measure for an object of measurement to hold the measurement outcome constant. Integrating a Rasch model with a substantive theory dictates the form and substance of permissible conjoint interventions. Rasch analysis absent construct theory and an associated specification equation is a black box in which understanding may be more illusory than not. The mere availability of numbers to analyze and statistics to report is often accepted as methodologically satisfactory in the social sciences, but falls far short of what is needed for a science.

1 Introduction

The vast majority of psychometric thought over the last century has had as its focus the item. Shortly after Spearman's (1904) original conception of reliability as whole instrument replication proved to be difficult when there existed little understanding

Joint International IMEKO TC1 + TC7 + TC13 Symposium. 2011.

A. J. Stenner (✉) · D. Burdick
MetaMetrics, Inc., Durham, NC, USA

M. Stone
Adler School of Professional Psychology, Chicago, IL, USA

D. Burdick
Department of Mathematics, Duke University, Durham, NC, USA

W. P. Fisher and P. J. Massengill (eds.), *Explanatory Models, Unit Standards, and Personalized Learning in Educational Measurement*,
https://doi.org/10.1007/978-981-19-3747-7_15

of what psychological instruments actually measured. The lack of substantive theory made it difficult indeed to "clone" an instrument—to make a genetic copy. In the absence of a substantive theory the instrument maker does not know what features of test items are essential to copy and what features are incidental and cosmetic (Irvine & Kyllonen, 2002). So, faced with the need to demonstrate the reliability of psychological instruments but lacking a substantive construct theory that would support instrument cloning early psychometrics took a fateful step inward. Spearman (1910) proposed estimating reliability as the correlation between sum scores on odd and even items of a single instrument. Thus was the instrument lost as a focus of psychometric study and the part score and inevitably the item became ascendant. The spawn of this inward misstep is literally thousands of instruments with non–exchangeable metrics populating a landscape devoid of unifying psychological theory. And, this is so because… "The route from theory or law to measurement can almost never be traveled backwards" (Kuhn, 1961).

There are two quotes that when taken at extreme face value open up a new paradigm for measurement in the social sciences:

It should be possible to omit several test questions at different levels of the scale without affecting the individuals [readers] score [measure] (Thurstone, 1926).

… a comparison between two individuals [readers] should be independent of which stimuli [test questions] within the class considered were instrumental for comparison; and it should also be independent of which other individuals were also compared, on the same or some other occasion (Rasch, 1961).

Both Thurstone and Rasch envisioned a measurement framework in which individual readers could be compared independent of which particular reading items were instrumental for the comparison. Taken to the extreme we can imagine a group of readers being invariantly ordered along a scale when there is not a single item in common. No two readers are exposed to the same item. This would presumably reflect the limit of "omitting" items and making comparisons "independent of the items" used to make the comparison. Compare a fully crossed data collection design in which each item is administered to every reader with a design in which items are nested in persons, i.e., items are unique to each person. Although easily conceived it is immediately clear that there is no data analysis method that can extract invariant reader comparisons from the second design type data. But is this not exactly the kind of data that is routinely generated say when parents report their child's weight on a doctor's office form? No two children (except for siblings) share the same bathroom scale nor potentially even the same underlying technology and yet we can consistently and invariantly order all children in terms of weight? What is different is that the same construct theory for weight has been engineered into each and every bathroom scale even though the specific mechanism (digitally recorded pressure vs. spring driven analog recording) may vary. In addition, the measurement unit (pounds or kilograms) has been consistently maintained from bathroom scale to bathroom scale. So, it is substantive theory and engineering specifications not data that is used to render comparable measurements from these disparate bathroom scales. We argue that this illustrates the dominant distinguishing feature between physical science and social science measurement. Social science measurement does not, as a rule, make use of substantive theory in the ways that the physical sciences do.

Validity theory and practice suffers from an egalitarian malaise, all correlations are considered part of the fabric of meaning and like so many threads each is treated equally. Because we live in a correlated world, correlations of 0.00 are rare, non-zero correlations abound and it is an easy task to collect a few statistically significant correlates between scores produced by virtually any human science instrument and other meaningful phenomena. All that is needed to complete our validity tale is a story about why so many phenomena are correlated with the instrument we are making. And so it goes hundreds and thousands of times per decade, dozens of new instruments are islands unto themselves accompanied by dozens of hints of connectivity whispered to us through dozens of middling correlations. This is the legacy of the nomological network (Cronbach & Meehl, 1955). May it rest in peace!

Validity, for us, is a simple straightforward concept with a narrow focus. It answers the question "What causes the variation detected by the instrument?" The instrument (a reading test) by design comes in contact with an object of measurement (a reader) and what is recorded is a measurement outcome (count correct). That count is then converted into a linear quantity (a reading ability). Why did we observe that particular count correct? What caused a count correct of 25/40 rather than 20/40 or 30/40? The answer (always provisional) takes the form of a specification equation (Stenner et al., 1983) with variables that when experimentally manipulated produce the changes in item behavior (empirical item difficulties) predicted by the theory. In this view validity is not about correlations or about graphical depictions of empirical item orderings called Wright maps (Wilson, 2004). It is about what is causing what? Is the construct well enough understood that its causal action can be specified? Clearly our expectation is unambiguous. There exist features of the stimuli (test or survey items) that if manipulated will cause changes in what the instrument records (what we observe). These features of the stimuli interact with the examinee and the instrument records the interaction (correct answer, strong agreement, tastes good etc.). The window onto the interaction between examinee and instrument is fogged up. We can't observe directly what goes on in the mind of the examinee but we can dissect and otherwise manipulate the item stimuli, or measurement mechanism, and observe changes in recorded behavior of the examinee (Stenner et al., 2009a). Some of the changes we make to the items will matter (radicals) to examinees and others will not (incidentals). Sorting out radicals (causes) from incidentals is the hard work of establishing the validity of an instrument (Irvine & Kyllonen, 2002). The specification equation is an instantiation of these causes (at best) or their proxies (at a minimum).

Typical applications of Rasch models to human science data are thin on substantive theory. Rarely is there an a priori specification of the item calibrations (i.e., constrained models). Instead the analyst estimates both person parameters and item parameters from the same data set. For Kuhn this practice is at odds with the function of measurement in the "hard" sciences in that almost never will substantive theory be revealed from measurement (Kuhn, 1961). Rather "the scientist often seems rather to be struggling with facts [e.g., raw scores], trying to force them to conformity with a theory he does not doubt" (Kuhn, 1961). Here Kuhn is talking about substantive theory not axioms. The scientist imagines a world and formalizes these imaginings as a theory and then makes measurements and checks for congruence between what is observed and what theory predicted: "Quantitative facts cease to seem simply the 'given'. They must be fought for and with, and in this fight the theory with which they are to be compared proves the most potent weapon". It's not just that unconstrained models are less potent; they fail to conform to the way science is practiced and most troubling they are least revealing of anomalies (Andrich, 2004).

Andrich (Andrich, 2004) makes the case that Rasch models are powerful tools precisely because they are prescriptive not descriptive and when model prescriptions meet data, anomalies arise (Andrich, 2004). Rasch models invert the traditional statistical data-model relationship. Rasch models state a set of requirements that data must meet if those data are to be useful in making measurements. These model requirements are independent of the data. It does not matter if the data are bar presses, counts correct on a reading test, or wine taste preferences, if these data are to be useful in making measures of rat perseverance, reading ability, or vintage quality all three sets of data must conform to the same invariance requirements. When data sets fail to meet the invariance requirements we do not respond by, say, relaxing the invariance requirements through addition of an item specific discrimination parameter to improve fit; rather, we examine the observation model and imagine changes to that model that would bring the data into conformity with the Rasch model requirements.

A causal Rasch model (item calibrations come from theory not the data) is doubly prescriptive (Stenner et al., 2009a). First, it is prescriptive regarding the data structures that must be present.

"The comparison between two stimuli [text passages] should be independent of which particular individuals [readers] were instrumental for the comparison; and it should also be independent of which other stimuli within the considered class [prose] were or might also have been compared. Symmetrically, a comparison between two individuals [readers] should be independent of which particular stimuli within the class considered [prose] were instrumental for [text passage] comparison; and it should also be independent of which other individuals were also compared, on the same or on some other occasion" (Rasch, 1961).

Second, Causal Rasch Models (CRM) (Burdick et al., 2006; Stenner et al., 2008) prescribe that item calibrations take the values imposed by the substantive theory. Thus, the data, to be useful in making measures, must conform to both Rasch model invariance requirements and substantive theory invariance requirements as represented in the theoretical item calibrations. When data meet both sets of requirements then those data are useful not just for making measures of some construct but are useful for making measures of that precise construct specified by the equation that produced the theoretical item calibrations. We note again that these dual invariance requirements come into stark relief in the extreme case of no connectivity across stimuli or examinees. How, for example, are two readers to be measured on the same scale if they share no common text passages or items? If you read a Harry Potter novel and answer questions and I read a Lord of the Rings novel and answer questions, how is it possible that from these disparate experiences an invariant comparison of our reading abilities is realizable? How is it possible that you can be found to read 250L better than I and, furthermore, that you had 95% comprehension and I had 75% comprehension of our respective books. Given that seemingly nothing is in common between the two experiences it seems that invariant comparisons are impossible, but, recall our bathroom scale example, different instruments qua experiences underlie every child's parent reported weight. Why are we so quick to accept that you weigh 50lbs less than I do and yet find claims about our relative reading abilities (based on measurements from two different books) inexplicable. The answer lies in well developed construct theory, instrument engineering and metrological conventions.

Clearly, each of us has had ample confirmation that the construct WEIGHT denominated in pounds and kilograms can be well measured by any carefully calibrated bathroom scale. Experience with diverse bathroom scales has convinced us that within a pound or two of error these instruments will produce not just invariant relative differences between two persons (as described in the Rasch quotes) but the more stringent expectation of invariant absolute magnitudes for each individual independent of instrument. Over centuries, instrument engineering has steadily improved to the point that for most purposes "uncertainty of measurement" usually reported as the standard deviation of a distribution of imagined or actual replications taken on a single person can be effectively ignored for most bathroom scale applications. Finally, by convention (i.e., the written or unwritten practice of a community) in the U.S. we denominate weight in pounds and ounces. The use of pounds and ounces is arbitrary as is evident from the fact that most of the world has gone metric, but what is decisive is that a unit is agreed to by the community and is slavishly maintained through consistent implementation, instrument manufacture, and reporting. At present READING ABILITY does not enjoy a commonly adhered to construct definition, nor a widely promulgated set of instrument specifications nor a conventionally accepted unit of measurement, although, the Lexile Framework for Reading (Stenner et al., 2006) promises to unify the measurement of READING in a manner precisely parallel to the way unification was achieved for LENGTH, TEMPERATURE, WEIGHT and dozens of other useful attributes (Stenner & Stone, 2010).

A causal (constrained) Rasch model (Stenner et al., 2009b) that fuses a substantive theory to a set of axioms for conjoint additive measurement affords a much richer context for the identification and interpretation of anomalies than does an unconstrained Rasch model. First, with the measurement model and the substantive theory fixed it is self evident that anomalies are to be understood as problems with the data ideally leading to improved observation models that reduce unintended dependencies in the data. Recall that The Duke of Tuscany put a top on some of the early thermometers thus reducing the contaminating influences of barometric pressure on the measurement of temperature. He did not propose parameterizing barometric pressure so that the boiling point of water at sea level would match the model expectations at 3,000 feet above sea level. Second, with both model and construct theory fixed it is obvious that our task is to produce measurement outcomes that fit the (aforementioned) dual invariance requirements. By analogy, not all fluids are ideal as thermometric fluids. Water, for example, is non-monotonic in its expansion with increasing temperature. Mercury, in contrast, has many useful properties as a thermometric fluid. Does the discovery that not all fluids are useful thermometric fluids invalidate the concept of temperature? No! In fact, a single fluid with the necessary properties would suffice to validate temperature as a useful construct. The existence of a persistent invariant framework makes it possible to identify anomalous behavior (water's strange behavior) and interpret it in an expanded theoretical framework. Analogously, finding that not all reading item types conform to the dual invariance requirements of a Rasch model and the Lexile theory does not invalidate either the axioms of conjoint measurement theory or the Lexile reading theory. Rather, anomalous behaviors of various item types are open invitations to expand the theory to account for these deviations from expectation. Notice here the subtle shift in perspective. We do not need to find 1000 unicorns; one will do to establish the reality of the class. The finding that reader behavior as a single class of reading tasks can be regularized by the joint actions of the Lexile theory and a Rasch model is sufficient evidence for the reality of the reading construct.

2   Model and Theory

Equation (1) is a causal Rasch model for dichotomous data, which sets a measurement outcome (raw score) equal to a sum of modeled probabilities

$$Expected\ raw\ score = \sum_i \frac{e^{(b_n - d_i)}}{1 + e^{(b_n - d_i)}} \tag{1}$$

The measurement **outcome** is the dependent variable and the measure (e.g., person parameter, b) and instrument (e.g., the parameters di pertaining to the difficulty d of item i) are independent variables. The measurement outcome (e.g., count correct on a reading test) is observed, whereas the measure and instrument parameters are not observed but can be estimated from the response data and substantive theory, respectively. When an interpretation invoking a predictive mechanism is imposed on the equation, the right-side variables are presumed to characterize the process that generates the measurement outcome on the left side. The symbol =: was proposed by Euler circa 1734 to distinguish an algebraic identity from a causal identity (right hand side causes the left hand side). The symbol =: exhumed by Judea Pearl can be read as *manipulation of the right hand side* via *experimental intervention will cause the prescribed change in the left hand side of the equation.*

A Rasch model combined with a substantive theory embodied in a specification equation provides a more or less complete explanation of how a measurement instrument works (Stenner et al., 2009a). A Rasch model in the absence of a specified measurement mechanism is merely a probability model. A probability model absent a theory may be useful for describing or summarizing a body of data, and for predicting the left side of the equation from the right side, but a Rasch model in which instrument calibrations come from a substantive theory that specifies how the instrument works is a causal model. That is, it enables prediction after intervention.

Causal models (assuming they are valid) are much more informative than probability models: "A joint distribution tells us how probable events are and how probabilities would change with subsequent observations, but a causal model also tells us how these probabilities would change as a result of external interventions... Such changes cannot be deduced from a joint distribution, even if fully specified." (Pearl, 2000).

A satisfying answer to the question of how an instrument works depends on understanding how to make changes that produce expected effects. Identically structured examples of two such narratives include (a) a thermometer designed to take human temperature and (b) a reading test.

2.1 The NexTemp® Thermometer

The NexTemp® thermometer is a small plastic strip pocked with multiple enclosed cavities. In the Fahrenheit version, 45 cavities arranged in a double matrix serve as the functioning end of the unit. Spaced at 0.2 °F intervals, the cavities cover a range from 96.0 to 104.8 °F. Each cavity contains three cholesteric liquid crystal compounds and a soluble additive. Together, this chemical composition provides discrete and repeatable change-of-state temperatures consistent with the device's numeric indicators. Change of state is displayed optically and is easily read.

2.2 The Lexile Framework for Reading®

Text complexity is predicted from a construct specification equation incorporating sentence length and word commonality components. The squared correlation of observed and predicted item calibrations across hundreds of tests and millions of students over the last 15 years averages about 0.93. Available technology for measuring reading ability employs computer-generated items built "on-the-fly" for any continuous prose text. Counts correct are converted into Lexile measures via a Rasch model estimation algorithm employing theory-based calibrations. The Lexile measure of the target text and the expected spread of the cloze items are given by theory and associated equations. Differences between two readers' measures can be traded off for a difference in Lexile text measures. When the item generation protocol is uniformly applied, the only active ingredient in the measurement mechanism is the choice of text complexity.

In the temperature example, if we uniformly increase or decrease the amount of soluble additive in each cavity, we change the correspondence table that links the number of cavities that turn black to degrees Fahrenheit. Similarly, if we increase or decrease the text demand (Lexile) of the passages used to build reading tests, we predictably alter the correspondence table that links count correct to Lexile reader measure. In the former case, a temperature theory that works in cooperation with a Guttman model produces temperature measures. In the latter case, a reading theory that works in cooperation with a Rasch model produces reader measures. In both cases, the measurement mechanism is well understood, and we exploit this understanding to address a vast array of counterfactuals (Woodward, 2003). If things had been different (with the instrument or object of measurement), we could still answer the question as to what then would have happened to what we observe (i.e., the measurement outcome). It is this kind of relation that illustrates the meaning of the expression, "there is nothing so practical as a good theory" (Lewin, 1951).

3   Distinguishing Features of Causal Rasch Models

Clearly the measurement model we have proposed for human sciences mimics key features of physical science measurement theory and practice. Below we highlight several such features.

1. The model is individual centered. The focus is on explaining variation within person over time.

Much has been written about the disadvantages of studying between person variation with the intent to understand within person causal mechanisms (Barlow et al., 2009; Grice, 2011). Molenaar (2004) has proven that only under severely restrictive conditions can such cross level inferences be sustained. In general in the human sciences we must build and test individual centered models and not rely on variable or group centered models (with attendant focus on between person variation) to inform our understanding of causal mechanisms. Causal Rasch models are individually centered measurement models. The measurement mechanism that transmits variation in the attribute (within person over time) to the measurement outcome (count correct on a reading test) is hypothesized to function the same way for every person (the second ergodicity condition of homogeneity) (Molenaar, 2004). Note, however, that the fact that there are different developmental pathways that led you to be taller than me and me to be a better reader than you does not mean that the attributes of height and reading ability are somehow necessarily different attributes for both of us.

2. In this framework the measurement mechanism is well specified and can be manipulated to produce predictable changes in measurement outcomes (e.g., percent correct).

For purposes of measurement theory we don't need a sophisticated philosophy of causal inference. For example, questions about the role of human agency in the intervention/manipulation based accounts of causal inference are not troublesome here. All we mean by the claim that the right hand side of Eq. 1 causes the left hand side is that experimental manipulation of each will have a predictable consequence for the measurement outcome (expected raw score). Stated more generally all we mean by x causes y is that an intervention on x yields a predictable change in y. The specification equation used to calibrate instruments/items is a recipe for altering just those features of the instrument/items that are causally implicated in the measurement outcome. We term this collection of causally relevant instrument features the "measurement mechanism". It is the "measurement mechanism" that transmits variation in the attribute (e.g., temperature, reading ability) to the measurement outcome (number of cavities that turn black or number of reading items answered correctly).

Two additional applications of the specification equation are: (1) the maintenance of the unit of measurement independent of any particular instrument or collection of instruments (Stenner & Burdick, 2011), and (2) bringing non-test behaviors (reading a Harry Potter novel, 980L) into the measurement frame of reference.

3. Item parameters are supplied by substantive theory and, thus, person parameter estimates are generated without reference to or use of any data on other persons or populations.

It is a feature of the Rasch model that differences between person parameters are invariant to changes in item parameters, and differences between item parameters are invariant to change in person parameters. These invariances are necessarily expressed in terms of differences because of the one degree of freedom over parameterization of the Rash model, i.e., locational indeterminacy. There is no locational indeterminacy in a causal Rasch model in which item parameters have been specified by theory.

4. The quantitivity hypothesis (Michell, 1999) can be experimentally tested by evaluating the trade-off property for the individual case. A change in the person parameter can be off-set or traded-off for a compensating change in text complexity to hold comprehension constant. The trade-off is not just about the algebra in Eq. 1. It is about the consequences of simultaneous intervention on the attribute (reader ability) and measurement mechanism (text complexity). Careful thinking about quantitivity makes the distinction between "an attribute" and "an attribute as measured." The attribute "hardness" as measured on the Mohs scale is not quantitative but as measured on the Vickers scale (1923) it is quantitative. So, it is confusing to talk about whether an attribute, in and of itself, is quantitative or not. If an attribute "as measured" is quantitative then it can always be represented as merely ordinal. But the obverse is not true. twenty-first century science still uses the Mohs scratch test which produces more-than-less-than statements about the "hardness" of materials. Pre 1923 it would have been inaccurate to claim that hardness "as measured" was a quantitative attribute because no measurement procedure had yet been invented that produced meaningful differences (the Mohs scratch test produces meaningful orders but not meaningful differences). The idea of dropping, with a specified force, a small hammer on a material and measuring the volume of the resulting indentation opened the door to testing the quantitivity hypothesis for the attribute "hardness". "Hardness" as measured by the falling hammer passed the test for quantitivity and correspondence tables now exist for re expressing mere order (Mohs) as quantity (Vickers).

Michell (1999) states "Because measurement involves a commitment to the existence of quantitative attributes, quantification entails an empirical issue: is the attribute involved really quantitative or not? If it is, then quantification can sensibly proceed. If it is not, then attempts at quantification are misguided. A science that aspires to be quantitative will ignore this fact at its peril. It is pointless to invest energies and resources in an enterprise of quantification if the attribute involved is not really quantitative. The logically prior task in this enterprise is that of addressing this empirical issue (p. 75)."

As we have just seen we cannot know whether an attribute is quantitative independent of attempts to measure it. If Vickers' company had Michell's book available to them in 1923 then they would have looked at the ordinal data produced by the Mohs scratch test and concluded that the "hardness" attribute was not quantitative and, thus, it would have been "misguided" and "wasteful" to pursue his hammer test. Instead Vickers and his contemporaries dared to imagine that "hardness" could be measured by the hammer test and went on to confirm that "hardness as measured" was quantitative.

Successful point predictions under intervention necessitate quantitative predictors and outcomes. Concretely, if an intervention on the measurement mechanism (e.g., increase the text complexity of a reading passage by 250L) results in an accurate prediction of the measurement outcome (e.g., how many reading items the reader will answer correctly) and if this process can be successfully repeated up and down the scale then text complexity, reader ability and comprehension (success rate) are quantitative attributes of the text, person and reader/text encounter respectively. Note that, if say, text complexity was measured on an ordinal scale (think Mohs) then making successful point predictions about counts correct based on a reader/text difference would be impossible. Specifically, successful prediction from differences requires that what is being differenced has the same meaning up and down the respective scales. Differences on an ordinal scale are not meaningful (will lead to inconsistent predictions) precisely because "one more" means something different depending on where you are on the scale.

Note that in the Rasch model performance (count correct) is a function of an exponentiated difference between a person parameter and an instrument (item) parameter. In the Lexile Framework for Reading (LF) Eq. 1 is interpreted as:

Comprehension $=$ Reader Ability - Text Complexity (success rate)

The algebra in Eq. 1 dictates that a change in reader ability can be traded-off for an equal change in text complexity to hold comprehension constant. However, testing the "quantitivity hypothesis" requires more than the algebraic equivalence in a Rasch model. What is required is an experimental intervention/manipulation on either reader ability or text complexity or a conjoint intervention on both simultaneously that yields a successful prediction on the resultant measurement outcome (count correct). When manipulations of the sort just described are introduced for individual reader/text encounters and model predictions are consistent with what is observed the quantitivity hypothesis is sustained. We emphasize that the above account is individual centered as opposed to group centered. The LF purports to provide a causal model for what transpires when a reader reads a text. Nothing in the model precludes averaging over readers and texts to summarize evidence for the "quantitivity hypothesis" but the model can be tested at the individual level. So, just as pressure and volume can be traded off to hold temperature constant or volume and density can be traded off to hold mass constant so can reader ability and text complexity be traded off to hold comprehension constant. Following Michell (1999) we note that a trade-off between equal increases (or decrements) in text complexity and reader ability "identifies equal ratios directly" and "Identifying ratios directly via trade-offs results in the identification of multiplicative laws between quantitative attributes. This fact connects the theory of con- joint measurement with what Campbell called derived measurement" (Michell, 1999).

Garden variety Rasch models and IRT models are in their application purely descriptive. They become causal and law like when manipulations of the putative quantitative attributes produce changes (or not) in the measurement outcomes that are consistent with model predictions. If a fourth grade reader grows 100L in reading ability over one year and the text complexity of her fifth grade science textbook also increases by 100L over the fourth grade year textbook then the forecasted comprehension rate (whether 60%, 70%, or 90%) that that reader will enjoy in fifth grade science remains unchanged. Only if reader ability and text complexity are quantitative attributes will experimental findings coincide with these model predictions. We have tested several thousand students' comprehension of 719 articles averaging 1150 words. Total reading time was 9794 h and the total number of unique machine generated comprehension items was 1,349,608. The theory based expectation was 74.53% correct and the observed 74.27% correct.

4 Conclusion

This article has considered the distinction between a descriptive Rasch model and a causal Rasch model. We have argued for the importance of measurement mechanisms and specification equations. The measurement model proposed and illustrated (using NexTemp thermometers and the Lexile Framework for Reading) mimics in several important ways physical science measurement theory and practice. We plead guilty to "aping" the physical sciences and despite the protestations of Michell (Michell, 1999) and Markus and Boorsboom (Markus & Borsboom, 2011) do not view as tenable any of the competing go forward strategies for the field of human science measurement.

References

Andrich, D. (2004). Controversy and the Rasch model: A characteristic of incompatible paradigms? *Medical Care, 42*, 1–16.

Barlow, D. H., Nock, M. K., & Hersen, M. (2009). *Single case experimental designs* (3rd ed.). Pearson.

Burdick, D. S., Stone, M. H., & Stenner, A. J. (2006). The combined gas law and a Rasch reading law. *Rasch Measurement Transactions, 20*(2), 1059–1060.

Cronbach, L. J., & Meehl, P. E. (1955). Construct validity in psychological tests. *Psychological Bulletin, 52*, 281–302.

Grice, J. W. (2011). *Observation oriented modelling*. Elsevier.

Irvine, S. H., & Kyllonen, P. C. (2002). *Item generation for test development*. Lawrence Erlbaum Associates Inc.

Karabatsos, G. (2001). The Rasch Model, additive conjoint measurement, and new models of probabilistic measurement theory. *Journal of Applied Measurement, 2*, 389–423.

Kuhn, T. S. (1961). The function of measurement in modern physical science. *Isis, 52*(168), 161–193.

Lewin, K. (1951). *Field theory in social science: Selected theoretical papers*. Harper & Row.

Markus, K. A., Borsboom, D. (2011). Reflective measurement models, behavior domains, and common causes. *New Ideas in Psychology*, In Press.

Michell, J. (1999). *Measurement in psychology: A critical history of a methodological concept*. Cambridge University Press.

Molenaar, P. C. M. (2004). A manifesto on psychology as ideographic science: Bringing the person back into scientific psychology, this time forever. *Measurement: Interdisciplinary Research and Perspective, 2*, 201–218.

Pearl, J. (2000). *Causality: Models, reasoning, and inference*. Cambridge University Press.

Rasch G. (1961). On general laws and the meaning of measurement in psychology. In *Proceedings of the Fourth Berkeley Symposium on Mathematical Statistics and Probability, IV* (pp. 321–334). University of California Press.

Stenner, A. J., & Burdick, D. S. (2011). Can psychometricians learn to think like physicists? *Measurement, 9*, 62–63.

Stenner, A. J., & Stone, M. H. (2010). Generally objective measurement of human temperature and reading ability: Some corollaries. *Journal of Applied Measurement, 11*(3), 244–252.

Stenner, A. J., Smith, M., & Burdick, D. (1983). Toward a theory of construct definition. *Journal of Educational Measurement, 20*(4), 305–316.

Stenner, A. J., Burdick, H., Sanford, E., & Burdick, D. S. (2006). How accurate are Lexile text measures? *Journal of Applied Measurement, 7*(3), 307–322.

Stenner, A. J., Burdick, D. S., & Stone, M. H. (2008). Formative and reflective models: Can a Rasch analysis tell the difference? *Rasch Measurement Transactions, 22*(1), 1152–1153.

Stenner, A. J., Stone, M. H., & Burdick, D. (2009a). The Concept of a Measurement Mechanism. *Rasch Measurement Transactions, 23*(2), 1204–1206.

Stenner, A. J., Stone, M. H., & Burdick, D. S. (2009b). Indexing vs measuring. *Rasch Measurement Transactions, 22*(4), 1176–1177.

Thurstone, L. L. (1926). The scoring of individual performance. *Journal of Educational Psychology, 17*, 446–457.

Wilson, M. (2004). *Constructing measures: An item response modeling approach*. Lawrence Erlbaum Associates Inc.

Woodward, J. (2003). *Making things happen*. Oxford University Press.

Can Psychometricians Learn to Think Like Physicists?

A. Jackson Stenner and Donald S. Burdick

Abstract The last 50 years of human and social science measurement theory and practice have witnessed a steady retreat from physical science as the canonical model. Humphry (2011) unapologetically draws on metrology and physical science analogies to reformulate the relationship between discrimination and the unit. This brief note focuses on why this reformulation is important and on how these ideas can improve measurement theory and practice.

In principle, any characteristic of the instrument, objects of measurement, or measurement contexts that influences the change in probability of a modeled response (i.e., slope of the item response function) is a threat to unit invariance. Such influences produce instability in the unit. Progress in science requires stable unit specifications because it is only through such conventions that a unit is reproduced and shared. A large fraction of the metrology budget for more mature sciences is devoted to identifying and engineering around threats to unit stability.

A big step in realizing the metrology program outlined by Humphry is the abandonment of descriptive Rasch models in favor of an explicit causal interpretation of the regression of the response probability on the exponentiated difference between a person parameter and an instrument/item parameter. All that is meant by this causal claim is that an intervention on the person parameter can be traded off for an offsetting intervention on the item parameters to hold the probability of a correct response constant. If this trade-off property is experimentally verified throughout the range of the attribute and is invariant across task types, person characteristics, and measurement contexts, then, a stable reproducible unit for measuring persons and items has been specified and actualized. If invariance is lacking, say for a new task type, the first

Measurement: Interdisciplinary Research and Perspectives, 9. 2011.

A. J. Stenner (✉) · D. S. Burdick
MetaMetrics, Inc., Durham, NC, USA

D. S. Burdick
Department of Mathematics, Duke University, Durham, NC, USA

W. P. Fisher and P. J. Massengill (eds.), *Explanatory Models, Unit Standards, and Personalized Learning in Educational Measurement*,
https://doi.org/10.1007/978-981-19-3747-7_16

213

thing to check is whether the new task type purportedly measuring the same construct as the task type evidencing invariance has unexpected added easiness/hardness or is measuring in a differently sized unit. In writing research we have found that human and machine scoring of student writings need to be adjusted for differences in unit size, and once this adjustment is made the concordance between machine scoring and human ratings is striking.

A measurement instrument is built to detect variation of a kind. The specification equation answers the question what causes the variation the instrument detects? It is also clear that non-test behaviors (e.g., reading a Harry Potter novel) can be brought into a reading measurement frame of reference by imagining that the novel contains an ensemble of test items the distribution properties of which may be treated as known. The text complexity for the novel is then the reader ability required to correctly answer, say, 75% of the virtual items making up the novel. The specification equation is the tool used to calibrate these non-test behaviors. A third use of the specification equation is to calibrate actual test items making it possible to convert counts correct into quantities for, for example, computer-generated reading items. However, all of the above uses of the specification equation pale in relation to the role the specification equation can and should play in maintaining the unit of scale for an attribute. Once a specification equation for an attribute is "locked down" the unit origin and unit size are fixed. New task types and measurement contexts (e.g., machine vs. human scoring) can be linked back to the fixed unit. Tampering with the specification equation by changing the intercept or adjusting the regression weights alters the origin and unit size, respectively. Thus, the specification equation defines the unit, independent of any particular test form or linking study, and maintains that unit over widely varying instrumentation and measurement contexts.

It often happens that new task types, improved theory or improved technology necessitate changes to the specification equation that, if not taken into account, will result in a "new" unit. Adjustments to the specification equation can be made to ensure that the "old" unit and "new" unit are comparable as to origin and unit size. Typically some standard artifact (boiling point of normal water, platinum meter bar, or collection of empirically calibrated texts) is used to ensure unit stability over time.

Scale unification is a well-understood theme in the history of science. The obverse, scale proliferation, is a prominent feature of measurement theory and practice in the human and social sciences. Today there are dozens of scales for measuring every important attribute (anxiety, depression, reading ability, spatial reasoning). There is often debate about whether the fact that the task types vary in added easiness/hardness or unit size may signal that something different is being measured. Infrequently, attempts are made to document that the same attribute is being measured by the various instruments and then linking studies are launched that result in correspondence tables or equations that link the respective scales (similar to the equation that links the Fahrenheit and Celsius scales).

Humphry has sketched a metrology program for the human and social sciences to follow as we begin the arduous task of building a system of units. Although the far-term goal is a system of units for the human and social sciences, the near-term goal should be an invariant unit shared by a relevant community for a single attribute. We will learn much as these first attempts play out.

Reference

Humphry, S. M. (2011). The role of the unit in physics and psychometrics. *Measurement: Interdisciplinary Research and Perspectives, 9*(1), 1–24.

Metrology for the Social, Behavioral, and Economic Sciences

William P. Fisher Jr. and A. Jackson Stenner

Abstract A metrological infrastructure for the social, behavioral, and economic sciences has foundational and transformative potentials relating to education, health care, human and natural resource management, organizational performance assessment, and the economy at large. The traceability of universally uniform metrics to reference standard metrics is a taken-for-granted essential component of the infrastructure of the natural sciences and engineering. Advanced measurement methods and models capable of supporting similar metrics, standards, and traceability for intangible forms of capital have been available for decades but have yet to be implemented in ways that take full advantage of their capacities. The economy, education, health care reform, and the environment are all now top national priorities. There is nothing more essential to succeeding in these efforts than the quality of the measures we develop and deploy. Even so, few, if any, of these efforts are taking systematic advantage of longstanding, proven measurement technologies that may be crucial to the scientific and economic successes we seek. Bringing these technologies to the attention of the academic and business communities for use, further testing, and development in new directions is an area of critical national need.

1 The Technologies to Be Developed

Science underwrites the value of tangible forms of economic capital by ensuring that measures of commodities from minutes to kilowatts to barrels are universally

Abstract published on p. 38, National Science Foundation, Directorate for Social, Behavioral, and Economic Sciences. 2011. SBE 2020: White Papers; Titles, Authors, and Abstracts. Arlington, VA: National Science Foundation. https://www.nsf.gov/sbe/sbe_2020/Abstracts.pdf

W. P. Fisher Jr. (✉)
Living Capital Metrics LLC, Sausalito, CA, USA

Graduate School of Education, BEAR Center, University of California, Berkeley, CA, USA

A. J. Stenner
MetaMetrics, Inc., Durham, NC, USA

verifiable and comparable. Increasing interest in the economic value of the intangible forms of human and social capital studied in the social and behavioral sciences raises the question as to whether the quality of measurement in these areas might one day approximate the scientific rigor and practical convenience of measures in the natural sciences. Over 80 years of research strongly suggests that the answer to this question is yes (Fisher, 2009; Stenner et al., 2006).

There are two phases in the development of universally uniform metric systems. The first determines whether something exists in a persistent, stable and measurable state independent of the sample measured, the equipment and operator measuring, time, and space. In this first phase, things themselves act as agents compelling agreement among observers as to their separate and real status as objective phenomena in the world.

The second phase in the process of establishing metrological uniformity transforms this agent of agreement into a product of agreement. Now, given a measurable phenomenon, research technicians collaborate on the unit size and range, nomenclature, and terminology by which its quantitative and qualitative features will be communicated. Systems for calibrating instruments in the standard metric, and checking their traceability to it, are devised and implemented.

The first phase in the process of establishing metrological uniformity for intangible forms of capital measured via ability tests, surveys, assessments, and ratings has effectively been underway for at least 50 years, since the work of Rasch in the 1950s, and for more than 80 if Thurstone's pioneering work from the 1920s is included. Though virtually unknown outside psychometric circles, the facts of additive, independent, transitive, linear, ratio, and separable parameters for constructs measured in the human and social sciences are not controversial.

Much of this work has been conducted by researchers trained in the natural sciences who turned their attention to the social sciences, such as Thurstone (an electrical engineer), Rasch (a mathematician), and Wright (a physicist who worked as an assistant to Nobelists Townes and Mulliken before taking up psychometrics). A number of reviews of this work are available (for instance, Bezruczko, 2005).

There are, however, few signs of any research programs, funding, or practical demands targeting the second phase in the work required for universally available metrologic uniformity in the measurement of intangible forms of capital. The viability of such a goal is suggested repeatedly with every calibration of an instrument producing data that meets demands for separable, independent model parameters. New instrument calibrations are published every day, providing evidence of another construct that is a possible candidate for a standardized metric.

A critical national need exists for a widespread awareness of the viability and desirability of reference standard metrics for human, social, and natural capital. Barriers include:

- the huge cost of developing and deploying these metrics;
- a general lack of awareness of the decades of research proving the viability and special advantages of reference standards in the behavioral sciences;

- an underdeveloped public appreciation for both the high returns provided by investments in metrology and the vital role played by metrology in the history of science and capitalism;
- institutional orientations better able to serve the needs of existing paradigms than the emergence of new ones; and
- deeply rooted cultural presuppositions about the nature of number and the alleged limits of psychosocial measurement.

Over 80 years of research successfully refutes the assumption that measurement in the social sciences is epistemologically inferior to that of the natural sciences. We need research exploring possibilities for more rigorously defined metrics and the benefits that could be obtained from them. The primary results that could be obtained in an economic context informed by universally uniform, linear and ratio metrics for intangible forms of capital follow from the oft-repeated saying, "You manage what you measure." Most of the metrics currently used in the management of human, social, and natural capital are nonlinear and ordinal scores, ratings, and percentages. Because their unit magnitudes are dependent on locally variable score distributions, these alleged "metrics" are often uninformative, confusing, or deceptive. The incommensurability of these so-called measures effectively locks up human, social, and natural capital markets by making individual information transactions so expensive that decisions are made with no information, or with the wrong information.

Though advanced measurement applications have demonstrated highly desirable advantages and capabilities for decades, they have not become the mainstream paradigm. This may in part be due to the opinion held by many that improved measurement is an academic nicety, an end in itself, or is an expression of particular researchers' special theoretical investments.

Despite these opinions, and though we rarely stop to think about it, we all know that fair measures are essential to efficient markets. When different instruments measure in different units, market transactions are encumbered by the additional steps that must be taken to determine the value of what is being bought and sold. Health care and education are now so hobbled by myriad varieties of measures that common product definitions for the outcomes of these industries seem beyond reach.

As has been pointed out in a wide variety of works over the last several decades, we need to broaden the focus of business management beyond investments, factories, equipment, property, and labor. Instead of the traditional three forms of capital (land, labor, and manufactured equipment), we actually employ four (natural, human, manufactured equipment, and social). Land and labor are far more complex than a mere piece of ground and the functionality of a job description. These complexities are captured by the multifaceted concepts of natural and human capital, which have to include diverse and distinct dimensions of the resources brought to bear. And social capital is of such vital importance that capitalism itself could not have gotten off the ground without it.

In order to make capitalism live up to its own accounting principles, we need better measures fit into an accounting framework that redefines profit so that it is less a matter of liquidated capital, and more a matter of removing wasted resources from within a closed system of limited capacities. In order to learn how to reduce the waste and increase the stocks of human potential, community trust, and natural resources, we must better learn the truth of the maxim, we manage what we measure.

Calibrated instruments traceable to reference standards express value in universally uniform metrics that function as common currencies. New efficiencies for human, social, and natural capital markets come from the reduced friction in transactions, which are made meaningful and comparable via metrological networks not much different from the one connecting all the clocks. The intangibles of health care, education, social services, and human/natural resource management may not be forever doomed to locally dependent product definitions that defy pricing.

What we need are ways of extending the basic capitalist ethos into the domain of the intangibles. How can we set up markets so that the invisible hand efficiently promotes social and environmental ends unintended by individuals maximizing their own gains? How might we extend the free play of self-interest into more comprehensively determined returns for the global dividend? Better measurement will inevitably be of central concern in answering these questions.

In this context, the existence of an actual market of shared uniform information would coordinate the collective decisions of purchasers and providers to match supply and demand far more efficiently than could ever be the case in the current system of high-friction, ordinal, and locally dependent "metrics."

Innovation is increasingly seen as best conceived as a group effort. The wisdom of crowds phenomenon makes it possible for actors coordinated by shared information to accomplish in short order tasks that either could not be done by independent individuals at all, or only by using much more time and resources. The profit motive is an energy source of incredible power and potential. Creating an economic and social context in which innovation on the broadest scale could be brought to bear on issues of human, organizational, and environmental performance and management would be productively disruptive and transformative on the highest levels.

2   Justification for NSF SBE Attention

The societal challenges associated with the development and deployment of a metrological infrastructure for human, social, and natural capital are of a magnitude that prevent even the largest corporations or research institutes from undertaking the task alone. And piecemeal efforts often are not just inadequate to the task, they actually make things worse as uncoordinated and mutually contradicting efforts compete for attention and resources. We need broad efforts undertaken by society as a whole, with everyone's interests represented. No individual, small business, major corporation, or nonprofit foundation could ever hope to succeed alone in a task of this scale and scope.

Likely proposers to a funding competition in this area would include commercial agencies already making use of the available advanced measurement techniques. In education, these include the Northwest Evaluation Association; MetaMetrics, Inc.;

Educational Testing Service; ACT; Pearson; and many others. State departments of education with longstanding expertise in this area include those in Oregon, Michigan, Illinois, Vermont, and in many other states. In health care, research groups with relevant expertise include QualityMetric, FOTO, Inc., the Rehabilitation Institute of Chicago, PeaceHealth, and others.

Academic departments of psychology, education, sociology, public health, health systems management, and others conducting research and teaching in this area would also likely propose projects for funding. Academic departments with particularly high profiles in this area can be found at the University of California, Berkeley; University of Illinois, Chicago; Johns Hopkins University; University of Toledo (Ohio); Emory University; Northwestern University; Boston College; University of Denver; University of Michigan; the Chicago Medical School; and elsewhere around the world.

The societal challenges related to the improved measurement of human, social, and natural capital are not being addressed for a number of reasons. First, despite the longstanding existence of data, instruments, and theory to the contrary, it is widely and mistakenly believed that the fundamental measurement of constructs measured by way of ordinal observations is impossible. Second, it is also widely and mistakenly believed that all numbers are inherently and always effectively quantitative, or that supposed differences in kinds of numbers are purely academic and of no practical consequence.

Third, the metrological infrastructure is almost completely invisible to and taken for granted by the public, meaning that local efforts aimed at expanding it into a new domain are virtually meaningless and inevitably futile. Fourth, the existing system of incentives and rewards makes it very difficult, if not impossible, for individual researchers and teachers to have an impact on the behaviors and decisions of their clients and students, as these are culturally rooted in the familiar, albeit misunderstood and misapplied, ordinal and local systems. Fifth, even when an individual or organization does grasp the significance of the new measurement technologies, these very few isolated and uncoordinated instances depend too heavily on local leadership, and eventually starve for lack of sustenance from a larger networked metrological culture.

The sum meaning of these five conditions is that research and development of metrological infrastructure for intangible capital will not proceed at all without leadership at the national level and without public funding. But the nation's scientific frontiers and commercial frontiers would potentially be greatly stimulated simply by publicly introducing the idea that such an infrastructure could be possible, and by suggesting that considerable benefits relative to existing goals for improving education, health care reform, and environmental management could accrue from it.

References

Bezruczko, N. (Ed.). (2005). *Rasch measurement in health sciences*. JAM Press.

Fisher, W.P., Jr. (2009). Invariance and traceability for measures of human, social, and natural capital: Theory and application. *Measurement (Elsevier)*, *42*(9), 1278–1287.

Stenner, A. J., Burdick, H., Sanford, E., & Burdick, D. S. (2006). How accurate are Lexile text measures? *Journal of Applied Measurement*, *7*(3): 307–322.

Causal Rasch Models

A. Jackson Stenner, William P. Fisher Jr., Mark H. Stone,
and Donald Burdick

Abstract Rasch's unidimensional models for measurement show how to connect
object measures (e.g., reader abilities), measurement mechanisms (e.g., machine-
generated cloze reading items), and observational outcomes (e.g., counts correct on
reading instruments). Substantive theory shows what interventions or manipulations
to the measurement mechanism can be traded off against a change to the object
measure to hold the observed outcome constant. A Rasch model integrated with
a substantive theory dictates the form and substance of permissible interventions.
Rasch analysis, absent construct theory and an associated specification equation, is a
black box in which understanding may be more illusory than not. Finally, the quantitative hypothesis can be tested by comparing theory-based trade-off relations with
observed trade-off relations. Only quantitative variables (as measured) support such
trade-offs. Note that to test the quantitative hypothesis requires more than manipulation of the algebraic equivalencies in the Rasch model or descriptively fitting data to
the model. A causal Rasch model involves experimental intervention/manipulation
on either reader ability or text complexity or a conjoint intervention on both simultaneously to yield a successful prediction of the resultant observed outcome (count
correct). We conjecture that when this type of manipulation is introduced for individual reader text encounters and model predictions are consistent with observations,
the quantitative hypothesis is sustained.

Frontiers in Psychology, 4. 2013.

A. J. Stenner (✉) · D. Burdick
Metametrics, Inc., Durham, NC, USA

W. P. Fisher Jr.
Living Capital Metrics LLC, Sausalito, CA, USA

Graduate School of Education, BEAR Center, University of California, Berkeley, CA, USA

M. H. Stone
Adler School of Professional Psychology, Chicago, IL, USA

D. Burdick
Department of Mathematics, Duke University, Durham, NC, USA

> The thermometer, as it is at present construed, cannot be applied to point out the exact
> proportion of heat .... It is indeed generally thought that equal divisions of its scale represent
> equal tensions of caloric; but this opinion is not founded on any well decided fact.
>
> Gay-Lussac (1802)

Thirty years ago, three of the authors of this article introduced the concept of
the specification equation as a new model for measurement validity studies in the
human and social sciences (Stenner & Smith, 1982; Stenner et al., 1983; Stone &
Wright, 1983). The primary characteristic setting the concept of the specification
equation apart from other approaches to psychological and social measurement is its
decidedly mechanismic (causal) approach. In the years that have passed, the causal
notions associated with the specification equations have been variously endorsed
(Hobart et al., 2007) and ignored (Messick, 1989).

Perhaps the causal perspective on measurement would not be deserving of an
update were it not for the theoretical fruitfulness and widespread adoption of the
Lexile Framework for Reading. In this paper, the ideas presented in the initial papers
are situated in the contemporary psychometric context and some recent advances
are outlined. The framework for measurement in the human and social sciences
set out here adopts from 17th- and 18th-century physical science measurement a
focus on experimental manipulations of three variable systems (such as $F = MA$).
We conjecture that in three-variable systems of this kind, demonstrating that all
three variables are quantitative requires manipulations of one variable to result in
predictable changes to a second variable when the third is held constant. When this
manipulation works up and down the scale no matter which variable is held constant,
then all three variables are quantitative.

For instance, given that a change in mass can be offset by a change in acceleration
such that force is left unchanged, and this trade-off operates the same way at every
point along the mass scale and along the acceleration scale, then mass, acceleration,
and force are quantitative attributes. In what follows, we make some strong claims
regarding the possibility that human and social science measurement might find
foundations in causal relationships akin to those obtained in the physical sciences.
Measurement is conceived in this context as a three-variable equation involving an
attribute measure, a measurement mechanism and a measurement outcome (often
a count). The primary conjecture is that a prescribed change in a mechanism (such
as text complexity) is offset by a change in an attribute measure (such as reading
ability) to hold a measurement outcome (percent correct) constant. Our unproven
conjecture is that, when this trade-off functions precisely the same way up and down
the scale, then the three variables are quantitative (equal interval) variables. We
follow Pearl (2000, 158) in taking "the stand that the value of structural equations
lies not in summarizing distribution functions but in encoding causal information for
predicting the effects of policies."

In other words, our purpose is to elaborate the intuition that if like differences
in differences have the same predictive outcomes wherever along the scale the
differences are experimentally induced, then the attributes are quantitative. A 100
L difference between text complexity and reader ability has the same implication

for predicted success rates on reading items wherever along the scale the 100 L difference is experimentally induced. If the predicted success rates are observed, then 100 L means the same thing up and down the scale in the substantive terms of real texts encountered by real readers. Is not this capacity to infer the repetition of a qualitatively meaningful constant amount precisely what we mean by "equal interval"?

1 Quantity Versus Heterogeneous Orders

Michell (1997, 2000, 2007) made the case for what measurement is and in his earlier writings, exhorted psychologists to adopt this "standard model" in their practice. Over time, this eminently sensible call morphed into a diagnosis of so-called pathological behavior on the part of the field of psychology and its high priests and priestesses, called *psychometricians*. Most recently, some researchers have asserted that no psychological attributes have been shown to be quantitative and have offered the explanation that none will be because these attributes are actually heterogeneous orders:

> Scientists who care more about appearing to be quantitative and the advantages that might accrue from that appearance, than they do about investigating fundamental scientific issues, put expedience before the truth. In this, they do not conform to the values of science and elevate non-scientific interests over those values, thereby threatening to bring science as a whole into disrepute. If the attributes that psychometricians aspire to measure are heterogeneous orders then psychometrics, as it exists at present, is fatally flawed and destined to join astrology, alchemy and phrenology in the dustbin of science. (Michell, 2012, 16)

Throughout this paper, we draw analogies to human temperature measurement via the NexTemp thermometer[1] and to English reading ability measurement via the

---

[1] The NexTemp thermometer is a thin, flexible, paddle-shaped plastic strip containing multiple cavities. In the Fahrenheit version, the 45 cavities are arranged in a double matrix at the functioning end of the unit. The columns are spaced 0.2 °F intervals covering the range of 96.0–104.8 °F ... Each cavity contains a chemical composition comprised of three cholesteric liquid crystal compounds and a varying concentration of soluble additive. These chemical compositions have discrete and repeatable change-of-state temperatures consistent with an empirically established formula to produce a series of change-of-state temperatures consistent with the indicated temperature points on the device. The chemicals are fully encapsulated by a clear polymetric film, which allows observation of the physical change but prevents any user contact with the chemicals. When the thermometer is placed in an environment within its measure range, such as 98.6 °F (37.0 °C), the chemicals in all of the cavities up to and including 98.6 °F (37.0 °C) change from a liquid crystal to an isotropic clear liquid state. This change of state is accompanied by an optical change that is easily viewed by a user. The green component of white light is reflected from the liquid crystal state but is transmitted through the isotropic liquid state and absorbed by the black background. As a result, those cavities containing compositions with threshold temperatures up to and including 98.6 °F (37.0 °C) appear black, whereas those with transition temperatures of 98.6 °F (37.0 °C) and higher continue to appear green (Medical Indicators, 2006, 1–2). Thus, the observed outcome is a count of cavities turned black. The measurement mechanism is an encased chemical compound that includes a varying soluble agent that changes optical properties according to changes in temperature. Amount

EdSphere™ technology[2] (Hanlon et al., 2012). Both measurement systems share a basic three-part structure in common with many other measurement technologies (Hebra, 2010): (1) An observational outcome, often a count (number of NexTemp cavities changing from green to black reflecting temperature change or count of correct responses to EdSphere™ four-choice embedded cloze items reflecting change in reading ability), (2) a causal mechanism that transmits variation in the intended attribute (temperature or reading ability) to the observed outcome, and (3) an attribute measure denominated in some unit (such as degrees Celsius or Lexiles). As Michell (2012) correctly observed, "Only quantitative attributes can be measured because only they possess the necessary kind of homogeneity" (7). The crucial question is: How does one test for the necessary homogeneity that distinguishes quantity from mere order?

The history of science reveals a developmental course traveled by every attribute that figures in advanced scientific discourse. For some of these attributes, the full course took centuries and for others many decades, but in no case was an attribute born quantitative. Most quantities that are now recognized began as qualitative distinctions (hot, cold, or good reader, poor reader), were later understood to be ordinal and grew to adulthood as a quantity that admits of homogeneous differences. Historical and philosophical treatments that ignore this developmental pathway and the struggles along the way, whether purposeful or not, only confuse (Chang, 2004; Sherry, 2011). In ancient times, only extensive attributes (time, volume, weight, etc.) were considered measurable. Attributes such as temperature, density, and electromagnetism were in their infancy and were specifically not thought to be measurable (Heilbron, 1993; Roche, 1998).

It took hundreds of years of science and a solution to the particularly knotty problem of forsaking human sense impression as the final arbiter of disputes concerning quantitative attributes. In the case of temperature, degrees were obviously not homogeneous because *one more* had such ridiculously discontinuous consequences (e.g., a nominal one-degree change makes water freeze or boil). The lesson here is that perceived qualitative distinctions are a poor guide as to whether an attribute is quantitative as measured. Magnitudes of the same quantitative attribute appear to differ in dozens of ways. In the infancy and adolescence of many attributes

---

of soluble agent can be traded off for change in human temperature to hold number of black cavities constant.

[2] The EdSphere™ technology for measuring English language reading ability employs computer generated, four-option, multiple choice cloze items built on the fly for any prose text. Counts correct on these items are converted into Lexile measures via an applicable Rasch model. Individual cloze items are one off and disposable; an item is used only once. The cloze and foil selection protocol ensures that the correct answer (cloze) and incorrect answers (foils) match the vocabulary demands of the target text. The Lexile text complexity measure and the expected spread of the cloze items are given by a proprietary text theory and associated equations. Thus, the observed outcome is a count of correct answers. The measurement mechanism is a text with a specified Lexile text complexity and an item generation protocol consistent with that text complexity measure. The text complexity measure can be traded off for a change in reading ability to hold constant the number of items answered correctly.

that today are considered quantitative, *quantitative homogeneity* was impure. Quantitative measurement has been a hard-fought and dearly won battle; it has never been and never will be a given or something to be decided solely by speculation about perceived qualitative distinctions and heterogeneous orders. The historical record includes many attributes thought at one time to indicate mere order and later rendered as quantities and is void of confirmed quantitative attributes that are later found to be merely ordinal. This does not mean that every ordered attribute will, given time, be rendered quantitative, but science has taught us to leave the door ajar because, over and over, science has found ways to render some of today's *orders* as tomorrow's *quantities.*

Yet another argument is due to Sherry (2011): if treating an attribute such as text complexity or reading ability as quantitative allows one to employ mathematical models to bring order to the data, "then a plausible explanation for this success is that the attribute is approximately quantitative" (523).

So, if speculation about a hierarchy of states, stages, discontinuities, heterogeneous orders, and conceptual inclusions are unworthy guides to whether or not an attribute is quantitative, how does one proceed? As we will show, there is a simple, powerful test for essential homogeneity: the trade-off property. In many three-variable systems, such as, for example, a causal Rasch model that relates observed outcome (dependent variable) to a difference (or product) between the measurement mechanism (e.g., item difficulties) and the measurement attribute (e.g., reading ability), the trade-off property asserts that a manipulation on the measurement mechanism (e.g., increasing text complexity by 100 L) can be traded off for an off-setting manipulation on the measurement attribute (increasing reader ability by 100 L) to hold constant the observed outcome (percent correct).

We conjecture that when this trade-off property can be experimentally verified up and down the scale, all variables in the system are quantitative and differences (units) are monotonic, invariant, and homogeneous. As suggested by Trendler (pers. commun., 2011), and expanding on Burdick et al. (2006), the capacity to successfully predict differences from differences up and down a scale are an acid test for quantity. This trade-off property (in the service of successful substantive theory) is all that matters in demonstrating causality, deep understanding of the construct, and predictive control over interventions. There will always be extraneous qualities that can be identified in rhetorical arguments against a quantitative claim. Endless speculations can be made about why this or that feature of an attribute is a qualitative distinction worthy of notice (such as variation in the freezing point of water, relative to temperature), thus, appearing to render the unit non-homogeneous.

But such speculations are unable to offer distinctions that make a difference when evidence supports the trade-off property. This does not, however, mean that data fit to a descriptive (not explicitly causal) Rasch model implies that all three variables in the system are quantitative. Such fit may be suggestive in the same way that highly correlated data are a good place to look for causal relationships, but it is inconclusive as to the quantitative claim until experimental manipulation sustains the trade-off relation.

Perhaps an intuitively accessible illustration of the power of the trade-off relation would be helpful. Suppose Lexile reader measures and Lexile text complexity
measures were replaced with randomly spaced numbers that preserve mere order.
In this case, a text complexity difference between two articles might be increased
from +10 to 150 L and the adjacent 10 L article difference might be increased to
160 L. Thus, order is preserved but essential homogeneity is destroyed. Clearly,
the trade-off relation would be violated even though strict order was maintained.
Each trade off would produce wildly varying predicted observed outcomes, not the
constant observed outcome asserted under the trade-off property. Order is not sufficient; differences must be preserved and retained as invariant ranges up and down
the scale.

Note that there is a crucial difference between the trade-off property and order-
restricted inference tests such as double and triple cancelation (Kyngdon, 2008;
Michell, 1997). The trade-off relation is a claim about what will happen in the
individual case or a token test of a causal model. Most tests of double and triple
(and so on) cancelation with which we are familiar test between-person orders, not
within-person (intra-individual) differences (but see von Winterfeldt et al., 1997;
Luce, 1998). The trade-off property can and should be tested within persons over
time. We further develop this idea later in the paper.

Process talk localizes the active features of a measurement process within the
person or object of measurement. *Mechanism talk* conceptualizes the active ingredients as a tunable mechanism within the instrument. In temperature measurement, it
seems awkward to talk about a response process that functions within the person and
that interacts with the instrument to produce the cavity count on a NexTemp thermometer (see footnote 1). It is often more useful to reflect on the instrument features
(tunable mechanisms) that transmit variation in the attribute to the cavity count. When
the attribute *reader ability* is measured with a machine-generated, multiple choice,
cloze test, the tunable mechanism involves the text complexity of the passage and the
decision as to which words are "clozed" and how the foils are chosen. A causal Rasch
model may be seen as formalizing how a measurement mechanism and an attribute
measure cooperate to produce (cause) the observed outcome. The difference between
calibrated mechanism and measured attribute causes the observed outcome. When
viewed this way, it is clear that manipulations of the mechanism (e.g., added text
complexity) can be offset (traded off) by a manipulation of the attribute (more practice reading) to hold the observed outcome (success rate or comprehension) and the
measure based on it constant.

In this formulation of the measurement process, we require that the way the
mechanism works and the way it trades off against the attribute measure to cause
changes in the observed outcome must be invariant both within and between objects
of measurement (e.g., persons). Note that if the mechanism is well explicated and
functions invariantly across objects of measurement, it does not matter how the object
arrived at its position on the attribute scale. Process talk can confuse on this point
and may cause worry about the how. For example, two persons, Jane and Dave, may
both have a fever of 104 °F. For purposes of human temperature measurement (as
opposed to, say, treatment regime), it is not relevant that Jane has a bacterial infection

and Dave a viral infection. What is relevant is that temperature and the measurement mechanism cooperate in the same way for Jane and Dave, and for every other person, independent of what might have caused a deviation from normal human temperature. Similarly, two fourth-grade readers may both read at 800 L, but one got there with a particularly fortuitous genetic makeup (Castles et al., 1999) and the other because of 1 h of daily practice for 5 years. It is clarifying to maintain focus on exposing and explicating the measurement mechanism and not on distractions such as the myriad causes responsible for any specific attribute measure, be it temperature or reading ability.

In what we have termed theory-referenced measurement, instrument calibrations are provided by a construct theory and specifically not by data. For example, NexTemp thermometers and EdSphere[TM] reading tests[2] are calibrated via theory. Person-fit statistics become not just checks on how similar a person's response data are to the reference group's data but on how well each person conforms to theoretical expectations. What is intended to be measured (e.g., temperature or reading ability) is made explicit with the theory-based instrument calibrations, and the fit statistics confirm or disconfirm whether the respective observed outcomes and their associated measures are consistent with theoretical expectations in the individual case (Smith, 2000).

One way in which causal Rasch models differ from descriptive Rasch models is in the pattern of counterfactual dependencies inherent in the former. Woodward (2003) referred to these dependencies as "w questions" (11): What if things had been different with either the attribute or the measurement mechanism? What value would the observed outcome take? We document these counterfactual dependencies by manipulating the attribute or the measurement mechanism or conjointly manipulating them both and seeing whether the expected score outcome is in fact observed. The trade-off property is a special kind of counterfactual dependence that can be used to test the quantitative status of constructs, as shown below.

2 The Measurement Mechanism

Equation (2) is the familiar Rasch model for dichotomous data, which sets an observed outcome (raw score) equal to a sum of modeled probabilities. The observed outcome is the dependent variable and the measure (e.g., person parameter b) and instrument (e.g., item parameters d_i) are independent variables. The concrete outcome (e.g., count correct on a reading test) is observed, whereas the measures and instrument parameters are not observed but can be estimated from the response data. In Eq. (3), a mechanismic interpretation is imposed on the equation; the right-hand side (r.h.s.) variables are presumed to characterize the process that generates the observed outcome on the left-hand side (l.h.s.). An illustration of how such a mechanism can be exploited is given in Stone (2002). The item map for the Knox cube test (a test of short-term memory) revealed a 1 logit gap with no items. The specification equation was used to build an item that theory asserted would fill in the gap. Subsequent data analysis confirmed the theoretical prediction of the items' scale location:

Comprehension = Reading Ability − Text Complexity

Conceptual Rasch Model   (1)

$$\text{Raw Score} = \sum_i \frac{e^{(b_n - d_i)}}{1 + e^{(b_n - d_i)}}$$

Descriptive Rasch Model   (2)

$$\text{Raw Score} =: \sum_i \frac{e^{(b_n - d_i)}}{1 + e^{(b_n - d_i)}}$$

Causal Rasch Model   (3)

where raw score is the observed outcome, b is the attribute measure, and d_i's are mechanism calibrations. The observed outcome is thus, modeled as a sum of success probabilities. Typically, the item calibrations (d_i) are assumed to be known and the measure parameter is iterated until the equality is realized (i.e., the sum of the modeled probabilities equals the observed outcome). How is this equality to be interpreted? Is something more happening than simply the algebra?

In an effort intended to clarify the practical value of the algebra, Freedman (1997) proposed three uses for regression equations like those above:

(1) To describe or summarize a body of data,
(2) To predict the l.h.s. from the r.h.s., and
(3) To predict the l.h.s. after manipulation or intervention on one or more r.h.s. variables (measure parameter and/or mechanism parameter.

Description and summarization possess a reducing property in that they abstract away incidentals to focus on what matters in a given context. In a rectangular persons-by-items data matrix (with no missing data), there are $N_p \times N_i$ observations. Equations like those above summarize the data using only $N_p + N_i - 1$ independent parameters. Description and summarization are local in focus. The relevant concept is the extant data matrix with no attempt to answer questions that might arise in the application realm about "what if things were different." Note that if interest centers only on the description and summarization of a specific data set, additional parameters can be added, as necessary, to account for or better describe the data (e.g., as in fitting a polynomial equation).

Prediction typically implies the use of the extant data to project into an as yet unobserved context/future in the application realm. For example, item calibrations from the extant data are used to compute a measure for a new person, or person parameters are used to predict how these persons will perform on a new set of items. Predictions like these rest on a set of invariance claims. New items and new persons are assumed to behave as persons and items behaved in the extant data set (Andrich, 1989). Rasch fit statistics (for persons and items) are available to test for certain violations of these assumptions of invariance (Smith, 2000).

To explain how an instrument works is to detail how it generates the count it produces (the observed outcome) and what characteristics of the measurement procedure affect that count. This kind of explanation is neither just statistical nor synonymous with prediction. Instead, the explanation entails prediction under intervention: If one wiggles this part of the mechanism, the observed outcome will imply a measure different by this amount. As noted by (Hedström, 2005), "Theories based on fictitious assumptions, even if they predict well, give incorrect answers to the question of why we observe what we observe" (108). Rasch models, absent a substantive theory capable of producing theory-based instrument calibrations, may predict how an instrument will perform with another subject sample (invariance) but can offer only speculation in answer to the question, "How does this instrument work?" Rasch models without theory are not predictive under intervention and thus, are not causal models.

In 1557, the Welshman Robert Recorde remarked that no two things could be more alike (i.e., more equivalent) than parallel lines, and thus, was born the equal sign, as in $3 + 4 = 7$. We propose that the distinction between *descriptive* Rasch models and *causal* Rasch models should be signaled by the use of a 250-year-old symbol (=:) attributable to Euler (circa 1730) and exhumed by Pearl (1999), which denotes that interventions/manipulations on the r.h.s. of the equation *causes* a change to the l.h.s. of the equation (Pearl, 1999; Stenner et al. 2009). Allowable manipulations include changes to just reading ability, just text complexity, or conjointly to both. These equations purport to predict what will happen in the individual case (token causation) to the observed outcome (raw score, count correct, or percent correct) if allowable manipulations are made. Manipulations of reader ability and/or text complexity presume that these two variables (attributes of persons and text respectively), are well enough understood that manipulation is possible. Our recommended symbol also suggests that we adopt the more precise and explicit definition, as indicated above, to avoid the metaphysical implications of using causation in any narrative without an explicit understanding of what is implied. The symbol and definition reins in speculation surrounding the so-called swamp of language.

Cook and Campbell (1979) observed, "The paradigmatic assertion in causal relationships is that manipulation of a cause will result in the manipulation of an effect.... Causation implies that by varying one factor [variable] I can make another vary" (36). Holland (1986) reduced this to an aphorism: "No causation without manipulation" and Freedman (1997) distinguished the merely descriptive from the causal:

> Causal inference is different, because a change in the system is contemplated: for example, there will be an intervention. Descriptive statistics tell you about correlations that happen to hold in the data: causal models claim to tell you what will happen to Y if you change X. (116)

The descriptive Rasch model above, which employs a simple equality relating raw score to an exponentiated difference between ability (b) and item difficulty (d_i), is a descriptive model (Eq. 2). Algebraic manipulation of the ability parameter or item difficulty parameter can be evaluated for consequences to the raw score, but this says nothing about what would result following an experimental manipulation of b or d or a conjoint intervention on both simultaneously. Following Woodward (2003), we require causal Rasch models to be modular in the sense that it is possible to intervene on (manipulate) one variable in the equation on the right side without affecting another right-hand variable. Specifically, a manipulation of text complexity (e.g., choosing a more difficult text) should not alter reader ability and vice versa. A causal Rasch model should expose the mechanism that transmits variation in the attribute to the observed outcome: "One would also like to have more detailed information about just which interventions on X [r.h.s. of the equation] will change Y [the score or comprehension rate] and in what circumstances and exactly how they will change Y" (Woodward, 2003, 66). Reflecting on the distinction between a descriptive use and a causal use of a Rasch equation, asymmetry is apparent. A causal model can always be used for merely descriptive purposes even though it can do more, whereas a descriptive (e.g., correlational) model can offer no predictions about what would happen if right-side variables were manipulated as in Eq. (3).

It is often argued in garden-variety Rasch applications that the best evidence that an instrument is doing what it is supposed to do is data fit to the Rasch model, which implies that the observed outcome (e.g., count correct on a reading test) is a sufficient statistic for the attribute measure and so exhausts the information in the data about the attribute measure. Conditional on the difference between attribute measure (e.g., temperature) and measurement mechanism (amount of additive in a NexTemp cavity) the residuals taken over persons and taken within person over time are uncorrelated. Unfortunately, nothing in the fact that data fit the model licenses the conclusion that what is being measured is "temperature" or "reading ability" (Maraun, 1998). Typical so-called science tests are often poorly disguised reading tests. More than good fit of data to the model are needed to decide among competing claims regarding what attribute an instrument measures. We speculate that this fact is poorly understood because it is believed that conditioning on a variable is the same thing as intervening to fix the value of that variable. The former is a statistical manipulation and the latter is an experimental manipulation: "When one conditions, one takes as given the probability distribution. When one intervenes, one changes the probability distribution" (Hausman, 1998, 233). This is an important difference between descriptive Rasch models and causal Rasch models.

Measurement mechanism is the name given to just those manipulable features of the instrument that cause invariant observed outcomes for objects of measurement that possess identical measures. A measurement mechanism explains by opening the black box and showing the cogs and wheels of the instrument's internal machinery. A measurement mechanism provides a continuous and contiguous chain of causal links between the encounter of the object of measurement and instrument and the resulting observed outcome (Elster, 1989). We say that the observed outcome (e.g., raw score) is explained by explicating the mechanism by which those outcomes are obtained. In this view, to respond with a recitation of the Rasch equation for converting counts into measures, to reference a person by item map, to describe the directions given to the test-taker, to describe an item-writing protocol, or simply to repeat the construct label more slowly and loudly (e.g., extroversion), provide non-answers to the question, "How does this instrument work?".

Measurement mechanisms as theoretical claims, made explicit as specification equations, make point predictions under intervention: When one changes (via manipulation or intervention) either the object measure (e.g., reader experiences growth over a year) or measurement mechanism (e.g., increase text complexity measure by 200 L), the result will be a predictable change in the observed outcome. Notice how this process is crucially different from the prediction of the change in the observed outcome based on the selection of another, previously calibrated instrument with known instrument calibrations. Selection is not intervention in the sense used here. Sampling from banks of previously calibrated items is likely to be completely atheoretical, relying, as it does, on empirically calibrated items/instruments. In contrast, to modify the measurement mechanism requires intimate knowledge of how the instrument works. A theoretical psychometrics is characterized by the aphorism "test the predictions, never the postulates" (Jasso, 1988, 4), whereas theory-referenced measurement, with its emphasis on measurement mechanisms, says test the postulates, never the predictions. Those who fail to appreciate this distinction will confuse invariant predictors with genuine causes of observed outcomes.

We assert that a Rasch model combined with a substantive theory embodied in a specification equation provides a more or less complete explanation of how a measurement instrument works (Stenner et al., 1983). A Rasch model in the absence of a specified measurement mechanism is merely a probability model; a probability model absent a theory may be useful for Freedman's (1) and (2), whereas a Rasch model in which instrument calibrations come from a substantive theory that specifies how the instrument works is a causal model; that is, it enables prediction after intervention [i.e., Freedman's (3)].

3   Distinguishing Features of Causal Rasch Models

Admittedly, the measurement model we have proposed for the human sciences mimics key features of physical science measurement theory and practice (Bond & Fox, 2007). Below we highlight several such features.

First, the model is individual-centered. The focus is on explaining variation within person over time. Much has been written about the disadvantages of studying between-person variation with the intent to understand within-person causal mechanisms (Molenaar, 2004; Molenaar & Newell, 2010). Molenaar proved that only under severely restrictive and generally untenable conditions can such cross-level inferences be sustained. In general, in the human sciences, it is necessary to build and test individual-centered models and not rely on variable- or group-centered models (with attendant focus on between person variation) to inform one's understanding of causal mechanisms. Causal Rasch models are individually centered measurement models. The measurement mechanism that transmits variation in the attribute (within-person over time) to the observed outcome (count correct on a reading test) is hypothesized to function the same way for every person (the second ergodicity condition of homogeneity; Molenaar & Newell, 2010).

Second, in this framework, the measurement mechanism is well specified and can be manipulated to produce predictable changes in observed outcomes (e.g., percentage correct). For purposes of measurement theory, a sophisticated philosophy of causal inference is not necessary. For example, questions about the role of human agency in the intervention- and manipulation-based accounts of causal inference are not troublesome here. All that is meant by the claim that the r.h.s. of Eq. (3) causes the l.h.s. is that experimental manipulation of each r.h.s. variable will have a predictable consequence for the observed outcome (expected raw score). Stated more generally, what is meant by x causes y is that an intervention on x yields a predictable change in y. The specification equation used to calibrate instruments/items is a recipe for altering just those features of the instrument/items that are causally implicated in the observed outcome. As noted, we term this collection of causally relevant instrument features the measurement mechanism, which transmits variation in the attribute (e.g., temperature, reading ability) to the observed outcome (number of cavities that turn black or number of reading items answered correctly). Two additional applications of the specification equation are: (a) the maintenance of the unit of measurement independent of any particular instrument or collection of instruments, and (b) bringing non-test behaviors (reading a Harry Potter novel, 980 L) into the measurement frame of reference (Stenner & Burdick, 2011).

Third, item parameters are supplied by substantive theory and thus, person-parameter estimates are generated without reference to or use of any data on other persons or populations. When data fit a Rasch model, a consequence is that item parameters are estimated independent of person parameters and person parameters are estimated independent of item parameters (Andrich, 1989, 2002; Rasch, 1960). In a causal Rasch model, item/instrument calibrations are supplied by a substantive theory and associated specification equation. In the former, the separation is statistical; in the latter it is experimental. Karabatsos (2001) commented on the impossibility of complete separation when the same response data are used to estimate both person measures and item calibrations. Effects of the examinee population are completely eliminated from consideration in the estimation of an individual's person parameter and, thus, no information on other persons is needed because the item/instrument calibrations come from theory.

Fourth, the quantitative hypothesis (Michell, 1999) can be experimentally tested by evaluating the trade-off property for the individual case. A change in the reader parameter can be offset or traded off for a compensating change in text complexity to hold comprehension constant. The trade off is not just about the algebra in Eqs. (1–3). It is about the consequences of simultaneous intervention on the attribute (reader ability) and measurement mechanism (text complexity).

Finally, we conjecture that successful point predictions under intervention necessitate quantitative predictors and outcomes. Concretely, if an intervention on the measurement mechanism (e.g., increase the text complexity of a reading passage by 250 L) results in an accurate prediction of the observed outcome (e.g., how many reading items the reader will answer correctly), and if this process of offsets can be successfully repeated up and down the scale, then text complexity, reader ability, and comprehension (success rate) are quantitative attributes of the text, person, and reader/text encounter, respectively. When text complexity is measured on an ordinal scale, successful point predictions about counts correct based on a reader/text difference are impossible to make. Specifically, successful prediction from differences requires that what is being differenced has the same meaning up and down the respective scales. Differences on an ordinal scale are not meaningful (will lead to inconsistent predictions) precisely because *one more* means something different depending on the location of the text or reader on the scale.

The algebra in Eq. (2) dictates that a change in reader ability can be traded off for an equal change in text complexity to hold comprehension constant. However, to test the quantitative hypothesis requires more than the algebraic equivalence in a Rasch model. Rather, what is required is an experimental intervention/manipulation on either reader ability or text complexity or a conjoint intervention on both simultaneously that yields a successful prediction on the resultant observed outcome (count correct). We maintain that when manipulations of the type just described are introduced for individual reader/text encounters and model predictions are consistent with what is observed, the quantitative hypothesis is sustained.

We emphasize that the above account is individual-centered as opposed to group-centered. The Lexile Framework for Reading purports to provide a causal model for one aspect of what transpires when a reader reads a text. Nothing in the model precludes averaging over readers and texts to summarize evidence for the quantitative hypothesis, but the model can be tested at the individual level. These individual-level tests follow through from Rasch's (1960, 110–115) care in structuring his models in the same mathematical form as Maxwell's analysis of force, mass, and acceleration. By deliberately requiring models of this form, Rasch employs Maxwell's own method of analogy (Nersessian, 2002; Turner, 1955) and enables us to apply it, as Maxwell did, in setting up individual-level tests of hypotheses about potential causal relations among the model parameters (Fisher, 2010).

Maxwell's use of the method of analogy in developing electromagnetic theory has been shown to extend and focus everyday thinking processes into generic scientific model-based reasoning processes (Nersessian, 2006, 2008). In these reasoning processes, formal and structural analogies do not in any way imply content-based analogies. Thus, Rasch's analogy from masses and forces to persons and test items has no psychophysical connotations implying a role for mass and force in the way people respond to assessment questions. The point here extends beyond the present concerns to concept formation in science generally: neither Rasch's models nor Newton's laws are conclusions drawn from observations, being based as they are in reasoning from geometry, other equations, and information (see Crease, 2004a, 2004b for lists of similar mathematical laws).

The analogy is purely formal, and could involve any number of similarly structured relations from a wide variety of other domains in which lawful causal patterns are found, such as chemistry or genetics. No special value beyond their familiarity is to be inferred from the choice of physical constructs in the analogies made. That said, just as pressure and volume can be traded off to hold temperature constant or mass and volume can be traded off to hold density constant, so can reader ability and text complexity be traded off to hold comprehension constant. Michell (1999) made the following remarks on this point: "Identifying ratios directly via trade-offs results in the identification of multiplicative laws between quantitative attributes. This fact connects the theory of conjoint measurement with what Campbell called derived measurement" (204). See Kyngdon (2008) for a particularly clear discussion of this connection.

Garden-variety Rasch models and IRT models are in their application purely descriptive. They become causal and law-like when manipulations of the putative quantitative attributes produce changes (or not) in the observed outcomes that are consistent with model predictions. If a fourth-grade reader grows 100 L in reading ability in 1 year and the text complexity of the student's fifth-grade science text-book also increases by 100 L over the fourth-grade textbook, then the forecasted comprehension rate (whether 60, 70, or 90%) that the reader will enjoy in fifth grade science remains unchanged from that experienced in fourth grade. We offer without proof that only if reader ability and text complexity are quantitative attributes will experimental findings coincide with model predictions such as these. We have tested several thousand students' comprehension of 719 articles that averaged 1150 words in length. Total reading time was 9794 h and the total number of unique machine-generated comprehension items was 1,349,608. The theory-based expectation was 74.53% correct and the observed was 74.27% correct.

A measurement instrument comprises two kinds of features: radicals and inci-dentals (Irvine & Kyllonen, 2002). The *radicals* are those features that transmit variation in the attribute to the observed outcome. Radicals are tunable and when intervened upon, will change what is observed (count correct or number of cavities turning black). *Incidentals* include all features of the measurement instrument that if manipulated will not change the observed outcome. To change sentence length and vocabulary level will alter a text's complexity and will change count correct on an embedded reading test (observed outcome), whereas to change the font style will not. Stenner et al. (1983) and Stone and Wright (1983) proposed that radicals could be organized into a specification equation and could be used to provide theory-based calibrations for items, instruments, and ensembles. A powerful demonstration that a measurement process is under control is provided by a tradeoff between an inter-vention on the attribute measure (e.g., reading ability) for an intervention on the measurement mechanism/specification equation (e.g., text complexity) to hold the observed outcome (relative raw score, percentage correct) constant. Only when an instrument can be tuned to produce a desired change in the observed outcome when holding the attribute measure constant is the measurement mechanism well under-stood. This is, of course, the feature that enables the manufacture of large numbers of instruments that share the same correspondence table linking observed outcome to attribute measure. We call such instruments *strictly parallel*.

We maintain that when data fit a descriptive Rasch model, there is an important sense in which we remain dissatisfied. The source of this dissatisfaction resides in the fact that scientists eschew theories and models that contain many "free parameters...

the values of which are not determined by the theory itself but rather must, as it is commonly expressed, 'be put in by hand'—introduced with no other rationale than they are required by the data" (Woodward, 1989, 364). In descriptive Rasch models (and all IRT models), item and instrument calibrations are estimated from data, whereas in a causal Rasch model, the item, ensemble, or instrument calibrations are provided by theory, thus, dramatically reducing the free parameters in the model and not so coincidentally reducing the sense of arbitrariness.

We cannot overstate the importance of describing the measurement mechanism when explaining why a particular scored outcome was observed or how it came to be. If a Rasch or IRT study is submitted for publication and it makes no attempt to explicate the mechanism or active ingredients that transmit variation in the attribute to the observed outcome, then a truth-in-advertising disclaimer such as, "unfortunately, no mechanism is known to underlay the Rasch equation that we use," should accompany the report.

The simple fact that data fit a Rasch model where the dependent variable is count correct on a test and the predictor is a difference between a person parameter and an instrument parameter does not elicit understanding of the mechanism at work and thus, does not explain: "Because to explain is to exhibit or assume a (lawful) mechanism. This is the process—whether causal, random, or mixed that makes the system work the way it does…. This kind of explanation is usually called mechanistic. I prefer to call it mechanismic, because most mechanisms are not mechanical" (Bunge, 2004, 203). Without a mechanism (modeled as a specification equation), a Rasch model is unsatisfyingly functional and descriptive rather than mechanismic and explanatory. If editors embraced this position, attention might shift to explications of the mechanisms that underlie the human and social science instrumentarium.

The role of causal inference in human science measurement theory has been underdeveloped in part because causal inference is philosophically complex and more specifically the so-called mechanismic interpretation of measurement (Stenner et al., 2009) has lived in the shadow of sampling-based frameworks such as facet theory (Guttman, 1971), generalizability theory (Brennan, 2011), true score theory (Lord & Novick, 1968) and behavior domain theory (McDonald, 2009).

4 Illustrating the Tradeoff Property

In adopting the tradeoff property as a useful test for quantity in the human and social sciences, we reasoned that the symmetry of a Rasch model lends itself to thought about offsetting manipulations on the r.h.s. of Eq. (3) producing no change to the l.h.s. Concretely, a manipulation that increases text complexity by 200 L, if offset by an increase of 200 L in reader ability, should yield no change in observed comprehension rate. So, offsetting manipulations in two distinct attributes, text complexity and reader ability, can be experimentally shown to hold constant a third attribute (comprehension). If this tradeoff property holds up and down the scales for all three variables in the Lexile equation, the attributes (text complexity, reader ability, and comprehension) we conjecture must be quantitative attributes of text, reader, and the reader-text encounter, respectively. As we have seen, a particularly attractive feature of this approach to testing the quantitative hypothesis is that one can perform the test within-person with no reference to any between-person relations. Specifically, over a 13-year period (grades K-12), a reader is growing in reading ability and the computer trades off growth for new texts that have just the right amount of added complexity to hold the comprehension rate constant. Over the 13 years, the reader may read thousands of articles and millions of words, but the whole history can be summarized by the expected minus observed count correct. We assert without proof that only if text complexity, reader ability, and comprehension are quantitative, will we consistently observe a close correspondence between expected (under the theory) and observed count correct on machine-generated cloze items. Because one always wants the quantitative hypothesis to be sustained within-person over time, it seems best to test the hypothesis at the individual level and not resort to cross-level inferences (e.g., attempts to infer from between-person relationships something about within-person processes), which often find dubious rationales (Molenaar & Newell, 2010).

Figure 1 is an individual-centered growth trajectory for reading ability denominated in Lexiles. Student 1528 is a seventh grade male who read 347 articles (138,695 words) between May 2007 and April 2011. Each solid dot corresponds to a monthly average Lexile measure. The growth trajectory fits the data quite well, and this young man is forecasted (big dot on the far right of the figure) to be a college-ready reader when he graduates from high school. The open dots distributed around O on the horizontal axis are the expected performance minus observed performance for each month. Expected performance is computed using the Rasch model and inputs for text complexity and the reader's ability measure. Given these inputs, the apparatus forecasts a percentage correct. The observed performance is the observed percentage correct for the month. The difference between what the substantive theory (Lexile Reading Framework) in cooperation with the Rasch model expects and what is actually observed is plotted by month. The upper left-hand corner of the graphic summarizes the expected percentage correct (73.5%) and observed percentage correct (71.7%) across the 3342 items taken by this reader during 4 years. What may not be immediately obvious is that the apparatus is dynamically matching text complexity to the developing reader's ability to hold comprehension (percentage correct) at 75%. So, this graphic describes a within-person (intra-individual) test of the quantitative hypothesis: Can the apparatus trade off a change in reader ability for a change in text complexity to hold constant the success rate (comprehension)?

When a wide range of data fit a causal Rasch model, does the model assume the status of a law such as $F = MA$? Although the lines of demarcation between invariant generalizations and laws are difficult to draw, one feature seems paramount. How phenomena relate to one another must be independent of the particular mechanism used to measure each phenomenon. This independence presumes that multiple mechanisms exist for measuring each phenomenon that figures in the law. $F = MA$ would not be a law if it mattered how exactly mass were measured (what kind of weighing apparatus was employed) and similarly for acceleration and force. In fact, the very

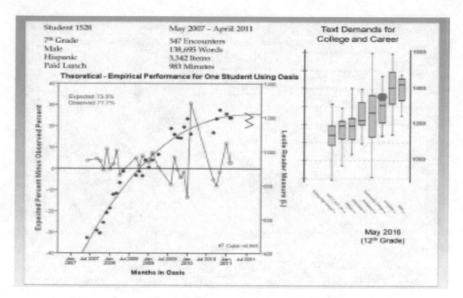

Fig. 1 Growth in reading ability relative to the reading demands of adulthood

claim for existence of a phenomenon (reader ability or text complexity) depends upon the fact that *different calibrated mechanisms return the same amounts.* Without demonstrated invariance of a phenomenon over measurement mechanisms, a counterclaim remains open that the hypothesized phenomenon is artifactually dependent on one particular instrument or measurement mechanism. We conjecture that laws are laws, in part, because they are invariant under changes in measurement mechanism(s).

So, what does it mean to claim that the r.h.s. of Eq. (3) is causal on the l.h.s. and thus, that the causal operator (=:) should be used in place of (=). Borrowing from Woodward (2003), we mean that an intervention or manipulation of the measurement mechanism and/or the attribute measure changes the observed outcome. The mathematical model explains how observed outcomes (counts correct on a reading test) are dependent on the measurement mechanism (text complexity and task type) and measured attribute (reader ability). The substantive theory specifies precisely what kinds of interventions on the object of measurement and measurement mechanism will change the observed outcome and, by omission, what interventions should have no effect on the observed outcome. In this sense, measurement as envisioned in the standard measurement model is about manipulation and control and not about correlation, description, and classification.

Figure 2 presents the results of a 5-year study of the relationship between theoretical text complexity as provided by a computer based text complexity engine (Lexile Analyzer) and empirical text complexity as provided by the EdSphere™ platform. The Lexile Analyzer computes the semantic demand of a text proxied by the log transformed frequency of each word's appearance in a multibillion word corpus of published text and the syntactic demand of each text proxied by log transformed mean

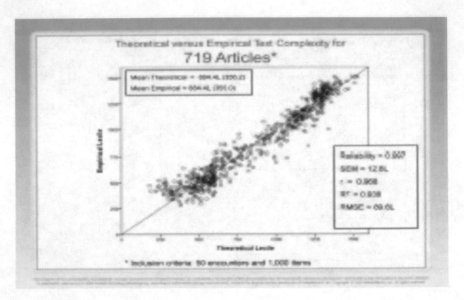

Fig. 2 Predicted versus observed text complexity measure

sentence length. The text preprocessing, what constitutes a *word* and what constitutes a sentence ending, involves thousands of lines of code. Modern computing enables the measurement of Tolstoy's *War and Peace* in a couple of seconds. The Lexile Analyzer has taken 25 years to build and optimize and is freely available for non-commercial use (Lexile.com).

The EdSphere™ platform enables students to select articles of their choosing from a vast range of content. Selected articles are targeted to ±100 L of each student's developing reading ability. Thus, as students' reading ability grows, the machine adjusts the text complexity of the articles from which the student chooses the next reading. The target success rate is 75%. The machine generates a reading comprehension item on the fly about every 70 words such that two students sitting side by side at computers and reading the same article will respond to different items.

Because only very rarely will more than one student take any particular item, a new kind of Rasch model was needed. The *ensemble* Rasch model exploits the raw score sufficiency property of Rasch models to convert counts correct on unique sets of items into Lexiles (Lattanzio et al., 2012; Stenner et al., 2006). The text complexity measure for the present article is 1480 L, which also is the mean difficulty of all the allowable cloze reading items that could be machine constructed for this article. Given this ensemble mean and a theory-specified ensemble variance, counts correct can be converted into a Lexile measure by sampling item calibrations from the ensemble and using these item calibrations to solve the Rasch equation. The key intuition that makes this possible is that it does not matter which particular sampled item calibrations are attached to which particular machine-generated items because a consequence of raw score sufficiency is that there is no information in the pattern of rights and wrongs about the reader parameter; thus, it does not matter how the sampled item calibrations are attached to individual items.

The 719 articles in this study were chosen because they were the first articles to meet the dual requirements of at least 50 readers and at least 1000 item responses. Well-estimated reader measures were available prior to an encounter between an article and a reader. The ensemble Rasch equation (not given) was rearranged to solve for text complexity given a count correct, a reader ability, and an ensemble variance. Thus, each of the 719 articles had a theoretical text complexity from the Lexile Analyzer and an empirical text complexity from EdSphere™. The correlation between theoretical text complexity and empirical text complexity is $r = 0.968$ ($r^2 = 0.938$). Reliability of the empirical text complexity measures is $r_{tt} = 0.997$ and the $RMSE = 89.6$ L. The RMSE is the square root of the average squared difference between the empirical and theoretical text complexities. Work has been ongoing for two decades to isolate new linguistic variables that might account for the 6% of uncontrolled variance. More recent work has shifted to a focus on possible artifacts (measurement error, range restriction, text preprocessor variation, and task specificity) that might account for the small remaining uncontrolled variation.

When data fit a descriptive Rasch model, relative differences between objects of measurement on the attribute of interest are independent of the instrument. When data fit a causal Rasch model in which instrument calibrations (item difficulty estimates) come from theory, then absolute measures are independent of instrument. Rasch called the former *specific objectivity*, and we called the latter *general objectivity* (Stenner, 1994). Both the temperature and reading attributes used as illustrations in this paper have available, well-developed, generally objective measurement systems. Equation (3) is an individual-centered measurement model (token causal model) and can be distinguished from SEM models and much philosophical discussion on causation that focuses on type causal models.

There is an unfortunate tendency in reading research and psychometrics to equate changes in measurement mechanism (e.g., task type) with change in the attribute being measured. For example, it is asserted that multiple-choice tests of reading must measure something other than short-answer, constructed responses because the former measure superficial, fact-based understanding and the latter measure higher level inferences. Thirty years ago, reading part scores were reported for subtests that mistakenly were thought to measure something else because the questions focused on different levels of Bloom's taxonomy. The idea that a reader constructs a meaning model and that all reading items interrogate that meaning model (Kintsch, 1974) and that different item types might measure the same attribute (reader ability) but with different added easiness or difficulty relative to some standard task type was a foreign notion. Adoption of the Lexile Reading Framework was resisted in part because it appeared to be too simple. The assertion that reading ability, like temperature, is a unidimensional attribute that can be measured with many mechanisms that appear different has been slow to take hold. That said, some measurement mechanisms may confound two or more attributes. Mercury in a tube thermometer without a top confounds the measurement of temperature with the measurement of atmospheric pressure. To seal the top of the thermometer eliminates the confound. Similarly, to ask readers to summarize in writing what they have just read may confound reading ability and writing ability. Both attributes of language facility are important, but for purposes of measurement, it is important to have mechanisms that transmit only one kind of variation and not a confounded mixture of two or more distinct kinds of variation.

5 Instrument Validity and Instrument Validation

Consider the following statements about validity and ponder the theoretical and practical implications that might follow for the measurement, respectively, of temperature and reading ability:

> Validation of an instrument calls for an integration of many types of evidence. The varieties of investigation are not alternatives any one of which would be adequate. The investigations supplement one another.... For purposes of exposition, it is necessary to subdivide *what in the end must be a comprehensive, integrated evaluation of the test.* (Cronbach, 1971, 445; emphasis in original)

> [Validity is] an integrated evaluative judgement of the degree to which empirical evidence and theoretical rationales support and *adequacy* and *appropriateness* of *inferences* and *actions* based on test scores or other modes of assessment. (Messick, 1989, 13; emphasis in original)

> Validation involves the evaluation of the proposed interpretations and uses of measurements. The interpretive argument provides an explicit statement of the inferences and assumptions inherent in the proposed interpretations and uses. The validity argument provides an evaluation of the coherence of the interpretive argument and of the plausibility of its inferences and assumptions. It is not the test that is validated and it is not the test scores that are validated. It is the claims and decisions based on the test results that are validated. Therefore, for validation to go forward, it is necessary that the proposed interpretations and uses be clearly stated. (Kane, 2006, 59–60)

So, from the above, validity calls for synthesis or integration of diverse sources of evidence. It is useful, for purposes of exposition, to group these sources. Not mentioned in the quotes above but described elsewhere are dozens of validity categories [face, criterion, concurrent, convergent, discriminant, predictive, construct, and 200 more documented by Shaw and Newton (2012) and Newton and Shaw (2012)]. For Cronbach, Messick, and Kane validity does not apply to the instrument (thermometer or reading test) nor just to the numbers produced by these instruments but rather to the claims and decisions that are made by users based on these numbers.

Even a casual reading of the history of science reveals a stunning disconnect between the quotations above and how instruments are invented, improved, and justified in the physical sciences. What holds center stage in physical science measurement is substantive theory (Maraun, 1998; Sherry, 2011) followed by intense focus on precision of measurement and to a lesser degree by an imperative for readable technologies (directly perceivable quantities for the latent variable being measured). Great care is taken in physics, for example, to separate what is being measured by an instrument from the uses to which measurements might be put and from the consequences of these uses. It is not that these consequential considerations are unimportant but that they have nothing to do with what an instrument measures. Validity, for us, is all about what an instrument measures.

Contrast the quotes from Cronbach, Messick, and Kane with those of Stenner et al., (1983) and Borsboom (2005):

> The notion that a test is valid if it measures what it purports to measure implies that we have some independent means of making this determination. Of course, we usually do not have an independent standard; consequently, validation efforts devolve into circular reasoning where the circle generally possesses an uncomfortably small circumference. Take, for example, Nunnally's (1978) statement, "A measuring instrument is valid if it does what it is intended to do" (86). How are we to represent intention independent of the test itself? In the past, educators and psychologists have been content to represent intentions very loosely, in many cases letting the construct label and its fuzzy connotations reconstruct the intentions of the test developers. Unfortunately, when intentions are loosely formulated, it is difficult to compare attainment with intention. This is the essence of the problem faced by classical approaches to validity. Until intentions can be stated in such a way that attainment can be explicitly tested, efforts to assess the adequacy of a measurement procedure will be necessarily characterized by post hoc procedures and interpretations. (Stenner et al., 1983, 31)

> Thus, validity theory has gradually come to treat every important test-related issue as relevant to the validity concept, and aims to integrate all these issues under a single header. In doing so, however, the theory fails to serve either the theoretically oriented psychologist or the practically inclined tester; the theoretically oriented are likely to get lost in the intricate subtleties of validity theory, while the practically oriented are unlikely to derive a workable conceptual scheme with practical implications from it. (Borsboom, 2005, 150)

If we follow this thread, where does it lead? First, let's observe that the past 400 years of successful physical science have managed with only the rudiments of a formal measurement theory. Progress in every physical science measurement armament has been realized from improved theory and improved engineering in the service of ever-increasing precision of measurement, and that is about it. Human science measurement, in contrast, has expended virtually no energy on substantive theory development, nor on increasing, say by an order of magnitude, the precision of ability, personality, or attitude measurement. Rather, we psychometricians expend much energy on mathematizing everything we do in a kind of caricature of physical science measurement without grasping the significance of just what physical science is accomplishing (Maraun, 1998; Sherry, 2011).

At present, much of the human science is too frequently actuarially driven rather than theory-driven. Consider the multitrillion dollar insurance industry as an analogy. The foundations of the industry are highly mathematized and predictions in the aggregate are highly precise, but there is no body of increasingly integrated substantive theory. Actuarial science has given us some wonderfully useful mathematics and in turn has found uses for some elegant pure mathematics. This sophisticated mathematization operates in the service of a simple criterion: prediction of how a population (humans, container ships, orange trees) will fare next year or next decade. As useful as this may be within the limits of the insurance industry's interests, theory-driven science optimizes a completely different kind of criterion.

The notion that validity is about the conformance of what an instrument actually measures to what the developers intended or what they *purport* to measure is vacuous unless *intent* can be independently formulated as, for example, in a specification equation. Too often in practice the instrument is offered as evidence both of intent and attainment of that intent. The specification equation breaks the relationship between intention and attainment by independently formalizing intent as an equation that can produce item calibrations or ensemble means that are in close correspondence with empirical estimates. Instruments are designed to detect variation of a kind. The specification equation specifies the kind and provides a means by which to test whether the variation detected by the instrument aligns with that specified by the construct theory:

> The construct specification equation affords a test of fit between instrument-generated observations and theory. Failure of a theory to account for variation in a set of empirical item scale values invalidates the instrument as an operationalization of the construct theory and limits the applicability of that theory. Competing construct theories and associated specification equations may be suggested to account for observed regularity in item scale values. Which construct theory emerges (for the time) victorious, depends upon essentially the same norms of validation that govern theory evaluation in other sciences. (Stenner et al., 1983, 307)

Everything important to instrument justification cannot be subsumed under the validity banner. Validity is not constituted by a bucket of correlations. It was these realizations among other deficiencies in the concept that led us to reformulate validity as a correspondence between intent and attainment where intent is formalized as a specification equation. Borsboom (2005) said it best:

> The argument to be presented is exceedingly simple, so simple, in fact, that it articulates an account of validity that may seem almost trivial. It is this: if something does not exist, then one cannot measure it. If it exists, but does not causally produce variations in the outcomes of the measurement procedure, then one is either measuring nothing at all or something different altogether. Thus, a test is valid for measuring an attribute if and only if (a) the attribute exists, and (b) variations in the attribute causally produce variations in the outcomes of the measurement procedure.
>
> The fact that the crucial ingredient of validity involves the causal effect of an attribute on the test scores implies that the locus of evidence for validity lies in the processes that convey this effect. This means that tables of correlations between test scores and other measures cannot provide more than circumstantial evidence for validity. What needs to be tested is not a theory about the relation between the attribute measured and other attributes, but a theory of response behavior. Somewhere in the chain of events that occurs between item administration and item response, the measured attribute must play a causal role in determining what value the observed outcomes will take; otherwise the test cannot be valid for measuring the attribute. Importantly, this implies that the problem of validity cannot be solved by psychometric techniques or models alone. On the contrary, it must be addressed by substantive theory. (151)

It is often difficult to conduct experimental manipulation (in the short term) of the latent variable (e.g., reader ability) of interest. We can and do, however, exploit the symmetry in the Rasch model and experimentally manipulate the items (e.g., text complexity) to produce the observed outcome (e.g., success rate or comprehension) expected under the construct theory. Manipulation of the text by the use of the specification equation results in changes consistent with theoretical expectations. The specification equation is one vehicle by which to introduce substantive theory into a psychometric model. The potential of this approach can perhaps be grasped most easily by setting up an equivalence relation. Suppose one wants to set the observed outcome (e.g., count correct on a reading test) equal for two readers who differ in reading ability. One could use the specification equation to build and calibrate items for the better reader that are just enough more difficult than the easier items given to the less able reader to cancel the way the first reader's ability surpasses that of the second reader. This trade off, or cancellation property, characterizes additive conjoint measurement frameworks (Kyngdon, 2008; Michell, 2005). In the Lexile Framework for Reading, a difference in reader ability of 200 L can be traded off for a difference in text readability of 200 L to hold constant the comprehension rate (Burdick et al., 2006). Relying only on substantive theory, the specification equation enables these causal manipulations.

In summary, validity is a straightforward concept with a specific meaning; it is not an amalgamation of dozens of kinds of evidence. Validity does not reduce to the scientific method and there is nothing in need of unification in the concept. The usefulness of measures and the consequences (intended and unintended) of their use, although important, have nothing to do with validity. In fact, "validity is not a judgment at all. It is the property being judged" (Borsboom, 2005, 154). When asserting that an instrument is a valid measure of an attribute, one makes an ontological claim that the attribute exists (a realist stance), that the instrument detects variation in the attribute, and that experimental manipulation of the attribute or the instrument (mechanism) will cause theoretically expected changes in the observed outcome (count correct).

If a measurement instrument's primary purpose is to detect variation of a kind then the paramount validity question is what causes the variation detected by the instrument. The specification equation provides an answer to this question by formalizing what it means to move up and down a scale for an attribute.

How well empirical item difficulties or ensemble means align with theoretical calibrations is a matter of degree (Stenner et al., 2006). Alternative construct theories and their attendant formalizations as specification equations compete in accounting for the variation detected by an instrument. As such, we adopt a causal rather than a correlational view of validity. We accentuate the role of substantive theory in our approach and have advanced the term *theory-referenced measurement* to label the method (Stenner et al., 2009).

6   Conclusion

Causal Rasch models expose the mechanism that transmits attribute variation to the observed outcome. Specification equations are one useful way to expose and express a mechanism's action. Although a specification begins its life as a local descriptive account of specifically objective item/instrument variation on a single form of a test, it can evolve into a universal causal account of the behavior of items/instruments previously used to measure an attribute and all items and instruments that might yet be manufactured. The specification equation bridges the here and now of the test in hand and the infinity of possible instruments that can be engineered and manufactured for the measurement of an attribute. The equation provides not only a specification for instrument manufacture but also yields the calibrations that can be used to convert counts into quantities.

So, an early use of the specification equation is to operationalize Rasch's (1960) notion of a frame of reference in a way that extends the frame beyond the specific objectivity obtained in the context of one or two tests for an attribute to an indefinitely large collection of actual and imagined instruments. Theory-based instrument calibration eliminates the need to use data both to calibrate instruments and to measure persons. The payoff of using theory instead of data to calibrate instruments is large and immediate. When data fit a Rasch model, differences among person measures are free of dependencies on other facets of the measurement context (i.e., differences are specifically objective). When data fit a causal Rasch model, absolute person measures are free of the conditions of measurement (items, occasions, contexts) and thus, absolute measures are objective. Under a causal Rasch model, attribute measures (temperature or reading ability) are individually centered statistics precisely because no reference to other person(s) data figures in their estimation.

Applications of descriptive Rasch models have (over the last 50 years) identified thousands of data sets that fit. For only a handful of these has there been an attempt to explicate the mechanism that transmits attribute variation to the observed outcome. We do not, however, want to miss the real possibility that there exist mechanisms for a non-negligible subset of these applications and that some mechanisms will explain observed outcomes across multiple applications. A starting point for this search might begin with the question, "How does this instrument work?" What are the characteristics of the items/instrument that, if experimentally manipulated, could be expected to change the observed outcome? Even tentative answers to these questions will move psychometrics into a closer alignment with the standard measurement model (Fisher & Stenner, 2011, 2012).

There has been a groundswell of negativity regarding the potential of psychology to realize *measurement* as that term is understood in the physical sciences (Barrett, 2008; Cliff, 1992; Grice, 2011; Michell, 2000; Schönemann, 1994). More recently, Trendler (2009) concluded, "Psychological phenomena are not sufficiently manageable. That is, they are neither manipulable nor are they controllable to the extent necessary for an empirically meaningful application of measurement theory. Hence they are not measurable" (592). For Trendler, psychology cannot have measurement because it cannot manipulate and control its constructs. This paper offers a demonstration that psychological measurement might be possible. Furthermore, a roadmap is presented for the realization of measurement for those attributes that, like early conceptions of temperature, are today merely ordinal.

Acknowledgements We are grateful to the Bill and Melinda Gates Foundation for its support of this work.

References

Andrich, D. (1989). Distinctions between assumptions and requirements in measurement in the social sciences. In J. A. Keats, R. Taft, R. A. Heath, & S. H. Lovibond (Eds.), *Mathematical and Theoretical Systems: Proceedings of the 24th International Congress of Psychology of the International Union of Psychological Science* (Vol. 4, pp. 7–16). Elsevier Science Publishers.

Andrich, D. (2002). Understanding resistance to the data-model relationship in Rasch's paradigm: A reflection for the next generation. *Journal of Applied Measurement, 3*, 325–359.

Barrett, P. (2008). The consequence of sustaining a pathology: Scientific stagnation. A commentary on the target article, "Is psychometrics a pathological science?" by Joel Michell. *Measurement, 6*, 78–123.

Bond, T. G., & Fox, C. M. (2007). *Applying the Rasch model*. Routledge.

Borsboom, D. (2005). *Measuring the mind: Conceptual issues in contemporary psychometrics*. Cambridge University Press. https://doi.org/10.1017/CBO9780511490026

Brennan, R. L. (2011). *Generalizability theory*. Springer.

Bunge, M. (2004). How does it work. *Philosophy of the Social Sciences, 34*, 182–210. https://doi.org/10.1177/0048393103262550

Burdick, D. S., Stone, M. H., & Stenner, A. J. (2006). The combined gas law and a Rasch reading law. *Rasch Measurement Transactions, 20*, 1059–1060.

Castles, A., Datta, H., Gayan, J., & Olson, R. K. (1999). Varieties of developmental reading disorder: Genetic and environmental influences. *Journal of Experimental Child Psychology, 72*, 73–94. https://doi.org/10.1006/jecp.1998.2482

Chang, H. (2004). *Inventing temperature*. Oxford University Press. https://doi.org/10.1093/019517 1276.001.0001

Cliff, N. (1992). Abstract measurement theory and the revolution that never happened. *Psychological Science, 3*, 186–190. https://doi.org/10.1111/j.1467-9280.1992.tb00024.x

Cook, T., & Campbell, D. (1979). *Quasi-experimentation: Design and analysis issues for field settings*. Houghton Mifflin.

Crease, R. (2004a). The greatest equations ever. *Physics World, 17*, 14.

Crease, R. (2004b). The greatest equations ever (Part 2). *Physics World, 17*, 19.

Cronbach, L. J. (1971). Test validation in educational measurement. In R. L. Thorndike (Ed.), *American Council on Education* (2nd ed., pp. 443–507). American Council on Education.

Elster, J. (1989). Social norms and economic theory. *Journal of Economic Perspective, 3*, 99–117. https://doi.org/10.1257/jep.3.4.99

Fisher, W. P., Jr. (2010). The standard model in the history of the natural sciences, econometrics, and the social sciences. *Journal of Physics: Conference Series, 238*. http://iopscience.iop.org/ 1742-6596/238/1/012016/pdf/1742-6596_238_1_012016.pdf

Fisher, W. P., Jr., & Stenner, A. J. (2011). *A technology roadmap for intangible assets metrology*. Paper presented in session on Fundamentals of Measurement Science, International Measurement Confederation Joint Symposium (Jena). http://ssrn.com/abstract=1925817

Fisher, W. P., Jr., & Stenner, A. J. (2012). *Metrology for the social, behavioral, and economic sciences* (Social, Behavioral, and Economic Sciences White Paper Series). Retrieved January 10, 2013, from http://www.nsf.gov/sbe/sbe_2020/submission_detail.cfm?upld_id=36

Freedman, D. A. (1997). From association to causation via regression. In V. R. McKim & S. P. Turner (Eds.), *Causality in crisis?* (pp. 113–161). University of Notre Dame Press.

Gay-Lussac, J. L. (1802). Enquiries concerning the dilation of the gases and vapors [Nicholson's]. *Journal of Natural Philosophy, Chemistry, and the Arts, 3*, 207–216, 257–267.

Grice, J. W. (2011). *Observation oriented modeling*. Academic Press.

Guttman, L. (1971). Measurement as structural theory. *Psychometrika, 36*, 329–341. https://doi.org/10.1007/BF02291362

Hanlon, S. T., Swartz, C. W., Stenner, A. J., Burdick, H., & Burdick, D. S. (2012). *Learning oasis*. www.alearningoasis.com

Hausman, D. (1998). *Causal asymetrics*. Cambridge University Press. https://doi.org/10.1017/CBO 9780511663710

Hebra, A. (2010). *The physics of metrology: All about instruments: From trundle wheels to atomic clocks*. Springer Wien.

Hedström, P. (2005). *Dissecting the social: On the principles of analytic sociology*. Cambridge University Press. https://doi.org/10.1017/CBO9780511488801

Heilbron, J. L. (1993). *Weighing imponderables and other quantitative science around 1800*. University of California Press.

Hobart, J. C., Cano, S. J., Zajicek, J. P., & Thompson, A. J. (2007). Rating scales as outcome measures for clinical trials in neurology: Problems, solutions, and recommendations. *Lancet Neurology, 6*, 1094–1105. https://doi.org/10.1016/S1474-442270290-9

Holland, P. (1986). Statistics and causal inference. *Journal of American Statistical Association, 81*, 945–960. https://doi.org/10.1080/01621459.1986.10478354

Irvine, S. H., & Kyllonen, P. C. (2002). *Item generation for test development*. Lawrence Erlbaum Associates.

Jasso, G. (1988). Principles of theoretical analysis. *Sociology Theory, 6*, 1–20. https://doi.org/10. 2307/201910

Kane, M. T. (2006). Validation. In R. L. Brennan (Ed.), *Educational measurement* (4th ed., pp. 18–64). American Council on Education/Praeger.

Karabatsos, G. (2001). The Rasch model additive conjoint measurement and new models of probabilistic measurement theory. *Journal of Applied Measurement, 2*, 389–423.

Kintsch, W. (1974). *The representation of meaning in memory*. Erlbaum.

Kyngdon, A. (2008). The Rasch model from the perspective of the representational theory of measurement (with commentary). *Theory & Psychology, 18*, 89–109. https://doi.org/10.1177/ 0959354307086924

Lattanzio, S., Burdick, D., & Stenner, A. J. (2012). *The ensemble Rasch model*. MetaMetrics Paper Series.

Lord, F. M., & Novick, M. R. (1968). *Statistical theories of mental test scores*. Addison-Wesley.

Luce, R. D. (1998). Coalescing, event commutativity and theories of utility. *Journal of Risk and Uncertainty, 16*, 87–114. https://doi.org/10.1023/A:1007762425252

Maraun, M. D. (1998). Measurement as a normative practice: Implications of Wittgenstein's philosophy for measurement in psychology. *Theory & Psychology, 8*, 435–461. https://doi.org/10.1177/ 0959354398084001

McDonald, R. P. (2009). Behavior domains in theory and in practice. *Alberta Journal of Educational Research, 49*, 3.

Medical Indicators. (2006). www.medicalindicators.com/pdf/NT-FC-Tech-bulletin.pdf

Messick, S. (1989). Validity. In R. L. Linn (Ed.), *Educational measurement* (3rd ed., pp. 13–103). Collier Macmillan.

Michell, J. (1997). Quantitative science and the definition of measurement in psychology. *British Journal of Psychology, 88*, 355–383. https://doi.org/10.1111/j.2044-8295.1997.tb02641.x

Michell, J. (1999). *Measurement in psychology*. Cambridge University Press. https://doi.org/10. 1017/CBO9780511490040

Michell, J. (2000). Normal science, pathological science and psychometrics. *Theory & Psychology, 10*, 639–667. https://doi.org/10.1177/0959354300105004

Michell, J. (2005). The logic of measurement: A realist overview. *Measurement, 39*, 285–294. https://doi.org/10.1016/j.measurement.2005.09.004

Michell, J. (2007). *Bergson's and Bradley's version of the psychometricians' fallacy argument*. Paper presented at the First Joint Meeting of ESHHS and CHEIRON. University College Dublin.

Michell, J. (2012). Alfred Binet and the concept of heterogeneous orders. *Frontiers in Psychology, 3*, 261–273. https://doi.org/10.3389/fpsyg.2012.00261

Molenaar, P. C. M. (2004). A manifesto on psychology as ideographic science: Bringing the person back into scientific psychology, this time forever. *Measurement: Interdisciplinary Research and Perspectives, 2*, 201–218. https://doi.org/10.1207/s15366359mea0204_1

Molenaar, P. C. M., & Newell, K. M. (2010). *Individual pathways of change*. American Psychological Association. https://doi.org/10.1037/12140-000

Nersessian, N. J. (2002). Maxwell and "the method of physical analogy": Model-based reasoning, generic abstraction, and conceptual change. In D. Malament (Ed.), *Essays in the history and philosophy of science and mathematics* (pp. 129–166). Open Court.

Nersessian, N. J. (2006). Model-based reasoning in distributed cognitive systems. *Philosophy in Science, 73*, 699–709. https://doi.org/10.1086/518771

Nersessian, N. J. (2008). *Creating scientific concepts*. MIT Press.

Newton, P. E., & Shaw, S. D. (2012). *We need to talk about validity*. Paper presented at the National Council on Measurement in Education, Annual Meeting.

Nunnally, J. C. (1978). *Psychometric theory*. McGraw-Hill.

Pearl, J. (1999). *Reasoning with cause and effect*. Paper presented at the International Joint Conference in Artificial Intelligence. https://bayes.cs.ucla.edu/IJCAI99/ijcai-99.pdf

Pearl, J. (2000). *Causality: Models, reasoning, and inference*. Cambridge University Press.

Rasch, G. (1960). *Probabilistic models for some intelligence and attainment tests* (Reprint, with Foreword and Afterword by B. D. Wright, University of Chicago Press, 1980). Danmarks Paedogogiske Institut.

Roche, J. (1998). *The mathematics of measurement: A critical history*. The Athlone Press.

Schönemann, P. E. (1994). Measurement: The reasonable ineffectiveness of mathematics in the social sciences. In I. Borg & P. Mohler (Eds.), *Trends and perspectives in empirical research* (pp. 149–160). Walter de Gruyter. https://doi.org/10.1515/9783110887617.149

Shaw, S. D., & Newton, P. E. (2012). *Cracks in construct validity theory*. Paper presented at the National Council on Measurement in Education Annual Meeting.

Sherry, D. (2011). Thermoscopes, thermometers, and the foundations of measurement. *Studies in History and Philosophy of Science, 42*, 509–524. https://doi.org/10.1016/j.shpsa.2011.07.001

Smith, R. M. (2000). Fit analysis in latent trait measurement models. *Journal of Applied Measurement, 1*, 199–218.

Stenner, A. J. (1994). Specific objectivity: Local and general. *Rasch Measurement Transactions, 8*, 374–375.

Stenner, A. J., & Burdick, D. S. (2011). Can psychometricians learn to think like physicists? *Measurement: Interdisciplinary Research and Perspectives, 9*, 62–63. https://doi.org/10.1080/15366367.2011.558797

Stenner, A. J., & Smith, M., III. (1982). Testing construct theories. *Perceptual and Motor Skills, 55*, 415–426. https://doi.org/10.2466/pms.1982.55.2.415

Stenner, A. J., Smith, M. I. I. I., & Burdick, D. S. (1983). Toward a theory of construct definition. *Journal of Educational Measurement, 20*, 305–316. https://doi.org/10.1111/j.1745-3984.1983.tb00209.x

Stenner, A. J., Burdick, H., Sanford, E. E., & Burdick, D. S. (2006). How accurate are Lexile text measures? *Journal of Applied Measurement, 7*, 307–322.

Stenner, A. J., Stone, M. H., & Burdick, D. S. (2009). The concept of a measurement mechanism. *Rasch Measurement Transactions, 23*, 1204–1206.

Stone, M. H. (2002). *Knox's cube test-revised*. Stoelting.

Stone, M. H., & Wright, B. D. (1983). Measuring attending behavior and short-term memory with Knox's cube test. *Educational and Psychological Measurement, 43*, 803–814. https://doi.org/10.1177/001316448304300315

Trendler, G. (2009). Measurement theory, psychology and the revolution that cannot happen. *Theory & Psychology, 19*, 579–599. https://doi.org/10.1177/0959354309341926

Turner, J. (1955). Maxwell on the method of physical analogy. *The British Journal for the Philosophy of Science, 6*, 226–238. https://doi.org/10.1093/bjps/VI.23.226

von Winterfeldt, D., Chung, N. K., Luce, R. D., & Cho, Y. (1997). Tests of consequence: Monotonicity in decision making under uncertainty. *Journal of Experimental Psychology: Learning, Memory, and Cognition, 23*, 406–426. https://doi.org/10.1037/0278-7393.23.2.406

Woodward, J. (1989). The causal/mechanical model of explanation. In W. Salmon & P. Kitcher (Eds.), *Minnesota studies in the philosophy of science, Vol. 13 Scientific explanation* (pp. 359–383). University of Minnesota Press.

Woodward, J. (2003). *Making things happen*. Oxford Press.

Comparison Is Key

Mark H. Stone and A. Jackson Stenner

Abstract Several concepts from Georg Rasch's last papers are discussed. The key one is *comparison* because Rasch considered the method of comparison fundamental to science. From the role of comparison stems scientific inference made operational by a properly developed frame of reference producing *specific objectivity*. The exact specifications Rasch outlined for making comparisons are explicated from quotes, and the role of causality derived from making comparisons is also examined. Understanding causality has implications for what can and cannot be produced via Rasch measurement. His simple examples were instructive, but the implications are far reaching upon first establishing the key role of comparison.

Rasch (1977) addressed science and objectivity in his last published paper: *On Specific Objectivity: An Attempt at Formalizing the Request for Generality and Validity of Scientific Statements*. In Section III (p. 68), he rhetorically asked, "What is Science?" His answer specified two conditions. Science is:

1. Making comparisons, and
2. Making these comparisons objectively.

Rasch (1977) further explained these two conditions.

Two features seem indispensible in scientific statements. They deal with comparisons, and the comparisons should be objective. However, to complete these requirements I have to *specify the kind of comparisons and the precise meaning of objectivity.* (p. 68, our italics)

Rasch (1977) follows the above quote by another one taken from an earlier paper (Rasch, 1967) but with each part given an important heading:

Journal of Applied Measurement, 15(1). 2014.

M. H. Stone (✉)
Departments of Mathematics, Psychology, and Social Work at Aurora University, Chicago, USA

A. J. Stenner
MetaMetrics, Inc., Durham, NC, USA

© The Author(s) 2023
W. P. Fisher and P. J. Massengill (eds.), *Explanatory Models, Unit Standards, and Personalized Learning in Educational Measurement*,
https://doi.org/10.1007/978-981-19-3747-7_19

Specifying Comparisons. Consider a class of 'objects' to be mutually compared. The sense in which they should be compared is specified through a class of 'agents,' to each of which each object may be 'exposed.' On each exposure an 'observation'—quantitative or qualitative—is made. The whole set of such observations made when a finite number of objects of $O_1...,O_n$ are exposed to a finite number of agents $A_1..., A_k$ from the data from which comparisons of the O's...to such agents as the A's can be inferred reactions.

Specifying Objectivity. Now, within this framework, which I have taken from psychophysics, the 'objectivity' of a comparative statement on, say two objects, O_1 and O_2 is taken to mean that although being based upon the whole matrix of data it should be independent of which set of agents $A_1,...,A_k$ out of the available class were actually used for the comparative purposes, and also of which objects...other than O_1 and O_2 were also exposed to the set of agents chosen.

Specific objectivity, some general properties. In order to distinguish this type of objectivity from other use of the same word I shall call it *'specific objectivity,'* and in passing I beg you notice the relativity of this concept: it *refers only to the framework specified by the class of objects, the class of agents and the kind of observations which define the comparison.* (pp. 2–3, our italics)

Rasch (1967) adds this important qualification:

...the objects and/or agents are subject to comparison...the data themselves are not directly compared, they only serve as the instruments for the comparison aimed at. *The consequences of introducing these two concepts: (specific) comparisons and specific objectivity, completed by the requirements that a comparison is always possible and its result always unambiguous, are really overwhelming.* (pp. 2-3, our italics)

The above quotes serve to introduce what is required in the method of comparison as well as the consequences from making such comparisons. It is not the data, but what it stands for that is the goal of science. The essential conditions are two: (1) specify 'the kind of comparisons' and (2) specify the precise meaning of objectivity.'

These stipulations suggest the following paradigm:

Propositions: Specifications:
1. Scientific statements   by the kind of comparisons made.
2. Objective comparisons   by the precise meaning and achievement of specific objectivity.

Scientific statements are the consequences of specifying comparisons. Objective comparisons are the consequences emanating from comparisons exemplifying the precise meaning of specific objectivity. Rasch (1977) draws attention to the ubiquity of observations made by comparisons:

...comparisons form an essential part of our recognition of our surroundings: we are ceaselessly faced with different possibilities for action, among which we have to choose just one, a choice that requires that we compare them. This holds both in everyday life and in scientific studies. (p. 68)

These remarks appear general, obvious, and rather simple. Every day finds us comparing prices and products or deciding activities. Rasch's observations on making comparisons in everyday life or pursuing science are self-evident, but clearly essential. However, a careful exam nation of the exact role for making comparisons is required.

Rasch (1977) next responds to those who suggest measurement as primary in making comparisons or conducting science:

That science should require observations to be measurable quantities is a mistake, of course; even in physics observations may be qualitative—as in the last analysis they always are. (p. 2)

This statement is vitally important in the social sciences because it challenges two common beliefs.

First, Rasch dismisses the notion of measurement as a prerequisite for science to proceed. His quote echoes Louis Guttman's comments of a similar nature,

I have avoided the term 'measurement' in all my writings and teachings. I have found it neither useful nor necessary.... No fixed a priori collection of abstract, contentless techniques or principles can be universally appropriate for scientific progress. (1971, pp. 330–331)

Second, Rasch raises qualitative observations to the prominence rightly deserved. Guttman shared this view also.

The basic data of most mental tests are qualitative, yet no treatment is given of the theory of such qualitative data. (in Levy, 1994, p. 324)

Guttman (1971) argued that his approach to science is via "…hypothesis construction for aspects of a universe of observations recorded" (p. 333). He eschewed measurement while promoting better investigations driven by theory. Guttman's remarks further suggest interesting comparisons between these two iconoclasts of traditional measurement practice. Andrich (1982, 1985, 1988) has made important psychometric comparisons of Guttman and Rasch perspectives. Engelhard (2008) has examined both perspectives critical to invariant measurement.

A contemporary expression of many of these same issues has been given by Rein Taagepera (2008) whose book *Making Social Sciences More Scientific* pursues these same goals (among others) by re-focusing attention to *constructing* predictive models. The need to repeat calls to the goal of predicting serves to show how important these matters are, but how neglected in practice they have been.

The important issue developed from Rasch thus far is that only specific comparisons, grounded in substance/objects, produce generalities and laws. It is the properties of these generalities and laws that we employ for guidance and direction. Data and its analyses are only a means to this end, and should not be mistaken for the conclusion. In practice, this means that what is derived from data must reach beyond to become a generalized outcome. Whether it does or not is critical.

1 The Strategy of Comparison

Comparison is driven by theory and experiment. Basic statements about making comparisons may initially appear simplistic. What makes these elementary concerns worthy of further investigation? Why such attention to making comparisons? Rasch

Falling Distance

	H_1	H_2	H_3	H_4	H_5	H_6
Heavy ashtray	+	+	+	+	-	-
Light ashtray	+	+	-	-	-	-

Fig. 1 Ashtray breakage by falling distance

illustrates his points with two examples. The first example involves a comparison of ashtrays. His choice of this example is unfortunate for these times, but illustrative nevertheless. Data from this experiment is produced by dropping heavy and light ashtrays from six different heights. It is reported as shown in Fig. 1.

Survival intact is denoted +, and breakage is denoted -. While the experiment is simple, Rasch reminds us that qualitative comparisons are fundamental to science. He stresses the importance of recognizing qualitative methodology as fundamental in science. This example, likewise for any example, requires confronting each object with an action to produce a reaction—a tripartite condition. But objects and actions must be systematically arranged to allow a valid observation to be made. A theory is stimulated by a thought experiment, an intuitive insight, or a historical review of relevant literature.

The experiment consists of a tripartite condition of *object, agent,* and *resulting reaction.* In a psychometric application this triplet could be person, item, and response. Rasch describes the process:

1. To determine whether an object has a certain property one must do something to the object, confront it with some action or different actions liable to create one of a number of reactions.
2. If knowledge gained in this way is to be used in making a choice it must be obtained for several objects of the kind in question so that a comparison becomes possible (Rasch, 1977, p. 69).

Rasch (1977) introduces his second example showing the sequential development of gas equations (IV:l–IV:15) not discussed here. The essential points that Rasch draws from the more rigorous exposition of this second example (pp. 71–73) are the following:

1. Varying the two parameters of temperature and pressure introduces a stepwise series of systematic experiments,
2. The results show "linearity to be a pervading trait,"
3. This results in "a law of greater generality,"
4. The sequence progresses from observations of volume, pressure and temperature and passes through a series of stepwise comparisons leading to summary equations.

Rasch concluded:

This procedure, it seems, can be taken as a prototype of the experimental charting of a complex field…only through systematic comparisons—experimental or observational—is it possible to formulate empirical laws of sufficient generality to be—speaking frankly—of real value, whether for furthering theoretical knowledge or for practical purposes...I see *systematic comparisons as a central tool* in our investigation of the outer world. (p. 74, our italics)

Great importance is given by Rasch to building sequence of experiments. It is not a singular, critical experiment that typifies science although it has occurred at times. It is a series of them, each of which builds upon the other. Parameters may be experimentally manipulated to observe outcomes, and confirm or deny theoretical predictions. The way that Rasch demonstrates development of the gas law brings a theory to an encompassing state built from and supported by a succession of critical experiments whose hypotheses are sequential, stepwise, and linear. This process constitutes Rasch's *causal model.* We have earlier stipulated, "The Rasch model in concert with a substantive theory is a powerful tool for discovering and testing the adequacy of formulations" (Burdick et al., 2006, p. 1059), and specified, "It takes a substantive theory to unambiguously distinguish between a latent variable and an index." (Stenner et al., 2008, p. 1177). Rasch was clearly proposing that comparisons be made in the context of substantive theory driving experimental studies. Only in this context can claims against latent variables be made following the specification outlined above.

In *Comparisons and Specific Objectivity,* Rasch (1977) generalizes and formalizes the implications of his two examples (p. 75). He begins with explanations and equations more rigorously defining comparison and specific objectivity that are abstracted below but sequenced by Rasch's order of them:

...two collections of elements O and A denoted here objects and agents...single elements and indices O_v and A_1...enter into a well-defined contact C...every contact an outcome R.

...the three collections of elements...the frame of reference

$$F = [O, A, R] \tag{1}$$

the concept of comparison...contact C determines outcome R as a function of object O and agent A

$$R = r(O, A). \tag{2}$$

Within a specified determinate frame of reference a comparison between two objects O_1 and O_2—with regard to their reactions to containing the agent A—is defined as a statement about them which is based solely on those reactions.

$$R_1 = r(O_1, A), R_2 = r(O_2, A) \tag{3}$$

...this comparing function $u (R_1, R_2)$ form a collection U

...the elements...may be qualitative

...inserting function U in

$$U(R_1, R_2) = u(r(O_1, A), r(O_2, A)) \qquad (4)$$

and a statement about O_1 and O_2...will depend on A, the agent.

...the comparing function $U(R_1, R_2)$ as a function of O_1 and O_2 conditioned by A

$$U(r(O_1, A) r(O_2, A)) = u(O_1, O_2|A) \qquad (5)$$

as a comparator for O_1 and O_2, conditioned by agent A (in analogy to the concept of conditional probability).

Statements dependent on the agent [object] are said by Rasch to be local comparisons.

Clarity is given about what is not, and what is objective, and what is local or specific in the frame of reference:

Local comparisons are...useful as pointers...a comparing statement.

...for O_1 and O_2 to be more than locally valid, the comparator must be independent of which a_i from A has been used to produce the reaction

... if the condition is fulfilled ... denote the comparison between these two objects as for agents between these two objects:

$$U(R_1, R_2) = u(r(O_1, A), r(O_2, A)) u(O_1, O_2). \qquad (6)$$

...if this globality with A holds for any two objects O_1 and O_2 in O [then]

...the pairwise comparison defined by (4) is specifically objective within frame of reference F.

The term 'objectivity' refers to the fact that the result of any comparison of two objects within O is independent of the choice of the agent a_i within A and also of the other elements in the collection of objects O; in other words: independent of everything else within the frame of reference, than the two objects which are to be compared and their observed reactions.

...the qualification "specific" is added because the objectivity of these comparisons is restricted to the frame of reference

F in [1] ...denoted as the frame of reference for the specifically objective comparisons in question. (pp. 75–77)

Rasch makes very clear,

... specific objectivity is not an absolute concept, it is related to the specific frame of reference ...this definition concerns only comparisons of objects, but within the same frame of reference it can be applied to comparisons of agents as well. (p. 77)

The philosophic issues encompassing the history of objectivity appear circumspectly avoided in these statements. Rasch specifies,

> The concept has therefore not been carved out in a conceptual analysis, but on the contrary its necessity has appeared in my practical [statistical] activity. (p. 58)

Why avoid the philosophic issue of causation? Speculation suggests it was not in good taste at the time he was writing to speak of causation in general. The zeitgeist did not appear to support objectivity and causality in this context. The consequences of quantum mechanics and the influence of logical positivism may have kept Rasch from wanting his methods contaminated by any digression into philosophy, or to have metaphysical issues injected into his systematic discourse. The strategy seems very clear in light of the quotes given in the opening paragraphs of this paper. Rasch wants to make his case without contending with excessive philosophical baggage concerning causality and objectivity. To illustrate the zeitgeist Rasch was avoiding, consider the opening sentence to Waismann's chapter entitled "The Decline and Fall of Causality" Chapter V in *Turning Points in Physics* (1961).

> 1927 is a landmark in the evolution of physics—the year which saw the obsequies of the notion of causality. (p. 84)

Hoover (2004) offers another illustration:

> [in reference to causality] ... a dip of about twenty percent in the occurrence of the causal family from the 1950s [for 'causally', 'causality', or 'causation' in econometric literature]. (p. 152)

Rasch is by no means conservative regarding the promotion of objectivity via theory/experiments constructed with a view to exploring results based upon engineered methods. He employs the word "law" in a favorable connotation more than a dozen times throughout his 1977 paper. While Rasch does not venture into "causation" stated explicitly, he does so implicitly via employment of comparison as described above. It seems clear, except for evading the philosophic realm of causality, Rasch advocates a causal mode of investigation by means of comparisons founded upon hypotheses and data.

It is important to observe his distinction between "indicators" and "specific objectivity". Rasch (1977) specifies,

> ... if this globality within A holds for any two objects O_1 and O_2 in O ... the pairwise comparison defined by (4) is specifically objective within the frame of reference F.
>
> The term 'objectivity' refers to the fact that the result of any comparison of two objects within O is independent of the choice of the agent a_i within A and also of the other elements in the collection of objects O; in other words: independent of everything else within the frame of reference, than the two objects which are to be compared and their observed reactions. (p. 76)

The essential point rests upon the comparison of two objects (or agents) independent of agent (or object) in the collection of objects (or agents) and their observed reactions within a specified frame of reference. As indicated earlier in the quotes given above, not all comparisons meet the conditions of specific objectivity. A key issue is distinguishing "... those statements dependent on the agent (object)," specified by Rasch to be "local comparisons" from those which emanate from "specific objectivity."

Returning to Rasch's ashtray example, we find there are no differences observed for the two ashtrays dropped from heights H_1, and H_2 because both survive breakage. No difference results from heights H_5 and H_6, also because neither ashtray survives the fall. The first two heights, H_1 and H_2, allow no comparison to be made, and the same occurs from observing H_5 and H_6. But every result occurs from making comparisons without knowledge of the heights or the composition of the ashtrays. This is a subtle but critical point in Rasch's methodology of comparisons. Nothing is required beyond the gross descriptions of the ashtray's composition/construction, and a sequence of heights for these drops. Rasch (1977) concludes:

> Comparing ashtrays at H_3 and H_4 shows the two middle distances with the heavier ones surviving breakage while the lighter ones do not...objects type—2 elements (heavy, light); agent distances—6 elements (each one ordered above the other); and reaction elements—2 results (survived breakage, did not survive breakage). (p. 69)

These specified elements reiterate Rasch's earlier contention that it is the qualities which are engaged while measurement quantities are not required.

> What is required, ... is to do something specific to the object, confront it with some action or different actions liable to create one of a number of actions. If knowledge gained in this way is to be used in making a choice it must be obtained for several objects of the same kind in question so that a comparison becomes possible. (p. 69)

Rasch (1977) draws these inferences from his ashtray experiment.

> A possible comparing function could be the assertion 'No. 1 is more solid than No. 2,' defined operationally by the sequence of reactions + -, that is, the first one holds and the second one breaks. This comparison is not global, it has the value 'true' for the intermediate falling distances and 'false' for the others. (p. 77)
>
> Another comparing function is 'No. 1 is at least as solid as No. 2,' defined operationally by the observed reactions ++ or +-: either they both hold or they both break or only No. 1 holds. This comparison is global within the frame of reference of the described experiment and can even be expected to be global also if more ashtrays and more falling distances are included in the frame of reference. (pp. 77–78)

This brings about the ashtray conclusions:

> 1. Ashtray No.1 is not more solid than No. 2 across all heights. It is not global because it is true for some heights, but not all. A general statement cannot be made for the two types of ashtrays over all heights.
>
> 2. Ashtray No. 1 is as solid as No. 2 can be supported as global in the No. 1 (heavy) is equal to or exceeds No. 2 (light) in heights 1 to 4 although both break at heights 5 and 6.

Rasch next delineates specific objectivity as separate from consideration of general objectivity:

> ...specific objectivity is not to be expected from an arbitrarily chosen comparing function of $u(R_1, R_2)$

Fig. 2 Light vs. heavy
ashtray breakage

		Heavy	
		Survives	Breaks
Light	Survives	H_1, H_2	no data
	Breaks	H_3, H_4	H_5, H_6

...bifactorial frames of reference...[where] every reaction is characterized by a so-called *scalar parameter*...characteristic of object, agent or reaction. [Given...]

...parameters O_v, A_i, and R_{vi}...denoted by ω_v, α_i, ξ_{vi} .

...the reaction is assumed to be uniquely determined by object and agent [in]...

$\xi_{vi} = \varrho(\omega_v, \alpha_i)$ the parametric reaction function.

The condition corresponding to equation (6) for specific objectivity of comparisons of objects... ω_λ and ω_v is:

$$u(\rho(\omega_\lambda, \alpha), \rho(\omega_v, \alpha)) = v(\rho(\omega_\lambda, \omega_v))$$

Under these conditions a decisive statement can be made on the properties of the reaction function $\varrho(\omega_v, \alpha_i)$ that are necessary for establishing specific objectivity of comparisons of objects within the framework F. (p. 78)

Rasch advocates strategies of comparison designed to tease out knowledge. His first example builds the case. The outcomes from the ashtray experiment can be arranged in a 2×2 table as shown in Fig. 2.

We already know enough about ashtrays and height to dismiss his simple example as unnecessary, but then we would miss the point of his generalizations that follow. Rasch's experiment establishes these major points.

1. Systematic comparisons made under specified conditions produce stochastically consistent results.
2. It becomes insightful science to systematically arrange each encounter between an object and an agent, and then observe the result, the classic experiment for determining an outcome.
3. Given enough such experiments, wisely contrived, we can often predict outcomes whenever we gain an understanding of the matter under study, and our theory is sound. But sometimes our predictions are surprising, and wrong!

There is a clear difference between the heavy and light ashtrays from the results for H_3 and H_4 because the heavy one survives, and the light one does not. This establishes Rasch's critical point illustrated by this experiment; comparisons are fundamental to measurement. Rasch also reminds us the results were produced by qualitative observations, and he indicates this is frequently the case with many scientific investigations in the physical sciences. There has been no requirement or need for prior quantitative measures in order to produce these results. We might wish to bring further clarification and more sophistication to this experiment by refining the conditions (introducing different heights with different ashtrays) and observing the results, but specific units for height and mass are not required. Order suffices.

2 Constructing Experimental Comparisons

Rasch argues that the process and strategy of constructing comparisons is the essence of scientific methodology. In his ashtray experiment every object comes in contact with every agent via a frame of reference. Each interaction produces some outcome resulting from this intersecting occurrence of agents with objects. Outcomes are recorded by one of two qualitative values in a dichotomous frame of reference defined by ashtrays and heights.

A 2×2 frame of reference permits these comparisons to be ordered in a systematic way. The outcome may be qualitatively the same or different as in the ashtray example. Further comparisons are made possible by progressing through a larger data frame, and subsequently aggregating all such useful comparisons regarding height and ashtrays. Order remains fundamental regardless of how complex the comparison framework grows.

Rasch (1977) characterizes pair-wise comparisons of objects (or agents) as *"specifically objective within the frame of reference"* (p. 77, our emphasis), and he delineates the process:

The term 'objectivity' refers to the fact that the result of any comparison of two objects within O is independent of the choice of the agent a_i, within A and also of the other elements in the collection of objects O, in other words: independent of everything else within the frame of reference than the two objects which are to be compared and their observed reactions. And the qualification 'specific' is added because the objectivity of these comparisons is restricted to the frame of reference F. This is therefore denoted as the frame of reference for the specifically objective comparisons in question. This also makes clear that *specific objectivity in not an absolute concept, it is related to the specified frame of reference.* (p. 77, our emphasis)

Designating ω_v, α_i, and ξ_{vi} as parameters for O_v, A_i, and R_{vi} gives $\xi_{vi} = \varrho(\omega_v, \alpha_i)$ as the parametric reaction function. This condition for the specific objectivity of comparisons for objects ω_λ and ω_v is:

$$u(\rho(\omega_\lambda, \alpha), \rho(\omega_v, \alpha)) = v(\rho(\omega_\lambda, \omega_v))$$

Rasch then formulates his main theorem of specific objectivity:

Let objects and agents in the bifactorial determinate frame of reference $\mathscr{F}$ be characterizable by scalar parameters ω and α and reactions by a scalar reaction function of 'convenient' mathematical properties:

$$\xi = \wp(\omega, \alpha)$$

with three monotonic functions:

$$\omega' = \phi(\omega), \alpha' = \psi(\alpha), \xi' = \chi(\xi)$$

transforming the scalar reaction function to an additive one:

$$\xi' = \omega' + \alpha'$$

...a necessary and sufficient condition for specifically objective comparability of objects as well as agents. (p. 79)

The model now becomes the source of objective measurement and not the details of the data. Hence, the sometimes used phrase, "When data fit the Rasch model ..." might be better expressed "The Rasch model has (1) identified a fit of data to the model across a frame of reference implied by this experiment, or (2) identified a lack of fit between the same."

Rasch (1960) had earlier alluded to this same condition:

It is tempting, therefore, in the cases with deviations of one sort or other to ask *whether it is the model or the test that has gone wrong*...the question is meaningful...the applicability of the model must have *something to do with the construction of the* test. (p. 51, our emphasis)

The model confirms or identifies inconsistencies by making experimental comparisons of agents, objects, and outcomes from the data. Inconsistent comparisons in an experiment suggest theory, data or both are suspect. Rasch's penchant for constructing data plots as a "check on the model" indicates his awareness of the importance of quality control. We further draw attention to the role of construction or engineering as critical. Every experiment is a fabrication in the sense of manufacturing a desired outcome from theory and data. To the degree that we succeed we know what is required to produce the desired outcome. Failure or deviations indicate full knowledge is lacking.

In such instances, theory, data or both require further investigation. This is an important point for applying the model to data. To use an illustration from physics, as Rasch often did, Newton's law of motion rests upon his model and not on data. It is the law and not the data to which we attend. Furthermore, we recognize data is fraught with contamination from many sources of error. Hence, we generalize from Newton's model of force/mass—his abstraction produced from contaminated data. Continuous validation has supported his theory until the advent of quantum physics required a new viewpoint, but even this evolutionary change has not obviated Newton's law.

Rasch's example began with six ashtrays, but only two ashtrays provided unique and key information. With a supply of additional ashtrays made of different types (shapes, composition, etc.) we can proceed to make every two-way comparison of different ashtrays dropped from various heights until we have exhausted all the two-way (height by ashtray) comparisons useful for this crash test. Summarizing the findings provides types of ashtrays arranged by their capacity to withstand breakage according to height employed. From a simple comparison of two similar ashtrays from different heights (or two different ones from the same height) this simple comparison may be extended as far as desired. We do not need to physically order the ashtrays, although we could do so, because a record of success and failure is sufficient. A summary of all the two-by-two comparisons will produce an ordered arrangement of all the ashtrays by all the heights employed. This process results in a durability/survival variable identifying ashtrays from the most fragile to the most durable, and from the lowest to highest heights in the frame of reference deemed useful for the experiment. We can also move from a two-way comparison to a multi-variable frame if desired.

Any expansion embodies the essence of experimentation guided by theory. It produces an ever encompassing frame of reference for determining durability for a variety of ashtrays dropped from a variety of heights. We can confirm previous predictions from theory as well as make further predictions about unexamined ashtrays and unexamined heights. As we proceed, it may be possible to derive other predictive hypotheses regarding what is expected to occur. A theory of ashtray composition dropped from various heights will probably produce additional predictive insight. Outcomes may be increasingly predictive compared to what was known initially. Theory regarding ashtray composition and heights may become increasingly better understood. Essential to the process is confirmation by cross-validation. If desired, we can describe the results by numeric values noting once more that such "measures" follow comparison.

Campbell (1921) offers a relevant remark,

> If measurement is really to mean anything, there must be some important resemblance between the property measured, on the one hand, and the numerals assigned to represent it on the other hand. In fundamental measurement this resemblance (or the most important part of it) arises from the fact that the property that is susceptible to addition is following the same rules as that of numbers. There is left resemblance in respect of "order." ...*Order then is characteristic of numerals; it is also characteristic of the properties represented by numerals. This is the feature which makes "measurement" significant.* (p. 126–127, our emphasis)

The key to measurement in psychometrics is (1) making systematic comparisons (items, persons, judges, etc.) and (2) making these comparisons objectively. Order becomes paramount. Assigning numerals in such cases is really an after-thought stemming from the initial pair-wise comparisons of ashtrays so as to systematize the findings and quantify the results. The numerals/numbers assigned and utilized become the summary of the experimental findings. They follow the experiment, but do not dictate it. Only when the results are valid and confirmed can we successfully apply/substantiate numeric/algebraic abstractions, and not the other way around. We construct measurement from results of the theory/experiment. We have Celsius and Fahrenheit scales, but one and the same temperature at a simultaneous measurement.

In Rasch's words (1977),

"Objectivity is achieved when a comparison of any two objects is independent of everything else within the frame of reference other than the two objects which are to be compared and their observed reactions." (p. 77)

Rasch (1977) also shares his emphasis upon comparison with the English philosopher Hume (1949) who wrote,

All kinds of reasoning consist in nothing but a comparison, and a discovery of those relations, either constant or inconstant, which two or more objects bear to each other. (p. 77, italics in original)

Making systematic comparisons constitutes the fundamental process by which we determine essential differences. Systematic comparisons when possessing transitivity produce order. Measures considered to be quantitative actually result from qualitative ordering. Comparison first makes clear how any two objects/agents relate to each other by whether one is "more" than the other. This is the ground of measurement. Hume stated the importance of comparison as a logico-philosophical deductive principle. Rasch specified it using a simple inductive example to deliver a mathematical generalization. Generalization from order via comparison produces a ground to measurement. Systematic comparisons produce order and understanding without the necessity of any a priori measurement scheme. Ricoeur (1977) speaks to their value and power:

... the power of making things visible, alive, actual is inseparable from either a logical relation of proportion or a comparison... Thus one and the same strategy of discourse puts into play the logical force of analogy and of comparison—the power to set things before the eyes, the power to speak of the inanimate as if alive, ultimately the capacity to signify active reality. (pp. 34–35)

The two-way frame is the data analysis procedure by which to determine order from the results observed by systematic comparisons produced from the tripartite elements of agent, object and outcome. Comparisons derived from this two-way frame are objective whenever "they are independent of everything else within the frame of reference." We restate this principle by some alternate expressions:

1. The frame of reference provides the basis for making objective comparisons.
2. Comparisons produce order in a frame of reference guided by theory.
3. Order (for agents, objects and results) is demonstrable (or not) by the frame of reference.
4. Comparisons are specifically objective when made within the frame of reference (2 and 3).

In a deterministic framework the consequences of comparison remain categorical (Guttman). In a stochastic framework the consequences of comparison are probabilistic (Rasch). Not every drop of a specified ashtray from a specified height produces the exact same outcome. There will be a distribution of errors surrounding each comparing event. But if the experimental conditions are the same, then a probabilistic result provides the answer. Residual analysis and misfit analysis become the important tools by which to evaluate every outcome resulting from every comparison. Rasch did not address quality control directly in his 1977 paper, but he did give careful consideration to always confirming the model to data as shown by his constant attention to making plots and graphs. He relied on data plots not correlations to show confirmation or identify deviance. This is good advice for today also.

The conclusions to be drawn from Rasch's exposition are as follows:

1. We construct science by making comparisons. These comparisons must be made by following a procedure leading to specific objectivity. Theory guides this process, but experimentation determines the outcome of hypotheses. This strategy assures clarity in the process and allows replications to confirm or refute the results.
2. The two-way frame of reference specifies the agent, object and resultant outcomes in a predictive guise.
3. The two-way frame of reference arranges and subsequently summarizes the comparisons. These comparisons are fundamental to what Rasch designated as specific objectivity.
4. Measurement follows from the results of qualitative comparisons that have been constructed in a systematic way using order as the fundamental characteristic.

3   Additive Conjoint Measurement

Rasch's experimental conditions demonstrate the first of two specifications essential for additivity, i.e., the rows and columns of a data matrix are monotonic for the data (Krantz et al., 1971). The independent variables of ashtrays; composition, height and result, were ordered in the ashtray experiment by a two-way frame of reference. Continued experimentation could sustain and expand upon the range thereby extending the frame of reference through additional experimentation.

A second property, double cancellation, identifies departures from additivity (Krantz et al., 1971). Luce and Tukey (1964, p. 3) show a simple way to graph the effects of two factors for demonstrating this property. Boorsboom (2005) sees double cancellation the consequence of additivity which "brings out the similarity between additive conjoint measurement and the Rasch model." He indicates that the condition of independence is "similar to what Rasch called parameter separation" (p. 97).

Perline et al. (1979) presented data to argue Rasch's psychometric model is a special case of additive conjoint measurement. Additivity is ascertained by examining these two necessary ordinal properties. Coombs et al. (1970) declare all independent variables are measured on an interval scale with a common unit when additivity exists.

The essential point of Perline et al. (1979) in their analysis of two studies was that Rasch model estimates (when data are orderly) correspond to the conditions for additive conjoint measurement, but data fit does not have to be ideal to be demonstrable.

Boorsboom (2005) argues,

...the Rasch model has little to do with fundamental measurement. In fact, the only things that conjoint measurement and Rasch models have in common, in this interpretation is additivity. (p. 132)

Indirectly, this simply confirms the comparability of Rasch estimates of person and item parameters to additive conjoint measurement. The key property of additive conjoint measurement rests upon order determined from making comparisons which returns us once more to the points made from Rasch's ashtray example.

Newby et al. (2002) provide a clear statement of the mathematical relationship between additive conjoint measurement (ACM) and the Rasch model saying,

... it is not the case that, given any data, the Rasch model will provide a natural numerical representation of the data that is comparable with ACM. (p. 350)

Why should the Rasch model address any data except to show if the conclusions drawn from making constructive comparisons are justified or not within the frame of reference. Our point is that a so-called Rasch model analysis does not cleanse data. Nor does it not signify the conclusion of data analysis. Further analysis must be conducted in the context of theory with clearly stated hypotheses Removed from the context of theory, no analysis sui generis provides substantive answers except incidentally (Stenner et al., 2008). For example; no statistical transformation of so-called raw scores to IQ values, etc., makes any improvement to a substantive understanding of intelligence or cognition.

Comparisons are the experimental results/consequence of an interaction of agents and objects guided by theory/hypotheses (Brogden, 1977). Comparison remains the key to any investigation. Content and substance are embodied in the selected comparisons made.

Are the results obtained from any Rasch analyses descriptive, or are they predictive? Does the ashtray frame of reference only describe, or can it predict? Does what we learn from dropping a variety of ashtrays from a variety of heights yield merely descriptive observations? Can these results predict stochastically suggested outcomes? Rasch answered these questions by the quotes we provided earlier.

Concerning parole data from Perline et al. (1979), we ask whether or not causality can be ascribed to the outcome emanating from this Rasch frame of reference? What do we learn from the parole data that provides a prediction? Is it a truly causal matter? Can we ascertain any information from an experimental manipulation of the tripartite object, agent, and outcome in a frame of reference to produce expected outcomes thought to be causal? If so, the variable moves beyond mere description to prediction and the consequences move to a higher plane (Stenner et al., 2008). But how do we truly ascertain prediction and causality in Rasch measurement?

Our answer ascribes the typical Rasch analysis as largely descriptive and lacking evidence without a causal connection or experimental manipulation. A Rasch analysis addresses order in the data, and may describe association or correlation, but not a direct instance of causal inference—suggestive, maybe, but not confirmatory. Lack of evidence suggests a crucial next step because (Stenner et al., 2008) argue,

There is no single piece of evidence more important to a construct's definition than the causal relationship between the construct and its indicators (p. 1153).

With respect to the parole data we ask, how can these nine items be placed in an experimental framework so as to examine their usefulness and predictive validity? The answer is to put these comparisons in a context for testing causality. The Rubin-Holland (Holland, 1980, 1986) framework for demonstrating causal inference specifies *no causation without manipulation*. Rasch analysis has produced considerable understanding of the variables exemplified in the resulting map of items, persons, and results from the two-way framework. The results remain descriptive unless contained within an experimental context guided by theory. Supporting any experiment should be a theory guiding inferences about what produced these outcomes. In the matter of parole, for example, how is granting or not granting parole to be experimentally controlled? How will outcome be determined? The hallmark of success is formulation of a specification equation to predict and validate successful manipulation of the variables required in a probabilistic causal model portrayed in the measuring variable (Pearl, 2000).

We therefore specify a process model for classifying Rasch investigations as shown in Fig. 3.

1. The comparison process begins with an idea, a conjecture, or history which develops into a thought experiment.

2. The experiment is designed and conducted:

   Input → Manipulation → Output

3. An equation may be formulated and specified:

   Domain → Function → Range

4. Data plots, the analysis of misfit and residuals provide information and data/model control.

5. From these strategies come the properties of comparison and order guided by theory which produce the essential ingredients for achieving specific objectivity.

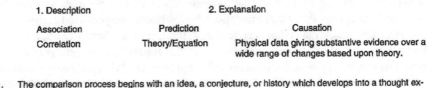

1. Description		2. Explanation
Association	Prediction	Causation
Correlation	Theory/Equation	Physical data giving substantive evidence over a wide range of changes based upon theory.

1. The comparison process begins with an idea, a conjecture, or history which develops into a thought experiment.

2. The experiment is designed and conducted:

   Input ⟶ Manipulation ⟶ Output

3. An equation may be formulated and specified:

   Domain ⟶ Function ⟶ Range

4. Data plots, the analysis of misfit and residuals provide information and data/model control.

5. From these strategies come the properties of comparison and order guided by theory which produce the essential ingredients for achieving specific objectivity.

Fig. 3 Strategies of comparison and order guided by theory leading to specific objectivity

References

Andrich, D. (1982). A index of person separation in latent trait theory, the traditional K.R.20 index, and the Guttman Scale response pattern. *Educational Research and Perspectives, 9*, 95–104.

Andrich, D. (1985). An elaboration of Guttman scaling with Rasch models for measurement. In N. Brandon-Tuman (Ed.), *Sociological methodology* (pp. 30–80). Jossey-Bass.

Andrich, D. (1988). *Rasch models for measurement.* Sage.

Borsboom, D. (2005). *Measuring the mind.* Cambridge University Press.

Brogden, H. E. (1977). The Rasch model, the law of comparative judgment and additive conjoint measurement. *Psychometrika, 42*, 631–634.

Burdick, D. S., Stone, M., & Stenner, A. J. (2006). The combined gas law and a Rasch reading law. *Rasch Measurement Transactions, 20*, 1059–1060.

Campbell, N. (1921). *What is science?* Methuen.

Coombs, C. H., Dawes, R. M., & Tversky, A. (1970). *Mathematical psychology: An elementary introduction.* Prentice-Hall.

Engelhard, G. (2008). Historical perspectives on invariant measurement: Guttman, Rasch, and Mokken. *Measurement: Interdisciplinary Research and Perspective, 6*, 155–189.

Guttman, L. (1971). Measurement as structural theory. *Psychometrika, 36*, 329–347.

Holland, P. W. (1980). Statistics and casual reference. *Journal of the American Statistical Association, 81*, 945–970.

Holland, P. W. (1986). Statistics and causal inference. *Journal of the American Statistical Association, 81*, 945–960.

Hoover, K. D. (2004). Lost causes. *Journal of the History of Economic Thought, 26*, 149–164.

Hume, D. (1949). *A treatise on human nature.* Dent and Sons (Original publication 1817).

Krantz, D. H., Luce, R. D., Suppes, P., & Tversky, A. (1971). *Foundations of measurement. Volume 1: Additive and polynomial representations.* Academic Press.

Levy, S. (1994). *Louis Guttman on theory and methodology: Selected writings.* Dartmouth.

Luce, R. D., & Tukey, J. W. (1964). Simultaneous conjoint measurement: A new kind of fundamental measurement. *Journal of Mathematical Psychology, 1*(1), 1–27.

Newby, V. A., Conner, G. R., Grant, C. P., & Bunderson, C. V. (2002). A formal proof that the Rasch Model is a special case of additive conjoint measurement. Paper presented at the International Objective Measurement Workshops, New Orleans, LA.

Pearl, J. (2000). *Causality: Models, reasoning, and inference.* Cambridge University Press.

Perline, R., Wright, B. D., & Wainer, H. (1979). The Rasch model as additive conjoint measurement. *Applied Psychological Measurement, 3*(2), 237–255.

Rasch, G. (1960). *Probabilistic models for some intelligence and attainment tests.* Danish Institute for Educational Research (Expanded edition, 1980, University of Chicago Press).

Rasch, G. (1967). An informal report of objectivity in comparisons. Psychological measurement theory. In *Proceedings of the NUFFIC International Summer Session in Science at "Het Onde Hof" at Den Hag July 14–28.*

Rasch, G. (1977). On specific objectivity: An attempt at formalizing the request for generality and validity of scientific statements. *Danish Yearbook of Philosophy, 14*, 58–94.

Ricoeur, P. (1977). *The rule of metaphor: Multi-disciplinary studies of the creation of meaning in language* (R. Czerny, Trans.). University of Toronto Press.

Stenner, A. J., Burdick, D. S., & Stone, M. H. (2008). Formative and reflective models: Can a Rasch analysis tell the difference? *Rasch Measurement Transactions, 22*(1), 1152–1153.

Taagepera, R. (2008). *Making social sciences more scientific: The need for predictive models.* Oxford University Press.

Waisman, F. (1961). *Turning points in physics.* Harper Torchbooks.

Individual-Centered Versus Group-Centered Measures

Mark H. Stone and A. Jackson Stenner

The Preface to *Probabilistic Models for Some Intelligence and Attainment Tests* (Rasch, 1960) cites Skinner (1956) and Zubin (1955a, 1995b). In an argument whereby, "… individual-centered statistical techniques require models in which each individual is characterized separately and from which, given adequate data, the individual parameters can be estimated" (Rasch, 1960, p. xx). The Skinner reference is easily located. The mimeographed work by Zubin has not been found, but we did find another Zubin paper given at the 1955 ETS Invitational Conference on testing problems in which he writes, "An Example of the application of individual-centered techniques which keeps the sights of the experimenter focused on the individual instead of on the group…" (p. 116) may have helped Rasch situate his thinking. Rasch goes on to state "… present day statistical methods are entirely group-centered so that there is a real need for developing individual-centered statistics" (p. xx). What constitutes the differences in these statistics?

While it is individual persons and groups of persons that are the focus of discussion, we begin with an even more simple illustration because human behavior is complex, and a single mechanical-like variable is a better illustration to one that is complex. We choose temperature for this illustration because measuring mechanisms (Stenner, Stone & Burdick, 2008) for temperature are well established and all report out in a common metric or degree (disregarding wind-chill, etc.). A measuring mechanism consists of (1) guiding substantive theory, (2) successful instrument fabrication, and (3) demonstrable data by which the instrument has established utility in the course of its developmental history.

Rasch Measurement Transactions, 29(1). 2015.

M. H. Stone
Departments of Mathematics, Psychology, and Social Work at Aurora University, Chicago, USA

A. J. Stenner (✉)
MetaMetrics, Inc., Durham, NC, USA

© The Author(s) 2023
W. P. Fisher and P. J. Massengill (eds.), *Explanatory Models, Unit Standards, and Personalized Learning in Educational Measurement*,
https://doi.org/10.1007/978-981-19-3747-7_20

269

Consider six mercury-tube outdoor thermometers that are placed appropriately in a local environment, but near each other. They all register approximately the same degree of temperature, independently verified by consulting NOAA for the temperature at this location. One by one each thermometer is placed in a compartment able to increase/decrease the prevailing temperature by at least ten degrees. Upon verifying the artificially induced temperature change for each thermometer, it is returned to its original location and checked to see if it returns to its previous value and agrees with the other five.

If each of the six thermometers measured a similar and consistent degree of temperature before and after the induced environmental intervention/manipulation, this consistency of instrument recording validates a deep understanding of the attribute "temperature" and its measurement. Each thermometer initially recorded the same temperature, and following a change to and from the artificial environment returned to the base degree of temperature. Furthermore, all the measurements agree.

Interestingly, the experimentally induced change of environment also produced what may be called *causal validity*, not unlike constructive validity (Cronbach & Meehl, 1955) inasmuch as the temperature was manipulated, fabricated, engineered, etc. via construction and use of the artificial environment. When measuring mechanism(s) such as outdoor thermometers are properly manufactured this result is to be expected, and this experimental outcome and its replication would be predicted prior to environmental manipulation from all we know about temperature and thermometers. This outcome might further be termed *validity as theoretical equivalence* (Lumsden & Ross, 1973) because the replications produced by all six thermometer recordings might be considered "one" temperature. Our theoretical prediction is expected as a consequence of the causal process produced by the experiment, and reported by all the instruments. *Causal validity* is a consequence of the successful theoretical predictions realized in the experiment. Its essence is "prediction under intervention." The manipulable characteristics of our experiment involving the base environment, change made by way of an artificial environment, and the final change of recorded temperature are the consequence of a well-functioning construct theory and measuring mechanism. Each of the six individual thermometers records a similarly induced experimental deviation and a return to the base state. Each thermometer constitutes an individual unit, and the six thermometers constitute a group albeit without variation, which is exactly what would be predicted.

Now consider a transition to human behavior. Height is the new outcome measure and the determination of height at a point in time can be obtained from another well-established measuring mechanism—the ruler, which provides a point-estimate for one individual measured at a single point in time. When this process is continued for the same individual over successive time periods we produce a trajectory of height for the person over time (purely individual centeresd as no reference to any other person(s) is required). From these values one may determine growth over time intervals as well as any observed plateaus and spurts well-known to occur in individual development. The individual's trajectory rate may also vary because of illness and old age, so we could discover different rates over certain time periods as well as determine a curvilinear average to describe the person's total trajectory. Growth in height is a function of time, and the human characteristics entailed in a person's overall development result from genetic and environmental makeup. These statistics are intra-individually determined. Such statistical analyses produce the "individual-centered statistics" that Rasch spoke about.

Aggregating individual measurements of height into a group or groups is a common method for producing "group-centered statistics" often employing some frequency model such as the normal curve. This is most common when generalizing the characteristics of human growth in overall height based upon a large number of individuals. The difference between measuring a group of individuals compared to our first illustration using a group of thermometers is that while we expected no deviation among the thermometers, we do not expect all individuals to gain the same height over time, but rather to register individual differences. Hence, we resort to descriptive statistics to understand the central trend, and the amount of variation found in the group or groups. An obvious group-centered statistical analysis might aggregate by gender; comparing the typical height of females to males or provide norms tables.

The measurement of height is sstraightforward and the measurement mechanism has been established over several thousand years. The same cannot be said for measuring mental attributes occurring in psychological, health, and educational investigations. Determining the relevant characteristics for their measurement is more difficult although the procedures for their determination should follow those already discussed. The major statistical hurdle is moving from the ordering of a variable's units to its "measurement application." The measurement models of Georg Rasch have been instrumental in driving this process forward.

Do we know enough about the measurement of reading that we can manipulate the comprehension rate experienced by a reader in a way that mimics the above temperature example? In the Lexile Framework for Reading (LFR) the *difference* between text complexity of an article and the reading ability of a person is *causal* on the success rate (i.e., count correct). It is true that short term manipulation of a person's reading ability is, at present, not possible, but manipulation of text complexity is possible because we can select a new article that possesses the desired text complexity such that any difference value can be realized. Concretely, when a 700L reader encounters a 700L article the forecasted comprehension rate is 75%. Selecting an article at 900L results in a decrease in forecasted comprehension rate to 50%. Selecting an article at 500L results in a forecasted comprehension rate of 90%. Thus we can increase/decrease comprehension rate by judicious manipulation of texts, i.e., we can experimentally induce a change in comprehension rate for any reader and then return the reader to the "base" rate of 75%. Furthermore, successful theoretical predictions following such interventions are invariant over a wide range of environmental conditions including the demographics of the reader (male, adolescent, etc.) and the characteristics of text (length, topic/genre, etc.).

Many applications of Rasch models to human science data are thin on substantive theory. Rarely proposed is an a priori specification of the item calibrations (i.e., constrained models). Causal Rasch Models (Burdick et al., 2006; Stenner & Stone, 2010; Stenner et al., 2013; Stenner, Stone & Burdick, 2009a, 2009b) prescribe (via engineering and manufacturing quality control) that item calibrations take the values imposed by a substantive theory. For data to be useful in making measures, those data must conform to the invariance requirements of both the Rasch model and the substantive theory. Thus, Causal Rasch Models are *doubly prescriptive*. When data meet both sets of requirements; the data are useful not just for making measures of some construct, but are useful for making measures of that precise construct specified by the equation that produced the theoretical item calibrations.

A Causal (doubly constrained) Rasch Model that fuses a substantive theory to a set of axioms for conjoint additive measurement affords a much richer context for the identification and interpretation of anomalies than does an unconstrained descriptive Rasch model. First, with the measurement model and the substantive theory fixed it is self-evident that anomalies are to be understood as problems with the data ideally leading to improved observation models that reduce unintended dependencies in the data (Andrich, 2002). Second, with both model and construct theory fixed it is obvious that our task is to produce measurement outcomes that fit the (aforementioned) dual invariance requirements. An unconstrained model cannot distinguish whether it is the model, data, or both that are suspect.

Over centuries, instrument engineering has steadily improved to the point that for most purposes "uncertainty of measurement," usually reported as the standard deviation of a distribution of imagined or actual replications taken on a single person, can be effectively ignored. The practical outcome of such successful engineering is that the "problem" of measurement error is virtually non-existent; consider most bathroom scale applications. The use of pounds and ounces also becomes arbitrary as is evident from the fact that most of the world has gone metric although other standards remain. What is decisive is that a unit is agreed to by the community and is slavishly maintained through substantive theory together with consistent implementation, instrument manufacture, and reporting. We specify these stages:

Theory ➡ Engineering ➡ Manufacturing ➡ Quality Control

The doubly prescriptive Rasch model embodies this process.

Different instruments qua experiences underlie every measuring mechanism; environmental temperature, human temperature, children's reported weight on a bathroom scale, reading ability. From these illustrations and many more like them we determine point estimates and individual trajectories and group aggregations. This outcome lies in well-developed construct theory, instrument engineering and manufacturing conventions that we designate *measuring mechanisms*.

References

Andrich, D. (2002). Understanding resistance to the data-model relationship in Rasch's paradigm: A reflection for the next generation. *Journal of Applied Measurement, 3,* 325–359.

Burdick, D. S., Stone, M. H., & Stenner, A. J. (2006). The combined gas law and a Rasch reading law. *Rasch Measurement Transactions, 20*(2), 1059–1060.

Cronbach, L., & Meehl, P. (1955). Construct validity in psychological tests. *Psychological Bulletin, 52,* 281–302.

Lumsden, J., & Ross, J. (1973). Validity as theoretical equivalence. *Australian Journal of Psychology, 25*(3), 191–197.

Rasch, G. (1960). *Probabilistic models for some intelligence and attainment tests.* The University of Chicago Press.

Skinner, B. F. (1956). A case history in scientific method. *The American Psychologist, 11,* 221–233.

Stenner, A. J., Burdick, D. S., & Stone, M. H. (2008). Formative and reflective models: Can a Rasch analysis tell the difference? *Rasch Measurement Transactions, 22*(1), 1152–1153.

Stenner, A. J., Stone, M. H. & Burdick, D. S. (2009a). Indexing vs. measuring. *Rasch Measurement Transactions, 22*(4), 1176–1177.

Stenner, A. J., Stone, M., & Burdick, D. (2009b). The concept of a measurement mechanism. *Rasch Measurement Transactions, 23*(2), 1204–1206.

Stenner, A. J., & Stone, M. H. (2010). Generally objective measurement of human temperature and reading ability: Some corollaries. *Journal of Applied Measurement, 11*(3), 244–252.

Stenner, A. J., Fisher, W. P., Stone, M. H., & Burdick, D. S. (2013). Causal Rasch models. *Frontiers in Psychology, 4,* 536. https://doi.org/10.3389/tpsyg.2013.00536

Zubin, J. (1955a). *Experimental abnormal psychology (mimeographed).* Columbia University Store.

Zubin, J. (1955b). Clinical vs. actuarial prediction: A pseudo-problem. In *Proceedings of the 1955b Invitational Conference on Testing Problems* (pp. 107–128). Princeton: Educational Testing Service.

Theory-Based Metrological Traceability in Education: A Reading Measurement Network

William P. Fisher Jr. and A. Jackson Stenner

Abstract Huge resources are invested in metrology and standards in the natural sciences, engineering, and across a wide range of commercial technologies. Significant positive returns of human, social, environmental, and economic value on these investments have been sustained for decades. Proven methods for calibrating test and survey instruments in linear units are readily available, as are data- and theory-based methods for equating those instruments to a shared unit. Using these methods, metrological traceability is obtained in a variety of commercially available elementary and secondary English and Spanish language reading education programs in the U.S., Canada, Mexico, and Australia. Given established historical patterns, widespread routine reproduction of predicted text-based and instructional effects expressed in a common language and shared frame of reference may lead to significant developments in theory and practice. Opportunities for systematic implementations of teacher-driven lean thinking and continuous quality improvement methods may be of particular interest and value.

1 Introduction

Metrology connects measurement applications across industrial, scientific, and practical tasks separated by space and time. Significant fractions of many nations' economic productivity are invested in ensuring traceability to standards for various units of measurement. The human, social, environmental, and economic value of the returns on these investments depends on the transparency of the measures and their integration into a wide range of decision processes at multiple organizational levels. Huge resources are required to create and maintain technologically produced effects,

W. P. Fisher Jr. (✉)
Living Capital Metrics LLC, Sausalito, CA, USA

Graduate School of Education, BEAR Center, University of California, Berkeley, CA, USA

A. J. Stenner
Metametrics, Inc., Durham, NC, USA

© The Author(s) 2023
W. P. Fisher and P. J. Massengill (eds.), *Explanatory Models, Unit Standards, and Personalized Learning in Educational Measurement*,
https://doi.org/10.1007/978-981-19-3747-7_21

such as volts, seconds, or meters, with the primary return on those resources being the illusion that the effects seem to be products of nothing but completely natural processes occurring with no human intervention.

New insights into how cognitive, social and technological resources aid in creating shared cultural frames of reference have emerged from close critical study of historical and contemporary scientific modelling and metrological practices. From this perspective, science is not qualitatively different from everyday ways of thinking and relating, except in more deliberately extending laboratory processes into the world as distributed cognitive systems supporting a range of associated problem-solving methods (Hutchins, 1995, 2012; Latour, 1987, 2005; Nersessian, 2012). Of particular interest here is the linking of specific ways in which organizations align and coordinate their processes and relationships relative to technical developments and expectations. A positive result of adopting this point of view is recognition of the value of previously obscured accomplishments in, and opportunities for, advancing the quality of research and practice in psychology and the social sciences. An illustrative example is found in the scientific modelling and metrological practices informing integrated reading assessment and instruction in education.

1.1 Transparent Instruments, Invisible Production

By definition, metrologists are doing their jobs best when no one knows they are there. Experimental scientists, for instance, may take little notice of their instrumentation until it breaks down or does not conform to expected standards. The general public and researchers in psychology and the social sciences are, then, also largely unaware of the resource-intensive work involved in establishing uniform unit standards and traceability to them (Latour, 1987, 2005; Schaffer et al., 1992; Wise, 1995).

The uniformity of the various phenomena described by natural laws allows scientists the convenient efficiency of not needing to specify scale units in statements of laws. Force equals mass times acceleration in kilograms, Newtons, and meters just as well as in pounds, poundals, and feet. The ability to skip over uniform details supports a division of labour in science that separates theoretical work from the calibration of instruments and both of these from the use of theory and instruments in experiments (Galison, 1997).

The convenience of separating theoretical, experimental and instrumental concerns has its drawbacks, too. Not knowing when or how reference standard units are established reinforces unexamined metaphysical assumptions—such as the idea that the universe or nature is inherently and innately numerical, quantitative, or mathematical—that rarely become explicit objects of attention.

The effect of these presuppositions is significant. Huge social, industrial, and economic efficiencies are gained by universal consensus on the facts of complex phenomena like electricity, temperature, distance, mass, and time. Though the dynamics of that consensus are complex and sometimes counterintuitive (Galison, 1997), making quantities seem natural is a cultural achievement of the highest order.

The advancement of science is put at risk when the historic and historical mathematical understanding of scientific objects is reified as unquestioned and unquestionable. Two questions emerge here: (1) how did the natural sciences succeed in making quant ties seem so thoroughly natural (Dear, 2012; Latour, 1987, 2005; O'Connell, 1993; Shapin, 1989; Sundberg, 2011), and (2) how might the social sciences learn from those successes? Recent advances in reading measurement embody important lessons in this regard for the social sciences.

1.2 Shortsightedly Focusing Attention on the Local Measurement Outcome

The technical processes of measurement were historically cut out of the picture of science by the positivist focus on empirical observation, as well as by the later antipositivist focus on theoretical constraints on observation (Galison, 1997). Sometimes this omission was literal and deliberate, as when a woodcut of a laboratory scene printed in its entirety in one place is trimmed in a later publication to exclude the means by which a technical effect was produced (Shapin, 1989). Other times the omission was metaphorical, as when technical processes were illustrated in summary form by angelic cherubs producing effects by means of divine intervention (Shapin, 1989).

Transparency in measurement is a two-edged sword. Wide access to comparable measures is achieved only to the extent that technical complexities can be ignored. This point was emphasized by Whitehead (1911), who observed that "Civilization advances by extending the number of important operations which we can perform without thinking about them" (p. 61). But what happens when those making these advances do not record—or do not themselves fully understand—*how* they extended the number of important operations that can be performed by persons unversed in their technicalities?

In his study of the geometric assumptions Galileo employed in his physics, Husserl (1970) was sensitive to the ways in which a hidden agenda set priorities. Like Galileo, we find ourselves in a situation, in accord with the philosophical problems attending measurement, in general, where

> Metrology has not often been granted much historical significance...Intellectualist condescension distracts our attention from these everyday practices, from their technical staff, and from the work which makes results count outside laboratory walls (Schaffer et al., 1992).

Researchers in the natural sciences make use of commercially available precision tools calibrated to universally uniform reference standards, standards capitalizing on the value of invariant laws. Transparent measures communicated in a network sharing common values situates metrology's often unrecognized historical significance in a complex overall context offering important lessons for psychology and the social sciences (Dear, 2012; Galison, 1997; Hutchins, 1995, 2012; Latour, 1987, 2005; Nersessian, 2012; O'Connell, 1993; Schaffer et al., 1992; Shapin, 1989; Sundberg,

2011; Wise, 1995). The culture of science rewards a mix of convergent, divergent, and reflective thinking in ways that have proven their productivity and inform a vital culture of ongoing innovation (Dear, 2012; Galison, 1997; Shapin, 1989; Sundberg, 2011; Woolley & Fuchs, 2011).

1.3 Consequences for Psychology and the Social Sciences

But in the social sciences, the lack of metrological institutions, methods, and traditions, and the associated absence of the intercalated disunity of distinct theoretical, experimental, and instrumental communities observed by Galison in the natural sciences (Galison, 1997), has been catastrophic. As social scientists have long recognized for themselves (Cohen, 1994; Salsburg, 1985; Wilson, 1971), mainstream research methods and statistical models employ scale-dependent ordinal data in a search for a kind of significance that is often irrelevant to and even antithetical to the production of new knowledge. Even when regularities akin to natural laws are sought and found in psychological and social phenomena (Burdick et al., 2006; Fisher, 2009; Luce, 1978; Luce & Tukey, 1964; Rasch, 1960), results are typically assessed in the language and methods of statistics rather than of measurement and metrology, meaning the focus is on data analysis and not on theory development or the calibration of instruments traceable to a standard unit. The human, social, economic, and scientific consequences of this failure to coordinate and balance convergent, divergent, and reflective field-defining activities are profound. Ideas on how such activities might be organized in education have recently been pro- posed (Fisher & Wilson, 2015).

The lack of institutions and traditions concerning metrological traceability and standards in psychology and the social sciences may have more to do with broad and deep cultural presuppositions than with an actual lack of a basis for them in evidence. After all, what systematic program of experimental evaluation has ever irrefutably established that uniform metrics based in lawful regularities are impossible in psychology and the social sciences? Evidence indicates that provisional possibilities exist in some circumstances (Burdick et al., 2006; Fisher, 2009; Fisher & Wilson, 2015; Luce, 1978; Luce & Tukey, 1964; Rasch, 1960).

2 Metrological Traceability for Reading Measurement

The longstanding need to provide students with reading challenges appropriate to their reading abilities is usually approached in terms of general curricular structures, and teacher training and experience. Theory has not been of significant interest (Engelhard, 2001; Sadoski & Paivio, 2007). Rasch's development of a new class of measurement models in the 1950s was an important step forward in improving the quantification of reading ability (Sadoski & Paivio, 2007). This research led to improvements in the matching of readers to text.

When Rasch's concept of specific objectivity (the modelled independence of the ability and difficulty parameters, as shown in Eq. (1)) as it was obtained in local measures was combined with a general predictive theory of English text complexity in the 1980s, following the work of Stenner and colleagues (Stenner & Smith, 1982; Stenner et al., 1983, 2006), the stage was set for the efficient creation of a network of reading measurement instruments calibrated in a common unit. By the late 1990, all of the major high stakes English reading tests in the U.S. had been brought into the system. These are today complemented by the hundreds of thousands of books, tens of millions of short articles and hundreds of millions of readers that have been brought into the system in the intervening years.

In this system, reader abilities and text complexities are measured in the same unit. The scale ranges from below 200 for beginning readers to over 1600 for very high level readers and texts. Knowing the text measure of a book and the reader's measure predicts the degree to which the book will be comprehensible to the student.

More than 30 million measures annually are reported in the U.S. from state and commercial assessments, and from classroom reading program assessments, in a common unit of measurement (Stenner et al., 2006). The 21 U.S. state departments of education that have formally adopted this unit for use are shown in the map in Fig. 1.

Traceability to the common unit is determined via both empirical, data-based equating studies and theory-based text analyses (Stenner et al., 2006). Additional features of the system include electronic tools integrating instruction and assessment for mass customized diagnosis (Rivero, 2011), and others charting growth in reading ability relative to college and career readiness (Williamson, 2008; Williamson et al., 2013). Establishing this network of comparable assessments required formal relationships with book and test publishers, teachers, schools and school districts, state departments of education, and psychometric researchers. Furthermore, a new array of material practices was needed to give all the parties involved ongoing and verifiable confidence in the theory. Though great efficiencies stood to be gained, credibility demanded a cautious approach to their implementation. Formal documentation of the birth of this traceability system would be a valuable contribution to the sociotechnical qualities of education.

3 Implications for Psychology and the Social Sciences

In 1965, the National Academy of Sciences published a report articulating common assumptions as to the sequence of events supposed to take place in the development of new instrumentation (Rabkin et al., 1992). Four stages were identified:

1. discovery of suitable means of observing some phenomenon,

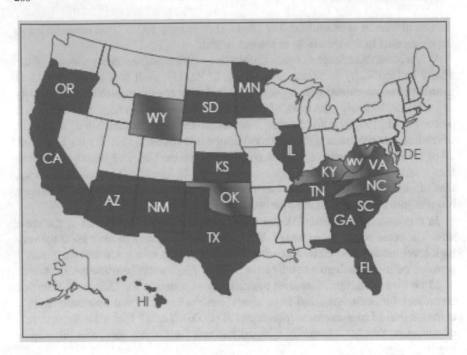

Fig. 1 Map of U.S. states employing a common unit of measurement (two tones indicate common units in use for both reading and mathematics)

2. exploration of this phenomenon with special, homemade instruments or commercial prototypes,
3. widespread use of commercial instruments,
4. routine applications of the instrument to control industrial production as well as research.

Textbook assumptions and presentations of this sequence have indoctrinated researchers in the human sciences to believe, mistakenly, that this is the normal sequence of events. Because hardly anyone is involved in every part of the process, unexamined assumptions cohere into a "just-so" narrative that says more about cultural expectations than about historical complexities. Scientists and non-scientists alike accept this story, against the grain of actual events. Rabkin (Rabkin et al., 1992) points out that

> this scheme seems to be at variance with much of the evidence in the history of science. It has been shown that the integration of instruments has been rarely due to the demand on the part of the researcher. Rather it occurs through vigorous supply of advanced instruments on the part of the industry. The company that proposes these four stages in the report has itself had experience when stages 3 and 4 occur in the reverse order and, moreover, stage 4 is by far the most decisive factor in the development of new instrumentation.

The "vigorous supply of advanced instruments", and not demand, also characterizes the introduction of popular electronic appliances. Just as Rabkin points out has been the case in research, there was little or no clamour among the public for telephones, televisions, faxes, the Internet, microwaves, blenders, or cell phones before they were developed and introduced.

Scientists and the public both tend to think of instrumentation only as tools employed in the service of the individuals who use them. This perspective is at odds with the historical evidence as well as with philosophers' observations, such as, for instance, Thoreau's realization that humanity has become the tool of its tools (Thoreau, 1854) and Nietzsche's insight that the victory of science is better cast as a victory of method over science (Nietzsche, 1967).

This alternative perspective is important because, in the history of science, theory follows from extensive experience with instruments more often than instruments are designed and built from theoretical projections. Standardized and commercially available instrumentation make possible the predictable and routine reproduction of scientific effects essential to the conduct of controlled experiments—and so also to the development of precise and accurate theoretical predictions. As stated by Price,

> Historically, we have almost no examples of an increase in understanding being applied to make new advances in technical competence, but we have many cases of advances in technology being puzzled out by theoreticians and resulting in the advancement of knowledge. It is not just a clever historical aphorism, but a general truth, that' thermodynamics owes much more to the steam engine than ever the steam engine owed to thermodynamics.' ...historically the arrow of causality is largely from the technology to the science (Price, 1986).

In the context of reading measurement, the repeated reproduction of consistent results following the work of Rasch and others led to the Anchor Test Study in the 1970s (Rentz & Bashaw, 1977). This study equated seven major reading tests in the U.S. and involved over 350,000 students in all 50 states. But the purely empirical basis of the calibration and the lack of predictive theory meant that the value of the common unit of measurement was lost as soon as new items were added to the tests, which was immediately.

A plain feature of the equated test results, however, was the similarity of the items from different tests that calibrated in the same locations. The stability of this phenomenon may not surprise anyone able to read, but its practical application in a predictive theory relating text complexity, comprehension rates, and reading ability was difficult to achieve (Stenner & Stone, 2010).

4   Theory for Reading Measurement

The ability to read is fundamental to education, and it is accordingly tested and measured more often than any other subject area. The index to the eighteenth edition of the Buros Mental Measurements Yearbook (Spies et al., 2010) includes over 140 tests with the word "reading" in their titles. This count does not include tests focused on vocabulary or word meaning, which are also numerous.

Though the issues are complex, literacy remains essential to productivity in the global economy (Hamilton & Pitt, 2011). The need for effective and efficient reading education will only intensify as communication, teamwork, and information management are increasingly demanded as basic skills (Neuman & Roskos, 2012).

And despite the longstanding fundamental importance of reading as the tool most essential to learning, reading research remained atheoretical until 1953, and interest in a unified theory of reading is a relatively new phenomenon (Engelhard, 2001; Sadoski & Paivio, 2007). Further, in the years since 1953, available reading theories have not generally been used to inform the design or interpretation of assessments of reading ability (Sadoski & Paivio, 2007).

Though it may seem counterintuitive, this failure to apply theory in the course of empirical measurement research is not unusual, nor is it restricted to reading research. On the contrary, measurement technologies in the natural sciences have historically been developed through socially-contextualized trial-and-error solutions to practical engineering problems, such as consistent, stable results, and not directly from theoretical principles (Bijker et al., 2012; Galison, 1997; Latour, 1987, 2005; Nersessian, 2012; Price, 1986; Rabkin et al., 1992; Schaffer et al., 1992; Wise, 1995). Theory generally comes later, after researchers have had the opportunity to employ standardized technologies in the routine and repeated reproduction of a controlled phenomenon. Only then do applicable general principles emerge as useful insights that can be fed back into technical refinements.

4.1 Syntactic and Semantic Elements

In the same way putting things in words reduces an infinite variety of ways an experience might be expressed into a particular set of words expressed in a particular language, science reduces the infinite variations that phenomena exhibit to simpler models. The truth of the models is less an issue than their usefulness (Box et al., 1979; Rasch, 1973). Simplification is usually achieved only in contexts that respect constraints and accept limited goals. The efficiency and power obtained when useful tools can be created, however, confers great value on a simplified process.

In the 1950s, Rasch's parameter separability theorem, concept of specific objectivity, and models useful in practical measurement applications combined in an important step forward in educational measurement (Loevinger, 1965). These developments were followed by Wright's introduction of improved estimation algorithms, model fit tests, and software in the 1960s, along with his vigorous championing of Rasch's ideas (Rasch, 1972). By the 1970s, enough data from reading tests had been successfully fit to Rasch models in the U.S. to support the viability of the Anchor Test Study (Rentz & Bashaw, 1977). Success in this large project and additional research predicting item difficulties on the Peabody Vocabulary Test and the Knox Cube Test (a measure of short term memory and attention span) (Stenner & Smith, 1982; Stenner et al., 1983), led to a new effort focused on developing explanatory theory for reading.

Reading theories build on the fact that all symbol systems share two features: a semantic component and a syntactic component. In language, the semantic units are words. Words are organized according to rules of syntax into sentences (Carver, 1974). Semantic units vary in familiarity and the syntactic structures vary in complexity. The readability of a text passage is dominated by the familiarity of the semantic units and by the complexity of the syntactic structures used in constructing the message. Many readability equations therefore use a two-variable equation to forecast text difficulty. The word-frequency and sentence-length measures combine to produce a regression equation, known as a construct specification equation (Stenner & Smith, 1982; Stenner et al., 1983). This equation provides a theoretical model evaluated in terms of the proportion of the variance of reading comprehension task difficulties (or, more recently, the means of specification-equivalent ensembles of item difficulties, following Gibbs (1981)) that can be explained as plausibly structured by causal relationships (Stenner et al., 2013).

4.2 The Specification Equation

One approach to such a specification equation first employs the mean of the logarithm of the frequencies with which words in a text appear in a 550-milion word corpus of K-16 texts. More specifically, the log frequency of the word family, which is more highly correlated with word difficulty, comprises one term in the equation. Word families include the stimulus word, all plurals, adverbial forms, comparatives, superlatives, verb forms, past participles, and adjectival forms. The frequencies of all words in the family are summed and the log of that sum is used in the specification equation.

The second term of the specification equation is the logarithm of the text's mean sentence length. This parameter is operationalized simply by counting and averaging the number of words in each sentence.

The theoretical logit is then a function of sentence length and word frequencies in the language stated in the specification equation:

$$\text{Reading difficulty (or readability)} = A * \log(MSL) - B * \overline{\log(WF)} + C \quad (1)$$

where MSL is the mean sentence length and WF is the word frequencies. Log(MSL) and the mean log(WF) are used as proxies for syntactic complexity and semantic demand, and the coefficients are drawn from the empirical regression study (Stenner & Burdick, 1997). Research is continuing into the decimal place significance of the coefficients and measurement uncertainty for the values of A (9.82247), B (2.14634), and C (a constant). The resulting logits are then scaled as follows:

$$(\text{logit} + 3.3) * 180 + 200 \quad (2)$$

The relationship of word frequency and sentence length to text readability was investigated in research that extended a previous study on semantic units (Stenner et al., 1983). The original study found success on items at about 3.3 logits as indicating the earliest reading ability, and set that level at 200. A practical top to the scale for the end of high school was at 2.3 logits, and this was set 1000 units higher, to 1200. There is no upper limit to the scale, but text measures above 1600L are rare.

In this unit, when student and text measures match, a 75% comprehension level is expected. A student with a measure of 500L is expected to answer correctly 75% of the questions on an assessment made from any text that also measures 500L, within the range of uncertainty. The 75% comprehension rate differs from the default rate of 50% comprehension usually associated with matching measures and calibrations. Though the lowest uncertainty is associated with the 50% rate, teachers find that instruction has a firmer basis in student confidence when success is more likely. For this reason, the relation of ability to difficulty was shifted from 50 to 75% comprehension.

The uncertainty (standard error) of the individual measures (Wright & Stone, 1979) is

$$SE = X * [L/(r(L - r))]^{(1/2)} \tag{3}$$

which is the square root of the test length L divided by the count correct r times the L − r count incorrect, times an expansion factor X that depends on test width. This logit is then converted to the standard unit. A standard unit uncertainty for a well targeted 36-item test measuring with an uncertainty of about 0.40 logits is the original logit range of 2.3 (−3.3) = 5.6 divided into the 1000L range, times 0.40, which comes to about 71L.

The analysis reported in the original study (Stenner et al., 1983) involved calculation of the mean word frequency and the log of the mean sentence length for each of the 66 reading comprehension passages on the Peabody Individual Achievement Test. The observed difficulty of each passage was the mean difficulty of the items associated with the passage (provided by the publisher) converted to the logit scale.

A regression analysis based on the word-frequency and sentence-length measures produced a regression equation that explained much of the variance found in the set of reading comprehension tasks. The resulting correlation between the observed logit difficulties and the theoretical calibrations was 0.97 after correction for range restriction and measurement error (Stenner & Burdick, 1997).

The regression equation was further refined based on its use in predicting the observed difficulty of the reading comprehension passages on eight other standardized tests (see Table 1). Repeated and ongoing comparisons of theoretically expected calibrations with data-based estimates produced from test data analysis provide continually updated validity evidence.

The regression equation links the syntactic and semantic features of text to the empirically determined difficulty of text. That link, in turn, is reproduced across thousands of test items and millions of examinees.

Table 1 Correlations of theory-based calibrations produced by the specification equation and data-based item difficulties

Test	# of questions	# of passages	r(OT)[a]	R(OT)[b]	R*(OT)[c]
SRA	235	46	0.95	0.97	1.00
CAT-E	418	74	0.91	0.95	0.97
Lexile	262	262	0.93	0.95	0.97
PIAT	66	66	0.93	0.94	0.97
CAT-C	253	43	0.83	0.93	0.96
CTBS-U	246	50	0.74	0.92	0.95
NAEP	189	70	0.65	0.92	0.94
Battery	26	26	0.88	0.84	0.87
Mastery	85	85	0.74	0.75	0.77
Totals Means	1780	722	0.84	0.91	0.93

[a] $r_{(OT)}$ = raw correlation between observed difficulties (O) and theory-based calibrations (T)
[b] $R_{(OT)}$ = correlation between observed difficulties (O) and theory-based (T) corrected for range restriction
[c] $R^*_{(OT)}$ = correlation between observed difficulties (O) and theory-based calibrations (T) corrected for range restriction and measurement error

In applications the consistent display of the link over time provides a basis for using the equation to perform theory-based calibrations of test items and texts, thus rendering empirical calibrations necessary only as checks on the system.

This specification equation joins together previously separated but analogous developments in measures of information. Hartley's (1928) log(N) measure of information content (the number of signs in a message), for instance, is akin to the sentence length parameter in Eq. (1). Similarly, the word frequency parameter is akin to Shannon's (1948) classic expression p*log(p), where more information is implied by a word's greater rarity in the language. Including Shannon's extra p (multiplying the log of the probability of observing a sign by that probability) indicates the entropy of the area under the curve in the logistic ogive (Linacre, 2006).

5 Benefits of Metrological Comparability

A wide range of applications for text measures have emerged in recent years (Coviello et al., 2014; Williamson, 2008; Williamson et al., 2013; Zhang et al., 2008). Measures of information content are taking a wide range of forms, many involving entropy. These statistical approaches tend to be dependent on particular data sets and algorithms. Little, if any, attention is put into identifying and implementing an invariant unit of measurement, or into designing and maintaining a metrological network of instruments traceable to such a unit.

The benefits of metrological comparability for measuring reading ability extend from the advancement of education science's basis in theory to practical quality improvement methods in schools and classrooms (Fisher, 2013). The natural sciences and the monetary economy both enjoy a degree of efficiency in their markets for the exchange of information and prices. This efficiency stems in large part from the existence of rules, roles, and responsibilities (Dear, 2012; Galison, 1997; Hutchins, 1995, 2012; Latour, 1987, 2005; Miller & O'Leary, 2007; Nersessian, 2012; O'Connell, 1993; Schaffer et al., 1992; Wise, 1995) associated with the institutionalization of common units of measurement, such as meters, grams, degrees Celsius, or dollars. Suppliers, manufacturers, marketers, accountants, advocates, and customers are able to better coordinate and align their investments in physical capital when information systems employ common languages. Similar kinds of coordinations can be expected to emerge as teachers, researchers, and psychometricians establish firmer expectations for educational outcomes and the exception that prove (in the sense of test) the rules. For instance, quality circles will facilitate the exchange of instructional outcome information across classrooms, grades, and schools in ways not possible with test scores reported in traditional percentages correct. Curriculum publishers are already developing individualized reading instruction modules that integrate assessment information in ways that make student learning trajectories portable across proprietary tests, schools, and countries.

6 Discussion

Projected comprehension rates should not be the only factor influencing text selection. To make the quantified measure the sole determinant of a curricular decision would be analogous to reducing a table to its physical dimensions when its colour, style, or sentimental or historical value might also be relevant.

Initial efforts at deploying the unit of measurement quickly encountered a chicken and egg question from book publishers: why should they adopt the unit as a means of indicating the text complexity of their books and articles if there were no schools or students prepared to take advantage of that information? Conversely, state departments of education and school districts asked, why should they be interested in a universally uniform measure of reading ability if there were no books or articles to match with students' ability measures?

The solution arose when one publisher incorporated the unit in their own system, involving both a reading curriculum and a reading assessment system. This coordinated reader-text matching made the link to the unit more attractive to testing agencies, who could now point to an additional use for their results; to book publishers, who now were assured of a population of students with measures to match with their books; and to state departments of education and school districts, who could now effectively put the matching system to work.

The English-based system is in use in the U.S., Canada and Australia (with applications emerging in New Zealand, South Africa, and England), and in ESL applications in Korea, Japan, Malaysia, Hong Kong, and elsewhere in Asia. A Spanish system for matching readers and texts in the same unit is in use in Mexico and the Philippines. Researchers in various parts of the world are exploring possibilities for expanding the reader-text matching system to Mandarin, French, Arabic, and other languages.

Educational textbook and curriculum publishers have developed online software applications for tracking individual student growth in reading ability. A report from one such system is shown is Fig. 2.

Figure 3 shows the relationship between expected and observed text complexity measures in the online system. This plot illustrates the power of theory. Traceability to the standard unit is achieved not only by estimating student reading ability measure s from data, but by gauging text complexity from its syntactic and semantic makeup. Given theory-based estimates of item difficulty, items can be adaptively selected for custom-tailored individualized administration, and those students' measures may then be estimated from their comprehension rates relative to the scale values of those items.

The specification equation operationalizes Rasch's notion of a frame of reference in a way that extends the frame beyond the specific objectivity obtained in the context of a particular test or set of equated tests to an indefinitely large collection of actual

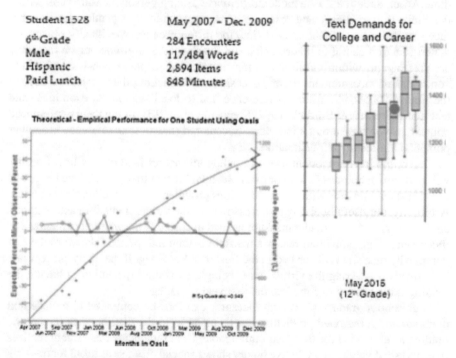

Fig. 2 Individual student online reading measure tracking system report, text domains from left to right are: High School (11–12), SAT 1 ACT AP, Military, Citizenship, Workplace, Community College, University, Graduate Record Exam

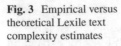

Fig. 3 Empirical versus theoretical Lexile text complexity estimates

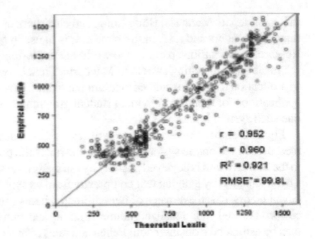

or virtual instruments, students, and texts. Theory-based instrument calibration eliminates the need to use data to both calibrate instruments and measure persons. The pay-off from using theory instead of data to calibrate instruments is large and immediate. When data fit a Rasch model, *differences* among person measures are, within the limits of uncertainty and response consistency, free of dependencies on other facets of the measurement context (i.e., the differences are specifically objective). When data fit a causal or theory-enhanced Rasch model, *absolute* person measures are free (again, within the limits of uncertainty and response consistency) of the conditions of measurement (items, occasions, etc.) making them objective beyond the limits of a specific frame of reference tied to local samples of examinees and test items (Stenner & Burdick, 1997; Stenner et al., 2013). In the theory-referenced context, person measures are individually-centered statistics; no reference to another person(s) figures in their estimation.

One of the most important uses of reading test scores is to predict how a reader will perform on non-test tasks. For example, imagine that first year college textbooks are virtual reading tests with item calibrations provided by the specification equation. Arbitrarily, but usefully, fixing a success rate on the virtual items for each textbook enables solving for the reader measure needed to correctly answer that percentage of those items. The individual reader's measure is then interpreted relative to the text complexity measure for each text in the freshman book bag. If the likely success rate in correctly answering the virtual items is high, so is the expectation of having the reading skills needed to complete the first year of college.

High school graduates' reading measures can thus be compared to college text demands and a reasoned prediction can be made as to the likelihood of having the reading level needed for first year completion. The efficiencies this system realizes from its use of validated predictive theory shows special promise as a tool for tracking reading readiness for post-secondary experiences in college, the work place, and the responsibilities of citizenship (Williamson, 2008; Williamson et al., 2013).

7 Conclusion

Historians of science have repeatedly documented the roles in theory development played by researchers with hands-on experience with instrumentation, as when Kuhn (1977) notes that seven of the nine pioneers in quantifying energy conversion processes were either trained as engineers or were working with engines when they made their contributions. Indeed, this attitude that an instrument can make a science was taken from physics into economics by both Stanley Jevons and Irving Fisher in their uses of the balance scale as a model of market equilibrium (Boumans et al., 2001; Maas et al., 2001).

But history shows that instruments alone are insufficient to the task of making a science. Furthermore, interestingly, equilibrium models have failed as guides to economic phenomena in large part because of problems in stochastic aggregation and variation in individual consumer behaviours (Ackerman, 2002). In specific circumstances (Dear, 2012; Galison, 1997; Hutchins, 1995, 2012; Latour, 1987, 2005; Nersessian, 2012; O'Connell, 1993; Schaffer et al., 1992; Shapin, 1989; Sundberg, 2011; Wise, 1995), however, instruments providing consistent information expressed in a common language throughout interconnected nodes of a network, as with the reading measurement system described here, may serve as a medium for coordinating spontaneous individual behaviours and decisions over time and space.

The historical success of science increasingly appears to stem from its embodiment of evolving ecologies of this kind of data-theory-instrument assemblage. Current conceptualizations and institutional systems prioritizing centralized design, data analysis, and policy formation stand in paradigmatic opposition to this ecologizing perspective (Arthur, 2014; Hayek, 1988; Hidalgo, 2015; Latour, 1995; Weitzel, 2004). How will cultures of decentralized innovation, complex self-organization, and authentic engagement with substantive, meaningful processes emerge in education and the social sciences? The organic integration of theory, data, and instruments in institutional contexts sensitive to ground-up self-organizing processes requires systematic conceptualizations of measurement as a distributed process, where scientific fields, markets, and societies operate as massively parallel stochastic computers (Arthur, 2014; Hidalgo, 2015). Recent comparisons of engineering and psychometric perspectives on the possibility of such systems in education suggest a viable basis for such conceptualizations (Mari Wilson, 2013; Pendrill, 2014; Pendrill & Fisher, 2013, 2015; Turetsky & Bashkansky, 2016; Wilson et al., 2015). Metrological traceability systems of this kind (Fisher & Wilson, 2015) will integrate qualitative progressions in learning defined by predictive theories of causal relations (Stenner et al., 2013), construct maps (Wilson, 2009), and associated item hierarchies in educational assessments generally. Systematically introduced infrastructural supports could effectively exploit the proven value of formative assessment (Hattie, 2008) in a hopeful development for broadly enhancing educational outcomes via research and local quality improvement efforts.

Acknowledgements This research was supported by funding from the Gates Foundation and Meta-Metrics, Inc. Thanks is extended to an anonymous reviewer for multiple suggestions that improved the quality of the article.

References

Ackerman, F. (2002). Still dead after all these years: interpreting the failure of general equilibrium theory. *Journal of Economic Methodology, 9*, 119–139.

Arthur, W. B. (2014). *Complexity and the economy*. Oxford University Press.

Bijker, W. E., Hughes, T. P., & Pinch, T. (Eds.). (2012). *The social construction of technological systems: New directions in the sociology and history of technology*. MIT Press.

Boumans, M. (2001). Fisher's instrumental approach to index numbers. In M. S. Morgan & J. Klein (Eds.), *The age of economic measurement* (pp. 313–344). Duke University Press.

Box, G. E. P. (1979). Robustness in the strategy of scientific model building. In R. L. Launer & G. N. Wilkinson (Eds.), *Robustness in statistics* (pp. 201–235). Academic Press.

Burdick, D. S., Stone, M. H., & Stenner, A. J. (2006). The Combined Gas Law and a Rasch Reading Law. *Rasch Measurement Transactions, 20*, 1059–1060.

Carver, R. P. (1974). Measuring the primary effect of reading: reading storage technique, understanding judgments and cloze. *Journal of Reading Behavior, 6*, 249–274.

Cohen, J. (1994). The earth is round (p < 0.05). *American Psychologist, 49*, 997–1003.

Coviello, L., Fowler, J. H., & Franceschetti, M. (2014). Words on the Web: noninvasive detection of emotional contagion in online social networks. *Proceedings of the IEEE, 102*, 1911–1921.

Dear, P. (2012). Science is dead; long live science. *Osiris, 27*, 37–55.

Engelhard, G. (2001). Historical view of the influences of measurement and reading theories on the assessment of reading. *Journal of Applied Measurement, 2*, 1–26.

Fisher, W. P. (2009). Invariance and traceability for measures of human, social, and natural capital: theory and application. *Measurement, 42*, 1278–1287.

Fisher, W. P., Jr. (2013). Imagining education tailored to assessment as, for, and of learning: theory, standards, and quality improvement. *Assessment and Learning, 2*, 6–22.

Fisher, W. P., Jr., & Wilson, M. (2015). Building a productive trading zone in educational assessment research and practice. *Pensamiento Educativo: Revista De Investigacion Educacional Latinoamericana, 52*, 55–78.

Galison, P. (1997). *Image and logic: A material culture of microphysics*. University of Chicago Press.

Gibbs, J. W. (1981). *Elementary principles in statistical mechanics*. Yale University Press.

Hamilton, M., & Pitt, K. (2011). Changing policy discourses: constructing literacy inequalities. *International Journal of Educational Development, 31*, 596–605.

Hartley, R. V. L. (1928). Transmission of information. *Bell System Technical Journal, 7*, 535–563.

Hattie, J. (2008). *Visible learning: A synthesis of over 800 meta-analyses relating to achievement*. Routledge.

Hayek, F. A. (1988). *The fatal conceit*. University of Chicago Press.

Hidalgo, C. (2015). *Why information grows: The evolution of order, from atoms to economies*. Basic Books.

Husserl, E. (1970). *The crisis of European sciences and transcendental phenomenology*. Northwestern University Press.

Hutchins, E. (1995). *Cognition in the Wild*. MIT Press.

Hutchins, E. (2012). Concepts in practice as sources of order. *Mind, Culture, Activity, 19*, 314–323.

Kuhn, T. S. (1977). *The essential tension: Selected studies in scientific tradition and change*. University of Chicago Press.

Latour, B. (1987). *Science in action: How to follow scientists and engineers through society.* Cambridge University Press.

Latour, B. (1995). Moderniser ou écologiser? A la recherche de la septième Cité. *Ecologie Politique, 13*, 5–27.

Latour, B. (2005). *Reassembling the social: An introduction to actor-network-theory.* Oxford University Press.

Linacre, J. M. (2006). Bernoulli, Fisher, Shannon and Rasch. *Rasch Measurement Transactions, 20*, 1062–1063.

Loevinger, J. (1965). Person and population as psychometric concepts. *Psychological Review, 72*, 143–155.

Luce, R. D. (1978). Dimensionally invariant numerical laws correspond to meaningful qualitative relations. *Philosophy of Science, 45*, 1–16.

Luce, R. D., & Tukey, J. W. (1964). Simultaneous conjoint measurement: A new kind of fundamental measurement. *Journal of Mathematical Psychology, 1*, 1–27.

Maas, H. (2001). An instrument can make a science: Jevons's balancing acts in economics. In M. S. Morgan & J. Klein (Eds.), *The age of economic measurement* (pp. 277–302). Duke University Press.

Mari, L., & Wilson, M. (2013). A gentle introduction to Rasch measurement models for metrologists [abstract]. *Journal of Physics: Conference Series, 459*, 012002.

Miller, P., & O'Leary, T. (2007). Mediating instruments and making markets: Capital budgeting, science and the economy. *Accounting, Organizations and Society, 32*, 701–734.

Nersessian, N. J. (2012). Engineering concepts: the interplay between concept formation and modeling practices in bioengineering sciences. *Mind, Culture, Activity, 19*, 222–239.

Neuman, S. B., & Roskos, K. (2012). Helping children become more knowledgeable through text. *Reading Teacher, 66*, 207–210.

Nietzsche, F. (1967). *The will to power.* Vintage.

O'Connell, J. (1993). Metrology: the creation of universality by the circulation of particulars. *Social Studies of Science, 23*, 129–173.

Pendrill, L. (2014). Man as a measurement instrument. *NCSLI Measure: Journal Measurement Science, 9*, 22–33.

Pendrill, L., & Fisher, W. P., Jr. (2013). Quantifying human response: linking metrological and psychometric characterisations of man as a measurement instrument. *Journal of Physics: Conference Series, 459*, 012057.

Pendrill, L., & Fisher, W. P., Jr. (2015). Counting and quantification: comparing psychometric and metrological perspectives on visual perceptions of number. *Measurement, 71*, 46–55.

Price, D. J. (1986). *In little science, big science—And beyond* (pp. 237–253). Columbia University Press.

Rabkin, Y. M. (1992). Rediscovering the instrument: research, industry, and education. In R. Bud & S. E. Cozzens (Eds.), *Invisible connections: Instruments, institutions, and science* (pp. 57–82). SPIE Optical Engineering Press.

Rasch, G. (1960). *Probabilistic models for some intelligence and attainment tests.* Danmarks Paedogogiske Institut (Reprint, with Foreword and Afterword by B.D. Wright, University of Chicago Press).

Rasch, G. (1972/1988). Review of the cooperation of Professor B. D. Wright, University of Chicago, and Professor G. Rasch, University of Copenhagen; letter of June 18, 1972. *Rasch Measurement Transactions, 2*, 19.

Rasch, G. (1973/2011). All statistical models are wrong! Comments on a paper presented by Per Martin-Löf, at the Conference on Foundational Questions in Statistical Inference, Aarhus, Denmark, May 7-12, 1973. *Rasch Measurement Transactions, 24*, 1309.

Rentz, R. R., & Bashaw, W. L. (1977). The National Reference Scale for Reading: an application of the Rasch model. *Journal of Educational Measurement, 14*, 161–179.

Rivero, V. (2011, June 10). Interview: Jack Stenner takes education beyond the metrics. *Edtech Digest*. Retrieved May 23, 2012 from http://edtechdigest.wordpress.com/2011/06/10/interview-jack-stenner-takes-education-beyond-the-metrics/

Sadoski, M., & Paivio, A. (2007). Toward a unified theory of reading. *Scientific Studies of Reading, 11*, 337–356.

Salsburg, D. S. (1985). The religion of statistics as practiced in medical journals. *American Statistician, 39*, 220–223.

Schaffer, S. (1992). Late Victorian metrology and its instrumentation: a manufactory of Ohms. In R. Bud & S. E. Cozzens (Eds.), *Invisible connections: Instruments, institutions, and science* (pp. 23–56). SPIE Optical Engineering Press.

Shannon, C. (1948). A mathematical theory of communication. *Bell System Technical Journal, 27*, 379–423.

Shapin, S. (1989). The invisible technician. *American Scientist, 77*, 554–563.

Spies, R. A., Carlson, J. F., & Geisinger, K. F. (Eds.). (2010). *The eighteenth mental measurements yearbook*. University of Nebraska Press.

Stenner, A. J., & Smith, M. (1982). Testing construct theories. *Perceptual and Motor Skills, 55*, 415–426.

Stenner, A. J., & Stone, M. (2010). Generally objective measurement of human temperature and reading ability: some corollaries. *Journal of Applied Measurement, 11*, 244–252.

Stenner, A. J., Smith, M., & Burdick, D. S. (1983). Toward a theory of construct definition. *Journal of Educational Measurement, 20*, 305–316.

Stenner, A. J., Burdick, H., Sanford, E. E., & Burdick, D. S. (2006). How accurate are Lexile text measures? *Journal of Applied Measurement, 7*, 307–322.

Stenner, A. J., Fisher, W. P., Jr., Stone, M. H., & Burdick, D. S. (2013). Causal Rasch models. *Frontiers in Psychology: Quantitative Psychology and Measurement, 4*, 1–14.

Stenner, A. J., & Burdick, D. S. (1997). *The objective measurement of reading comprehension*. MetaMetrics, Inc. www.lexile.com

Sundberg, M. (2011). The dynamics of coordinated comparisons: how simulationists in astrophysics, oceanography, and meteorology create standards for results. *Social Studies of Science, 41*, 107–125.

Thoreau, H. D. (1854). *Walden, or life in the woods*. Ticknor and Fields.

Turetsky, V., & Bashkansky, E. (2016). Testing and evaluating one-dimensional latent ability. *Measurement, 78*, 348–357.

Weitzel, T. (2004). *Economics of standards in information networks*. Physica-Verlag.

Whitehead, A. N. (1911). *An introduction to mathematics*. Henry Holt and Co.

Williamson, G. L. (2008). A text readability continuum for postsecondary readiness. *Journal of Advanced Academics, 19*, 602–632.

Williamson, G. L., Fitzgerald, J., & Stenner, A. J. (2013). The Common Core State Standards' quantitative text complexity trajectory: figuring out how much complexity is enough. *Educational Researcher, 42*, 59–69.

Wilson, M. R. (2009). Measuring progressions: assessment structures underlying a learning progression. *Journal of Research in Science Teaching, 46*, 716–730.

Wilson, M., Mari, L., Maul, A., & Torres Irribara, D. (2015). A comparison of measurement concepts across physical science and social science domains: instrument design, calibration, and measurement. *Journal of Physics: Conference Series, 588*, 012034.

Wilson, T. P. (1971). Critique of ordinal variables. *Social Forces, 49*, 432–444.

Wise, M. N. (1995). *The values of precision*. Princeton University Press.

Woolley, A. W., & Fuchs, E. (2011). Collective intelligence in the organization of science. *Organization Science, 22*, 1359–1367.

Wright, B. D., & Stone, M. H. (1979). *Best test design*. MESA Press.

Zhang, W., Yoshida, T., & Tang, X. J. (2008). Text classification based on multiword with support vector machine. *Knowledge-Based Systems, 21*, 879–886.

Towards an Alignment of Engineering and Psychometric Approaches to Uncertainty in Measurement: Consequences for the Future

William P. Fisher Jr. and A. Jackson Stenner

Abstract The International Vocabulary of Measurement (VIM) and the Guide to Uncertainty in Measurement (GUM) shift the terms and concepts of measurement information quality away from an Error Approach toward a model-based Uncertainty Approach. An analogous shift has taken place in psychometrics with the decreasing use of True Score Theory and increasing attention to probabilistic models for unidimensional measurement. These corresponding shifts emerge from shared roots in cognitive processes common across the sciences and they point toward new opportunities for an art and science of living complex adaptive systems. The psychology of model-based reasoning sets the stage for not just a new consensus on measurement and uncertainty, and not just for a new valuation of the scientific status of psychology and the social sciences, but for an appreciation of how to harness the energy of self-organizing processes in ways that harmonize human relationships.

1 Introduction

While only rarely ever thinking about it, we use language to manage the uncertainty of the future. Everyone benefits from being able to connect words, ideas, and things in the world—to communicate—but without knowing very much about how the language in use came to be, or why it is written, pronounced, or structured as it is. We have simply learned that we can rely on our words to mean about the same thing the next time we use them as they did the last time. We derive a great deal of security from knowing we can manage the uncertain future using words that have served us

---

18th International Congress of Metrology, 12,004. 2017.

W. P. Fisher Jr.
Living Capital Metrics LLC, Sausalito, CA, USA

Graduate School of Education, BEAR Center, University of California, Berkeley, CA, USA

A. J. Stenner (✉)
MetaMetrics, Inc., Durham, NC, USA

well in the past. If memory failed us all, or if it was impossible to link concepts and things with some material visual or auditory representation, we would have to invent new words for things in a constant process of reinvention. In that kind of world, life's unpredictability would make experience very different from what it has been for humanity.

Science extends and refines language in ways enabling the management of new, previously inaccessible tasks. Theoretical conceptualizations and experimental substantiations of new phenomena, such as disease-causing germs and electricity, are embodied in technologies like vaccines and appliances distributed throughout interconnected networks (Latour, 1988, 2005). Science systematically associates new words with new concepts in order to bring new things into the social world (Hutchins, 2014; Nersessian, 2012). These word-concept-thing assemblages are kept in close contact with standards by technicians trained in the creation and use of the relevant tools. But all of this effort is expended so that end users can employ the new words and ideas in the same way as any other words, which is to say, without understanding the technical details of exactly how specifically unforeseeable future connections with something real in the world were made predictable and manageable.

Though it is not often articulated in this way, uncertainty and its management are plainly a matter of central importance in science. Measurement is the crucial activity that brings together in a portable technology (an instrument) what was learned in the past from data. That learning hinges, first, on the data being explained well enough by theory to predict future observations, and second, on the reduction in uncertainty realized in the precision of the instrument calibration. The calibration of new classes of instruments intended to measure previously unknown phenomena is, then, fundamentally a process of bringing new things into language by establishing consistent and reproducible relationships between their properties and the theoretical ideas and words instrumental to their communication. Science goes beyond everyday language in ascertaining and systematically reducing the uncertainty of the number words representing quantities, and in so doing opens up new opportunities for the creation of shared meanings and communities.

2 Uncertainty in Metrology and Psychometrics

It is in this context that we come to the most recent editions of the International Vocabulary of Metrology (VIM) and the Guide to Uncertainty in Measurement (GUM), which document a shift in the terms and concepts used in communications on uncertainty and measurement information quality (Committee and for Guides in Metrology (JCGM, WG 1) Evaluation of measurement data–Guide to the expression of uncertainty in measurement 2008; Joint Committee for Guides in Metrology (JCGM/WG 2) International vocabulary of metrology: basic and general concepts and associated terms 2012). The change is away from an Error Approach (also known as the Traditional Approach or the True Value Approach) and in favour of an Uncertainty Approach. Instead of assuming that the goal of measurement is the closest possible estimation of an unknown true value (the Error Approach), metrology now holds that measurement information supports only an assignment of a range of values, given that no mistakes have been made. This range varies depending on what information is taken into account, leading to the development of uncertainty budgets (Bucher, 2012; Ratcliffe et al., 2015).

An analogous shift has taken place in psychometrics, where recent research presented at a series of symposia jointly sponsored by several International Measurement Confederation (IMEKO) technical committees (TC-1, TC-7, and TC-13) suggests provocative new practical and theoretical correspondences between metrology and psychometrics (Andrich, in press; Fisher & Stenner, 2016; Mari & Wilson, 2014; Pendrill & Fisher, 2015). In accord with those similarities, differences between the psychometric binomial model's True Score Theory (and associated Classical Test Theory) and measurement theoretical approaches to error/uncertainty (Andrich, in press) parallel aspects of the shift documented in the VIM and GUM.

First, both metrology's Error Approach and psychometrics' True Score Theory focus on an assumed distinction between random and systematic errors. In both metrology and psychometrics, these sources of error are always assumed distinguishable, should be treated differently, and cannot be combined to form a total error. Second, both metrology's Uncertainty Approach and developments in psychometrics involving the evaluation of uncertainty relative to interval units of measurement focus on mathematical treatments of measurement uncertainty by employing an explicit measurement model characterizing the measurand in terms of an essentially unique value.

The GUM and IEC documents provide guidance on the Uncertainty Approach to end users as to the case of a single reading of a calibrated instrument. No such guidance has yet been systematically available in psychometrics, in large part because instruments calibrated and traceable to uniform and universally available consensus unit standards are still unusual, though not unknown (Fisher & Stenner, 2016). Routine estimation of individual measurand uncertainties is, then, encumbered by widespread reliance on True Score Theory's scale- and sample-dependent ordinal score units, and associated statistical methods. True Score Theory's sampling approach to group-level uncertainty supports one motivation for the probabilistic form of statistical models evaluated via significance tests, while measurement theory is instead motivated by the response process itself in deriving a scientific model evaluated in terms of explanatory theory, meaningfulness, ethical criteria, and practical utility (Andrich, 1989; Duncan, 1992; Duncan & Stenbeck, 1988; Wilson, 2013a). The latter is then able to characterize unit traceability in terms of a unique value systematically qualified relative to the effects of local uncertainty, model fit, bias and DIF, range restriction, etc.

3   The Contrasting Shapes of Uncertainty

One especially salient contrast between True Score Theory's and psychometric measurement theory's approaches to uncertainty concerns their different U-shaped vs arch shaped error distributions (Fig. 1) (Linacre, 2007).

This contrast stems from the lack of expectations concerning the response structure in True Score Theory, and the explicit expectations for it in measurement theory (Wilson, 2013b). In True Score Theory, the motivation for making models probabilistic is rooted in statistical sampling. Items are assumed to be of equal difficulty. The normal distribution is assumed, meaning that the tendency toward the mean dominates model conceptualization. Uncertainty is contingent on the probability of a response being near the middle part of the distribution, where there are 50–50 odds of success, and not toward one extreme or the other, where scores of 0% or 100% have the lowest possible uncertainty.

In measurement, in contrast, variation in item difficulties is expected and modelled, with the goal of obtaining a sufficient statistic, one "that summarizes the whole of the relevant information," as Ronald Fisher put it in 1922 (Fisher, 1922). Statistical sufficiency formulated at the individual level, per Rasch (Rasch, 1960; Wright, 1997), is conceptually identical with invariance in measurement (Andersen, 1977; Hall et al., 1965).

The identification of invariant profiles in the difficulty or agreeability of test and survey items enables use of the item location hierarchy in qualitative, substantive interpretation and theory-building otherwise inaccessible to psychological and social measurement (Stenner et al., 2013; Wilson, 2005; Wright & Masters, 1982; Wright & Stone, 1979). A close match between a student's measure and a test item's difficulty,

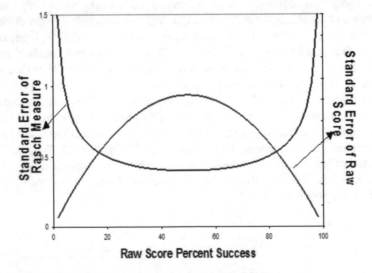

Fig. 1   , Standard errors of measures versus scores (Linacre, 2007)

for instance, indicates the border between the instructional materials the student has mastered (the easier items on which the student has a high probability of success), and those not yet mastered (the harder items on which the student has a low probability of success).

Uncertainty in this context becomes useful in allocating resources appropriate to the budgeting constraints of varying applications. Instructionally useful formative applications in the classroom, for instance, may be diagnostically valid for day to day use within wider confidence limits than can be tolerated for quality improvement, research, or accountability applications. Fewer test items and more uncertainty can be tolerated in individual applications given the background knowledge and familiarity with the student possessed by the teacher. High stakes assessments in which new interventions are being evaluated, or that are used as the basis for graduation, admissions, or other decisions, must necessarily obtain more and higher quality information with less uncertainty, typically by asking more questions in a more rigorously controlled environment.

The statistical sampling approach to uncertainty taken by True Score Theory, in contrast, is less applicable to instructional problems. Uncertainty is lowest when scores are at the extremes, when the student knows nothing or everything, but this minimal uncertainty does nothing to indicate either how much too difficult the test items are, or how much too easy they are.

4   The Role of Uncertainty in Unit Definition

Andrich (Andrich, in press) shows the role played by randomly unimodal error distributions with smooth density functions in the formulation and realization of measurement invariance using Rasch's probabilistic models. Ordered categories for scoring responses to test and survey items are required to partition a range of variation meaningfully to be useful as a basis for measurement. True Score Theory stops with these ordinal categories and scores, manipulating the numbers as though they stand for a qualitatively meaningful and substantive property (intelligence, achievement, health, etc.) that adds up like the numbers do. Rasch models interval properties in a way useful for experimentally testing ordinal data to assess its capacity for substantiating the hypothesis of an additive, invariant unit. Asserting the need for such a model, and for experimental tests of the existence of a unit of comparison, goes against the grain of the methods typically employed in psychology and the social sciences. Considerable resistance to the use of Rasch's models has come from those wedded to simpler and less demanding numerical methods, though the scientific defensibility and practical utility of Rasch-based results has led to the routine use of the models in many high-stakes domains of research and practice.

Broader scale applications of Rasch's models are likely to follow from the results described by Andrich (Andrich, in press). What he proposes should decisively resolve the "attenuation paradox" (Engelhard, 1993; Loevinger, 1954), namely, that higher reliability coefficients are not necessarily associated with better measurement. In

True Score Theory, a perfectly discriminating test, for instance, is one that separates all examinees into two groups, one with no correct responses and the other with all correct responses. This can be seen to follow from True Theory's peaked uncertainty distribution, where uncertainty is lowest at the extremes (Fig. 1).

The situation becomes more complex in the context of Guttman's deterministic approach to measurement. Here, the expectation of an ordered sequence of item difficulties contrasts with True Score Theory's supposition of equal item difficulties. But after an incorrect response is obtained, Guttman expects there to be no further correct responses to more difficult items. This rarely occurs in practice, so large percentages of data are often deemed unscalable in Guttman applications, leading to the method's rare application.

Rasch's models are basically a probabilistic formulation of Guttman's requirement of monotonic consistency in item responses (Andrich, 1982, 1985). The attenuation paradox takes a more specific form in the Guttman context than it does in True Score Theory. Guttman requires perfect reliability and discrimination in every score group, so that everyone with a score of 3 correct responses to 10 questions has exactly the same pattern. The end result, however, is that it becomes impossible to formulate any method for estimating how much more difficult one item is than another. This problem follows from the lack of information, the absence of any kind of a stochastic resonance, from which the magnitude of the difference might be estimated.

Duncan (Duncan, 1984) accordingly observes that "It is curious that the stochastic model of Rasch, which might be said to involve weaker assumptions than Guttman uses [in his deterministic models], actually leads to a stronger measurement model." Andrich's (Andrich, in press) explanation of the role randomly unimodal error distributions play in applications of Rasch's models may satisfy Duncan's curiosity. It seems to me that Andrich's account is likely to turn out to be an image of the deep structure of how certain kinds of noise-induced order (Gammaitoni et al., 1998; Matsumoto & Tsuda, 1983) come about. To what extent are the "sufficient conditions for a system to exhibit stochastic resonance" (Hess & Albano, 1998) the same as or different from the conditions sufficient for fit to a stochastic Rasch measurement model (Andersen, 1977; Andrich, 1989; Rasch, 1960; Wilson, 2013a; Wright, 1997)? How similar or different from the True Score Theory-Rasch or Guttman-Rasch contrasts is "the transition from chaotic behavior to ordered behavior induced by external noise…observed in a certain class of one-dimensional mappings" (Matsumoto & Tsuda, 1983)? To what extent could individual-based ecological models and agent-based economic models (Grimm & Railsback, 2013) be informed by Rasch's individual-level stochastic measurement models based in sufficient statistics?

This phenomenon of stochastic resonance provides a metaphor that meets the terms of Galison's (Galison, 1997) search for a way to talk about the unity-through-disunity observed in the social and conceptual discontinuities characterizing some complex systems, such as science (Fisher & Wilson, 2015; Fisher press, 1992, 2011a, 2011 in). Galison notes the insufficiency of Peirce's and Wittgenstein's multistrand cable and thread analogies as images of how communities of theoreticians, experimentalists, and instrumentarians interact. He suggests that instead of these homogenous images in which the whole is the sum of the parts, we need images of discontinuous structures, like amorphous semiconductors with disordered atomic properties, or laminated structural engineering materials. In these cases, microscopic failure and disorder provide otherwise-unattainable signal–noise ratios and structural integrity. A great deal more research and study is needed to see if the resonance properties exhibited in Rasch's stochastic measurement models' randomly unimodal error distributions (Andrich, in press) provide the combined harmonies and discordances needed to coordinate the varied material and symbolic processes creating the binding culture of science (Fisher & Wilson, 2015; Fisher press, 1992, 2011a, 2011 in; Galison, 1997).

5 Uncertainty Budgets

Andrich (Andrich, in press) notes that.

Measurement in the social sciences has not reached a level where the degree of precision is routinely stated in advance. To be able to do so, substantive research in the construction of ordered categories will require the same detailed empirical research that natural scientists carry out in constructing their instruments. The PRM [Polytomous Rasch Model] provides a basis for assessing the precision of measurement achieved.

Theory-informed empirical research on reading ability has begun to approximate the level of detail obtained in the construction of instruments in the natural sciences (Andrich, in press; Fisher & Stenner, 2016; Mari & Wilson, 2014; Pendrill & Fisher, 2015; Pendrill, 2014a; Stenner et al., 2013; Wright, 1997). Implications for uncertainty budgets and the practical interpretation of psychometric measures begin from a distinction between Type A and Type B uncertainties (Committee and for Guides in Metrology (JCGM, WG 1) Evaluation of measurement data–Guide to the expression of uncertainty in measurement 2008; Ratcliffe et al., 2015).

Type A uncertainties are statistical and random, whereas Type B uncertainties are usually systematic. The True Value Approach did not allow uncertainties to be combined, but the more recent Uncertainty Approach allows the estimation of a total uncertainty from the accumulation of Type A and Type B uncertainties. Uncertainty components are no longer classified as random or systematic because of this capacity to bring all uncertainty estimates together into a single frame of reference in which they are comparable.

What might an uncertainty budget for psychometric measures look like? Table 1 proposes some elements, following the model provided by Bucher (Bucher, 2012). Type A uncertainties may emerge primarily from sampling considerations. As to Type B, the usual modelled measurement uncertainty, estimated as a function of the number of items administered, is complemented by a version inflated by positive values of the mean square model fit statistic (a chi-square divided by the degrees

Table 1 A possible psychometric uncertainty budget [modelled from example in 7]: How to fill in the blanks?

Type A uncertainty	Uncertainty description	Uncertainty	Distribution	Divisor	Standard uncertainty	Variance
1	Repeatability					
2	Combined Type A Uncertainty					
	Type B Uncertainty					
	Uncertainty description	Uncertainty	Distribution	Divisor	Standard uncertainty	Variance
1	Modeled	$1 / (\sqrt{\sum}(Pni(1-Pni))$	Randomly unimodal			
2	Fit-inflated	Modeled * MnSq > 1				
3	Range restriction					
4	Bias					
	Combined Type B Uncertainty					

of freedom) (Wright, 1995). Additional sources of uncertainty introduced by range restriction or bias might be imagined.

Before uncertainty contributors can be combined (using a root sum square method), they must be random, independent, and normalized to a standard uncertainty using the divisor suited to the relevant distribution, as specified in the GUM (Committee and for Guides in Metrology (JCGM, WG 1) Evaluation of measurement data–Guide to the expression of uncertainty in measurement 2008).

6 Implications for a New Art and Science of Self-Organizing Complex Adaptive Systems

Chaitin (Chaitin, 1994) observes that the deterministic conceptualizations of Newtonian mechanics and the arithmetic of the natural numbers have given way to chaos and randomness. Reductionist approaches can no longer tenably posit that information on individuals will aggregate into reliable information on groups. The whole is no longer the sum of the parts, even in the areas of life where that seems most obviously true (Garfinkel & pp., 1991). Instead of fully unique, independent, and separable individual elements, even in physical phenomena like the combined gas law, structural presuppositions at the microlevel are unavoidable.

It seems as though a transition is occurring in fundamental concepts, a transition analogous to the difference between True Score Theory and Guttman, on the one hand, and Rasch models, on the other. True Score Theory and Guttman assume all individuals are interchangeable, that the whole is the sum of the parts, and that there is no need for structural presuppositions. Rasch, in contrast, shows that individuals are not entirely unique, that they exhibit interdependencies, that the whole is more than the sum of the parts, and that group-level structure consistently emerges whether or not one is looking for it. The question is how to set up media for the self-organized, bottom-up repeatable and reproducible display and communication of these complex adaptive multilevel structures (Arthur, 2014; Fisher, in press). New opportunities are opening up in this direction across the sciences. For instance, in a recent book on complexity economics, Arthur (Arthur, 2014) points out that.

> Science and mathematics are shedding their certainties and embracing openness...there is no reason to expect that economics will differ in this regard.

New agent-based models re-orient psychology and social science away from policies imposing mechanically aggregated statistical results from the top down toward bottom-up initiations of grassroots efforts. The overall effect is akin to the difference between quality control and continuous quality improvement methods in industry (Fisher, 2013). The former were characterized by "tail-chopping" methods that merely removed the undesired end of a quality distribution, just to have it filled again in the next iteration. The latter, "curve-shifting" methods empower everyone on the front line to act to improve the production system itself via bottom-up, organic, self-organizing approaches to creating genuine value (Arthur, 2014; Fisher, in press). Providing teachers, clinicians, and managers with the coherent information tools they need to manage their responsibilities across the discontinuities inherent to multilevel communications systems (Fisher, 2013; Fisher et al., in press; Star & Ruhleder, 1996) will likely result in quality revolutions in education, health care, human resource management, and other areas equal to or greater than that experienced in manufacturing (Fisher, 2009, 2011b, 2012). Realizing the practical value of measurements as objective bases for decision supports demands close attention to uncertainty (Fisher et al., 2010; Pendrill, 2014b). Communicating the value of educational, health, human resource, and other psychological and social measures in common units with known uncertainties coordinated across levels of complexity is essential to grasping the opportunities for improved outcomes we have within our reach.

References

Andersen, E. B. (1977). Sufficient statistics and latent trait models. *Psychometrika, 42*, 69.

Andrich, D. (1982). An index of person separation in Latent Trait Theory, the traditional KR-20 index, and the Guttman scale response pattern. *Education Research and Perspectives, 9*, 95.

Andrich, D. (1985). An elaboration of Guttman scaling with Rasch models for measurement. In N. B. Tuma (Ed.), *Sociological methodology* (p. 33–80). San Francisco:Jossey- Bass.

Andrich, D. (1989). Distinctions between assumptions and requirements in measurement in the social sciences. In Keats, J. A., Taft, R., Heath, R. A., & Lovibond, S. H. (Eds.), *Mathematical and Theoretical Systems* (pp. 7–16). North Holland: Elsevier Science Publishers.

Andrich, D. (2017). A law of ordinal random error: the Rasch measurement model and random error distributions of ordinal assessments. *Journal of Physics: Conference Series,* in Press.

Arthur, W. B. (2014). *Complexity and the economy.* Oxford University Press.

Bucher, J. L. (2012). *The metrology handbook.* ASQ Quality Press.

Chaitin, G. J. (1994). Randomness and complexity in pure mathematics. *International Journal of Bifurcation and Chaos, 4,* 3.

Duncan, O. D. (1984). *Notes on social measurement.* Russell Sage Foundation.

Duncan, O. D. (1992). What if? *Contemporary Sociology, 21,* 667.

Duncan, O. D., Stenbeck, M. (1988). Panels and cohorts: Design and model in the study of voting turnout. In C. C. Clogg (Ed.), *Sociological Methodology* (pp. 1–35). Washington, DC: American Sociological Association.

Engelhard, G. (1993). What is the attenuation paradox? *Rasch Measurement Transactions, 6,* 257.

Fisher, R. A. (1922). On the mathematical foundations of theoretical statistics. *Philosophical Transactions of the Royal Society of London, 222,* 309.

Fisher, W. P., Jr. (1992). Stochastic resonance and Rasch measurement. *Rasch Measurement Transactions, 5,* 186.

Fisher, W. P., Jr. (2009). Invariance and traceability for measures of human, social, and natural capital: Theory and application. *Measurement, 42,* 1278.

Fisher, W. P., Jr. (2011a). Stochastic and historical resonances of the unit in physics and psychometrics. *Measurement: Interdisciplinary Research and Perspectives, 9,* 46.

Fisher, W. P., Jr. (2011b). Bringing human, social, and natural capital to life: Practical consequences and opportunities. *Journal of Applied Measurement, 12,* 49.

Fisher, W. P., Jr. (2012). What the world needs now: a bold plan for new standards. *Standards Engineering, 64,* 1.

Fisher, W. P., Jr. (2013). Imagining education tailored to assessment as, for, and of learning: Theory, standards, and quality improvement. *Assessment and Learning, 2,* 6.

Fisher, W. P., Jr. (2017). A practical approach to modeling complex adaptive flows in psychology and social science. *Procedia Computer Science,* in press.

Fisher, W. P., Jr., Elbaum, B., & Coulter, W. A. (2010). Reliability, precision, and measurement in the context of data from ability tests, surveys, and assessments. *Journal of Physics Conference Series, 238,* 012036.

Fisher, W. P., Jr., Oon, E. P.-T., & Benson, S. (2017). Applying Design Thinking to systemic problems in educational assessment information management. *Journal of Physics Conference Series,* in press.

Fisher, W. P., Jr., & Stenner, A. J. (2016). Theory-based metrological traceability in education: a reading measurement network. *Measurement, 92,* 489.

Fisher, W. P., Jr., & Wilson, M. (2015). Building a productive trading zone in educational assessment research and practice. *Pensamiento Educativo, 52,* 55.

Galison, P. (1997). *Image and logic.* University of Chicago Press.

Gammaitoni, L., Hanggi, P., Jung, P., & Marchesoni, F. (1998). Stochastic resonance. *Reviews of Modern Physics, 70,* 223.

Garfinkel, A. (1991). Reductionism. In R. Boyd, P. Gasper, & J. D. Trout (Eds.), *Philosophy of science* (pp. 443–459). Cambridge, Massachusetts: MIT Press.

Grimm, V., & Railsback, S. F. (2013). *Individual-based modeling and ecology.* Princeton University Press.

Hall, W. J., Wijsman, R. A., & Ghosh, J. K. (1965). The relationship between sufficiency and invariance with applications in sequential analysis. *Annals of Mathematical Statistics, 36,* 575.

Hess, S. M., Albano, A. M. (1998). Minimum requirements for stochastic resonance in threshold systems. *International Journal of Bifurcation and Chaos, 8*, 395.

Hutchins, E. (2014). The cultural ecosystem of human cognition. *Philosophical Psychology, 27*, 34.

Joint Committee for Guides in Metrology (JCGM/WG 1) Evaluation of measurement data--Guide to the expression of uncertainty in measurement (International Bureau of Weights and Measures—BIPM, Sevres, France, 2008)

Joint Committee for Guides in Metrology (JCGM/WG 2) International vocabulary of metrology: basic and general concepts and associated terms, 3rd ed (with minor corrections) (International Bureau of Weights and Measures—BIPM, Sevres, France, 2012)

Latour, B. (1988). *The Pasteurization of France*. Harvard University Press.

Latour, B. (2005). *Reassembling the social*. Oxford University Press.

Linacre, J. M. (2007). Standard errors and reliabilities: Rasch and raw score. *Rasch Measurement Transactions, 20*, 1086.

Loevinger, J. (1954). The attenuation paradox in test theory. *Psychological Bulletin, 51*, 493.

Mari, L., & Wilson, M. (2014). An introduction to the Rasch measurement approach for metrologists. *Measurement, 51*, 315.

Matsumoto, K., & Tsuda, I. (1983). Noise-induced order. *Journal of Statistical Physics, 31*, 87.

Nersessian, N. (2012). Engineering concepts: The interplay between concept formation and modeling practices in bioengineering sciences. *Mind, Culture, and Activity, 19*, 222

Pendrill, L. R. (2014a). Man as a measurement instrument. *NCSLi Measure: The Journal of Measurement Science, 9*(4), 22–33.

Pendrill, L. (2014b). Using measurement uncertainty in decision-making and conformity assessment. *Metrologia, 51*, S206.

Pendrill, L., & Fisher, W. P., Jr. (2015). Counting and quantification: Comparing psychometric and metrological perspectives on visual perceptions of number. *Measurement, 71*, 46.

Rasch, G. (1960). *Probabilistic models for some intelligence and attainment tests*. Danmarks Paedogogiske Institut.

Ratcliffe, C., Ratcliffe, B. (2015). Expanded uncertainty of a measurement and an uncertainty budget for a single measurement. In *Doubt-free uncertainty in measurement* (pp. 33–37). Zurich, Switzerland: Springer International Publishing.

Star, S. L., & Ruhleder, K. (1996). Steps toward an ecology of infrastructure: Design and access for large information spaces. *Information Systems Research, 7*, 111.

Stenner, J., Fisher, W. P., Jr., Stone, M. H., Burdick, D. S. (2013). Causal Rasch models. *Frontiers in Psychology: Quantitative Psychology and Measurement, 4*(1).

Wright, B. D. (1995). Which standard error? *Rasch Measurement Transactions, 9*, 436.

Wright, B. D. (1997). A history of social science measurement. *Educational Measurement Issues and Practice, 16*, 33.

Wilson, M. (2005). *Constructing measures*. Lawrence Erlbaum.

Wilson, M. R. (2013a). Seeking a balance between the statistical and scientific elements in psychometrics. *Psychometrika, 78*, 211.

Wilson, M. R. (2013b). Using the concept of a measurement system to characterize measurement models used in psychometrics. *Measurement, 46*, 3766.

Wright, B. D., & Masters, G. N. (1982). *Rating scale analysis*. MESA Press.

Wright, B. D., & Stone, M. H. (1979). *Best test design*. MESA Press.

The Unreasonable Effectiveness of Theory Based Instrument Calibration in the Natural Sciences: What Can the Social Sciences Learn?

A. Jackson Stenner, Mark H. Stone, and William P. Fisher Jr.

Abstract In his classic paper entitled "The Unreasonable Effectiveness of Mathematics in the Natural Sciences," Eugene Wigner addresses the question of why the language of Mathematics should prove so remarkably effective in the physical [natural] sciences. He marvels that "the enormous usefulness of mathematics in the natural sciences is something bordering on the mysterious and that there is no rational explanation for it." We have been similarly struck by the outsized benefits that theory based instrument calibrations convey on the natural sciences, in contrast with the almost universal practice in the social sciences of using data to calibrate instrumentation.

1 Introduction

Why is mathematics so remarkably effective in the natural sciences, and what might the social sciences have to learn from the way it is used in those sciences? Is the effectiveness of mathematics in the natural sciences truly "unreasonable," as Wigner (1960) put it? Previous research on Maxwell's foundational contributions shows the effectiveness of mathematical model-based reasoning to be rooted in everyday thinking (Nersessian, 2002), and not in any special capacities associated with scientists or their objects of study. Rasch's (1960) adoption of Maxwell's method of analogy (Nersessian, 2002) set the stage for extending the effectiveness of mathematics into the social sciences (Fisher, 2010; Fisher & Stenner, 2013).

Journal of Physics: Conference Series 1044. 2018

A. J. Stenner (✉)
MetaMetrics, Inc., Durham, North Carolina, USA

M. H. Stone
Department of Psychology, Aurora University, Aurora, Illinois, USA

W. P. Fisher Jr.
BEAR Center, Graduate School of Education, University of California, Berkeley, California, USA

© The Author(s) 2023
W. P. Fisher and P. J. Massengill (eds.), *Explanatory Models, Unit Standards, and Personalized Learning in Educational Measurement,*
https://doi.org/10.1007/978-981-19-3747-7_23

In our ongoing explorations of the ways the natural sciences and social sciences invoke, define, and engage in measurement, we have identified a number of differences that are not as epistemologically necessary or predetermined as is popularly imagined (Stenner & Smith, 1982; Stenner et al., 1983, 2006, 2013; Williamson et al., 2013; Burdick et al., 2006; Stenner & Stone, 2010; Fisher, 2009; Fisher & Stenner, 2016). We have, to some benefit, contrasted human temperature thermometry (NexTemp™ thermometers Medical Indicators Inc (2006); see Appendix A) with the testing of mathematical ability and the measurement of English language reading ability. Although cataloging these differences has been useful, we now believe they are all traceable to a common cause. Physical science measurement virtually without exception takes place in the context of well-developed substantive theory, experimental evidence, and instruments calibrated to uniform unit standards. In the natural sciences, theories are not just compelling stories about the relationships between measurement outcomes (such as the count of cavities turning black on a NexTemp thermometer), unit standards (degrees Celsius), and measurement mechanisms (a chemical specification equation). They are instead sufficiently elaborated and precise in their specifications that they can be used to calibrate instrumentation.

In contrast, throughout the behavioral and social sciences, instrument calibration depends on data, is typically devoid of theory, and is not traceable to a unit standard. We hypothesize that most of the observed differences between behavioral and physical science measurement are traceable to these foundational differences. The absence of theory is the primary determinant of the need for data-based calibration and the lack of efficient methods for defining units and traceability to them. Further, we offer an example of a theory-referenced reading and text measurement system in the educational sciences that exhibits key theoretical, experimental, and instrumentation features analogous to those of human thermometry. Finally, we review the affordances shared by human thermometry and reading measurement (Cano et al., 2016).

2 A Reading Example

A consensus unit and systems for ensuring traceability to it are typical of most natural science measurement. Sometimes, as in temperature measurement, the unification process is not fully completed, but for the vast majority of natural science attributes (referred to as constructs in psychometrics), a unification process has resulted in diverse instrument makers sharing a unit of measure even when the measurement mechanisms vary from manufacturer to manufacturer. Mercury in glass tube thermometers for human temperature measurement differ substantively with NexTemp technology, but produce comparable results. Though the measurement mechanisms are drastically different they both report out in either Fahrenheit or Celsius units. In the case of NexTemp thermometry, a chemical specification equation calibrates the instrument in °C or °F. The chemical specification equation derived from experimental evidence enforces the unit, which is embodied in the instrument to a known degree of uncertainty. Similarly, in reading measurement, a text complexity specification equation enforces the unit and ensures that 100L of difference between two readers, two texts or a reader/text encounter is invariant over any of 100 + English reading tests that, at present, employ the unit (Fisher & Stenner, 2016).

Strictly parallel instruments are typical in the natural sciences. Such instruments share a common correspondence table that links a measurement outcome (count of cavities turning black on a NexTemp thermometer) to a °C or °F. The ability to manufacture essentially identical instruments in large quantities is a hallmark of natural science measurement. The specification equation is the recipe for manufacturing and calibrating clones of an instrument. The social sciences borrow the concept and talk about 'parallel' instruments or 'alternate forms' and advertise that say, form A and B produce exchangeable measures. But without a specification equation it is impossible to manufacture copies or clones that share the same correspondence table. The reading measurement specification equation can be used to build strictly parallel clones of any reading test (see Appendix B). No such capability exists, for example, for mathematics, and this is so precisely because, at present, there exists no specification equation for mathematical ability that can calibrate mathematics test items (see Appendix C). Different mathematics tests are empirically linked to a common scale through large scale, expensive field studies typically involving thousands of students.

Typical Rasch model applications in the social sciences are singly prescriptive. The major prescription that data must meet is non-intersecting item characteristic curves (ICCs), which relate the probability of a correct response to the difference between person ability and item difficulty. The data are used to estimate person and item parameters with no a priori constraints on the item parameters. Mathematics ability measurement is achieved in this way, as is typical of much social science measurement. Because there is no strong substantive theory for 'mathematical ability,' there is no specification equation and, thus, no potential for theoretically calibrating items/instruments. Instrument calibrations depend on sample data and a property of the Rasch model: when data fit the model *differences* between persons and *differences* between items are independent of items and persons, respectively.

Contrast this singly prescriptive measurement framework with the doubly prescriptive models underlying NexTemp human thermometry and the theoretical framework for reading. In both these cases strong substantive theory coupled with either a Guttman model or a causal Rasch model requires not just data fit to the model but also data fit to the theory specified item/instrument calibrations. For NexTemp a chemical specification equation is used as a recipe for the chemical compound that fills each cavity. By precisely varying the amount of additive the difference between any two adjacent cavities in sensitivity to the green component of light is precisely 0.2 degrees Fahrenheit. The chemical specification equation enforces this common unit difference for each of the 44 adjacent cavity differences across the 9 °F operating range for the instrument.

When data fit a doubly prescriptive Rasch model absolute person measures (not merely differences) are independent of items and instruments and are independent of person sample precisely because no person data figures in the instrument calibration process. Theory calibrated Rasch models are, thus, doubly prescriptive: prescriptive as to Rasch model requirements and prescriptive as to the substantive theory i.e., item/instrument calibrations. Person misfit to a doubly prescriptive model signals that the measurement mechanism that transmits variation in the attribute to the measurement outcome (often a count) is not working as intended for that individual. Frequent failures of theoretical invariance forces reexamination of the substantive theory, the measurement mechanism and instrument calibration procedures. Theoretical invariance can be tested within person over time (e.g. reading ability growth trajectories) and when intra individual theoretical invariance holds across persons then inter-individual theoretical invariance necessarily holds i.e., the attribute is homologous (Borsboom & Dolan, 2007; Borsboom et al., 2009b; Hamaker et al., 2007; Molenaar, 2004; Molenaar & Newell, 2010).

Molenaar (Hamaker et al., 2007; Molenaar, 2004; Molenaar & Newell, 2010) shows that inferences moving in the reverse direction, inferring from inter-individual factor structures something about intra-individual factor structures, is fraught with complications. The fact that so much of social and psychological measurement is based upon factor analysis of inter-individual variation prompted Molenaar (Molenaar, 2004; Molenaar & Newell, 2010) to call for a Kuhnian revolution, a paradigm shift in the concepts and methods of measurement in psychology. This paper is intended as another in a series of contributions to this revolution (Fisher, 2009, 2010; Stenner & Smith, 1982; Stenner et al., 1983).

3 Conclusion

Unification of measurement refers to a 200-year-old process whereby dozens if not hundreds of distinct scales for measuring a common attribute are, sometimes quickly and more often slowly, reduced to one, two or three exchangeable units of measure. The history of temperature measurement is a paradigmatic case (Chang, 2004; Sherry, 2011) that parallels many contemporary measurement movements in the social and behavioral sciences. Typically, an attribute (construct) captures the imagination of a community of scholars and engineers and different tests, instruments, mechanisms, and scales are proposed for measuring the attribute, and each is uniquely named. Once there is consensus that the selfsame attribute is being measured across these various devices small scale linking studies are undertaken to build conversion tables to re express one unit in one or more other units. More advanced linking studies reduce the link to an equation $°F = °C * 9/5 + 32$ making for quick and easy conversions. Since at this stage there is often not much to elevate one scale about the competition the market place takes over and '*unification*', with all its time and cost savings eventually prevails. Sometimes unification is swift and decisive but more often, particularly in the social sciences, metrology is poorly understood and unification plods along.

A useful case study of unification in the social sciences is the longstanding network of reading measures that has linked 100 + English language reading tests across the world, 250,000 book measures and 200 million article measures. The unification process is 27 years old and is accelerating but is far from complete. This effort drew inspiration and strategies from the history of the unification of temperature (Chang, 2004; Molenaar & Newell, 2010; Stenner et al., 2013).

Rather than using factor analysis of inter individual data to define an attribute structure and then asking if this structure obtains when examining intra individual data we suggest the use of substantive theory (in the form of specification/calibration equations) to establish the universality of attribute structure and measurement mechanism at the individual level. Once this is accomplished there is no puzzle about whether between person differences have the same structure as within person differences—of course they do. So, what this analysis reveals is that it is problematic to study between person variation at one point in time to glimpse truths about within person structures over time (Hamaker et al., 2007; Williamson et al., 2013). But the surprise is that if we start with within-person theory-referenced measurement, where in the extreme no two persons have any items in common over 5 years of measurement, then we would not stop for a moment to puzzle about the validity of the claim that at the end of year 1 Jane was higher than Bob but at the end of year 5 Bob was higher than Jane (i.e., a claim about inter-individual variation.) This is yet another benefit of theory based instrument calibration.

Several key features distinguishing physical science and behavioral science measurement systems can be traced to the absence of substantive theory sufficiently developed that said theory can be used to calibrate measurement instruments. Once such a calibration/specification equation is available most of these distinguishing features can quickly and easily be imported into the behavioral sciences.

Appendix A. The NexTemp Thermometer

"The NexTemp Thermometer is a thin, flexible, paddle-shaped plastic strip containing multiple cavities. In the Fahrenheit version, the 45 cavities are arranged in a double matrix at the functioning end of the unit. The columns are spaced 0.2^0 F intervals covering the range of 96^0 F to 104.8^0 F.... Each cavity contains a chemical composition comprised of three cholesteric liquid crystal compounds and a varying concentration of a soluble additive. These chemical compositions have discrete and repeatable change-of-state temperatures consistent with an empirically established formula to produce a series of change-of-state temperatures consistent with the indicated temperature points on the device. The chemicals are fully encapsulated by a clear polymeric film, which allows observation of the physical change but prevents any user contact with the chemicals. When the thermometer is placed in an environment within its measure range, such as 98.6^0 F (37.0^0 C), the chemicals in all of the cavities up to and including 98.6^0 F (37.0^0 C) change from a liquid crystal to an isotropic clear liquid state. This change of state is accompanied by an optical change that is easily viewed by a user. The green component of white light is reflected from the liquid crystal state but is transmitted through the isotropic liquid state and absorbed by the black background. As a result, those cavities containing compositions with threshold temperatures up to and including 98.6^0 F (37.0^0 C) appear black, whereas those with transition temperatures of 98.6^0 F (37.0^0 C) and higher continue to appear green" (Medical Indicators Inc., 2006). Thus, the observed outcome is a count of cavities turned black. The measurement mechanism is an encased chemical compound that includes a varying soluble agent that changes optical properties according to changes in temperature. Amount of soluble agent can be traded off for change in human temperature to hold number of black cavities constant.

Appendix B. Edsphere

The Edsphere™ technology for measuring English language reading ability employs computer-generated, four-option, multiple choice cloze items built on the fly for any prose text. Counts correct on these items are converted into reading measures in the standard unit via an applicable Rasch model. Individual cloze items are one-off creations and disposable; an item is used only once. The cloze and foil selection protocol ensures that the correct answer (cloze) and incorrect answers (foils) match the vocabulary demands of the target text. The text complexity measure and the expected spread of the cloze items are given by a proprietary text theory and associated equations. Thus, the observed outcome is a count of correct answers. The measurement mechanism is a text with a specified complexity and an item generation protocol consistent with that text complexity measure. The text complexity measure can be traded off for a change in reading ability to hold constant the number of items answered correctly.

Appendix C. Mathematics ability measurement

Mathematics ability measurement consists of a common supplemental metric that locates students relative to a taxonomy of mathematical skills, concepts, and applications. In order to develop the framework, several tasks were undertaken: (1) develop a structure of mathematics that spans the developmental continuum from first grade content through Algebra I, Geometry, and Algebra II content, (2) develop a bank of items that have been field tested, (3) develop the unit scale (multiplier and anchor point) based on the calibrations of the field-test items, (4) validate the measurement of mathematics ability as defined by the framework, and (5) link extant tests of mathematical ability to the scale. The process of scale unification for mathematics ability is well underway (Simpson, et al., 2015).

At present the attribute "mathematical ability" is unspecified; i.e. there is no specification equation and associated analyzer that can be used to locate 'math text' on the scale. Rather, data intensive methods are employed to calibrate instrumentation and human intensive qualitative analysis is employed to locate math text (e.g., a chapter on adding fractions with uncommon denominators) on the scale. The vast majority of social science attributes are similarly unspecified. By contrasting NexTemp thermometry, theory-based reading measurement, and data-based mathematics ability measurement we hope to illuminate the chasm of difference between instrumentation that employs strong substantive theory and that which does not. For the vast majority of measurement systems, it is the case "that the difference between any two points for one individual is qualitatively the same as a corresponding difference between two individuals at one time point" (Borsboom et al., 2009a); that is, the attribute is homologous. The same cannot be said for many measurement systems used in the social sciences. We propose that the routine adoption of theory-based instrument calibrations will pave the way for homologous attributes in the social sciences, thus assuring that the attribute on which I differ from myself over time is the same attribute on which I differ from my brother (Borsboom, 2005).

References

Borsboom, D. (2005). *Measuring the mind*. Cambridge University Press.

Borsboom, D., & Dolan, C. V. (2007). Commentary: Theoretical equivalence, measurement invariance, and the idiographic filter. *Measurement, 5*, 236–263.

Borsboom, D., Cramer, A. O., Kievit, R. A., Scholten, A. Z., & Franic, S. (2009a). The end of construct validity. In R. Lissitz (Ed.), *The concept of validity* (pp. 135–170). Information Age Publishing.

Borsboom, D., Kievit, R. A., Cervone, D., & Hood, S. B. (2009b). The two disciplines of scientific psychology, or: The disunity of psychology as a working hypothesis. In J. Valsiner, P. C. M. Molenaar, M. C. D. P. Lyra, & N. Chaudhary (Eds.), *Dynamic process methodology in the social and developmental sciences* (pp. 67–98). Springer.

Burdick, D. S., Stone, M. H., & Stenner, A. J. (2006). The Combined Gas Law and a Rasch Reading Law. *Rasch Measurement Transactions, 20*, 1059–1060.

Cano, S., Vosk, T., Pendrill, L., & Stenner, A. J. (2016). On trial: the compatibility of measurement in the physical and social sciences. *Journal of Physics: Conference Series, 772*, 012025.

Chang, H. (2004). *Inventing temperature*. Oxford University Press.

Fisher, W. P., Jr. (2009). Invariance and traceability for measures of human, social, and natural capital: Theory and application. *Measurement, 42*, 1278–1287.

Fisher, W. P., Jr. (2010). The standard model in the history of the natural sciences, econometrics, and the social sciences. *Journal of Physics: Conference Series, 238*, 012016.

Fisher, W. P., Jr., & Stenner, A. J. (2013). On the potential for improved measurement in the human and social sciences. n Q. Zhang & H. Yang (Eds.), *Pacific Rim Objective Measurement Symposium 2012 Conference Proceedings* (pp. 1–11). Springer.

Fisher, W. P., Jr., & Stenner, A. J. (2016). Theory-based metrological traceability in education: a reading measurement network. *Measurement, 92*, 489–496.

Hamaker, E. L., Nesselroade, J. R., & Molenaar, P. C. M. (2007). The integrated trait–state model. *Journal of Research in Personality, 41*, 295–315.

Medical Indicators Inc. (2006). *NexTemp*. http://medicalindicators.com/nextemp-2/

Molenaar, P. C. M. (2004). A manifesto on psychology as idiographic science: bringing the person back into scientific psychology, this time forever. *Measurement: Interdisciplinary Research and Perspective, 2*, 201–218.

Molenaar, P. C. M., & Newell, K. M. (2010). *Individual pathways of change*. American Psychological Association.

Nersessian, N. J. (2002). Maxwell and "the method of physical analogy"": Model-based reasoning, generic abstraction, and conceptual change. In D. Malament (Ed.), *Essays in the history and philosophy of science and mathematics* (pp. 129–166). Court.

Rasch, G. (1960). *Probabilistic models for some intelligence and attainment tests* (Reprint, University of Chicago Press, 1980). Danmarks Paedogogiske Institut.

Sherry, D. (2011). *Studies in History and Philosophy of Science, 42*, 509–524.

Simpson, M. A., Kosh, A., Elmore, J., Bickel, L., Stenner, A. J., Fisher, W. P., Jr., et al. (2015). A session presented at the Annual Meeting of the *Family group mathematics item generation theory: Large scale implementation*. National Council on Measurement in Education.

Stenner, A. J., Burdick, H., Sanford, E. E., & Burdick, D. S. (2006). How accurate are Lexile text measures? *Journal of Applied Measurement, 7*, 307–322.

Stenner, A. J., Fisher, W. P., Jr., Stone, M. H., & Burdick, D. S. (2013). Causal Rasch models. *Frontiers in Psychology | Quantitative Psychology and Measurement, 4*. https://doi.org/10.3389/fpsyg.2013.00536

Stenner, A. J., & Smith, M. (1982). Testing construct theories. *Perceptual and Motor Skills, 55*, 415–426.

Stenner, A. J., Smith, M., & Burdick, D. S. (1983). Toward a theory of construct definition. *Journal of Educational Measurement, 20*, 305–316.

Stenner, A. J., & Stone, M. (2010). Generally objective measurement of human temperature and reading ability: Some corollaries. *Journal of Applied Measurement, 11*, 244–252.

Wigner, E. P. (1960). The unreasonable effectiveness of mathematics in the natural sciences. *Communications on Pure and Applied Mathematics, 13*, 1–14.

Williamson, G. L., Fitzgerald, J., & Stenner, A. J. (2013). The Common Core State Standards' quantitative text complexity trajectory: Figuring out how much complexity is enough. *Educational Researcher, 42*, 59–69.

On the Complex Geometry of Individuality and Growth: Cook's 1914 "Curves of Life" and Reading Measurement

William P. Fisher Jr. and A. Jackson Stenner

Abstract Growth in reading ability varies across individuals in terms of starting points, velocities, and decelerations. Reading assessments vary in the texts they include, the questions asked about those texts, and in the way responses are scored. Complex conceptual and operational challenges must be addressed if we are to coherently assess reading ability, so that learning outcomes are comparable within students over time, across classrooms, and across formative, interim, and accountability assessments. A philosophical and historical context in which to situate the problems emerges via analogies from scientific, aesthetic, and democratic values. In a work now over 100 years old, Cook's study of the geometry of proportions in art, architecture, and nature focuses more on individual variation than on average general patterns. Cook anticipates the point made by Kuhn and Rasch that the goal of research is the discovery of anomalies—not the discovery of scientific laws. Bluecher extends Cook's points by drawing an analogy between the beauty of individual variations in the Parthenon's pillars and the democratic resilience of unique citizen soldiers in Pericles' Athenian army. Lessons for how to approach reading measurement follow from the beauty and strength of stochastically integrated variations and uniformities in architectural, natural, and democratic principles.

1 Growth in Reading Ability

Over the last several years in the U.S., new standards for reading education, termed the Common Core State Standards (CCSS), have been criticized for tendencies to focus too narrowly on quantitative representations of text complexity. This focus is

Journal of Physics: Conference Series 1065. 2018.

W. P. Fisher Jr. (✉)
BEAR Center, Graduate School of Education, University of California, Berkeley, California 94702, USA

A. J. Stenner
MetaMetrics, Inc., Durham, North Carolina 27713, USA

W. P. Fisher and P. J. Massengill (eds.), *Explanatory Models, Unit Standards, and Personalized Learning in Educational Measurement*,
https://doi.org/10.1007/978-981-19-3747-7_24

said to come at the expense of broader qualitative aspects of what makes reading easy or difficult. Many fear that inordinate concern with raising the complexity of text may displace legitimate concerns with other problems affecting education outcomes (Gamson et al., 2013; Hiebert, 2012). Debates have been contentious for the early grades, as these set the stage for later growth but lack clear guidelines for evaluating text complexity (Hiebert, 2013). Suggestions for such guidelines (Fitzgerald et al., 2015) are augmented by Williamson (2018), who notes individual differences in reading growth, marked by variation in starting points, rates of change, and decelerations as learning reaches the diminishing returns of higher levels.

Williamson (2018) also points out that few individuals grow in reading ability along the smooth path defined by the average trajectory. Any given end point might be achieved by one student who started from a higher level, showed slow, steady growth, and barely decelerated through middle and high school. Another student starting from a lower level of reading ability might progress very rapidly, with a moderate amount of deceleration, and could reach the same twelfth-grade reading ability as the first student, but via different individual growth curve.

Addressing these key features of growth (status, velocity, and deceleration) in educational policy and pedagogy could lead to new approaches to improving reading outcomes (Williamson et al., 2014). Initial status might be elevated on broad scales via early-intervention reading programs, velocity could be increased by means of intensive practice, and deceleration might be counter-acted by creative approaches to summer school.

These practical suggestions fit well, perhaps unexpectedly, in a larger context of aesthetic, scientific, and democratic values. This expanded perspective is opened by Theodore Cook's 1914 book, *The Curves of Life* (Cook, 1914/1979). In it, Cook anticipates a number of later developments in the philosophy of science, measurement modelling, and political philosophy that hinge in large part on understanding the relation of individual differences to larger population-based proportions and patterns. Cook anticipates fractal and chaotic concepts concerning the role irreducible randomness plays even in arithmetic and Newtonian physics Chaitin (1994). Cook's perspective on the roles of deviations and exceptions to growth in nature and its adoption in art and architecture is related to Kuhn's (1977) and Rasch's (1980) emphases on the role of anomalies in measurement, and to Bluecher's (1968) analogy between the uniqueness of the Parthenon's pillars and the individuality of Periclean citizen-soldiers. Improving growth in reading might find an enlarged base of support in a context foregrounding these aesthetic, scientific, and democratic values.

2 Aesthetics

Cook (1914/1979, p. 325) points out that "The 'straight lines' of the Parthenon...are in reality subtle curves." Details of the many delicate irregularities in the Parthenon (see Fig. 1) include the facts that

Fig. 1 Exaggerated view of unique Parthenon pillars (Pollitt, 1972)

No two neighbouring capitals correspond in size, diameters of columns are unequal, inter-columnar spaces are irregular, the metope spaces are of varying width, none of the apparently vertical lines are true perpendiculars, the columns all lean towards the centre of the building, as do the side walls, the antae at the angles lean forward, the architrave and frieze lean backward, the main horizontal lines of construction are in curves which rise in vertical planes to the centre of each side, and these curves do not form parallels (Cook, 1914/1979, pp. 328–329).

These variations from perfection in the Parthenon tend to be explained using one of three approaches: the compensation theory, the exaggeration theory, and the tension theory (Pollitt, 1972, pp. 75–76). Compensation suggests that the deviations are built into the structure in order to correct the deficiencies of perception, to make things look more proportionate than they would look if they were actually as proportionate as possible. The exaggeration theory takes the view that the intention was to make the building look more impressive or larger than it really was. The tension theory suggests that the deviations combine into wholes greater than the sum of the parts, and so give the structure a vitality it would otherwise lack.

Pollitt (1972, p. 78) suggests that the tension theory "seems to reflect most naturally the intellectual experience of the age," since the measured variations

are intentional deviations from 'regularity' for the purpose of creating a tension in the mind of the viewer between what he expects to see and what he actually does see. The mind looks for a regular geometric paradigm of a temple with true horizontals, right angles etc., but the eye sees a complex aggregate of curves and variant dimensions. As a result, the mind struggles to reconcile what it knows with what the eye sees, and from this struggle arises a tension and fascination which make the structure seem vibrant, alive, and continually interesting (Pollitt, 1972, p. 76).

Cook (1914/1979) similarly asserts that "...phenomena of life and beauty are always accompanied by deviations from any simple mathematical expression we can at present formulate" (p. 434).

3 Science

The task of science was long considered discovery of scientific laws, and scientific method was a set of routinely repeatable steps in a process sure to lead to new knowledge. But questions about the roles of language and technology in theory development and data acquisition led to new ideas about scientific change. One result was the realization, as Kuhn (1977, p. 219) put it, that

> The road from scientific law to scientific measurement can rarely be travelled in the reverse direction. To discover quantitative regularity one must normally know what regularity one is seeking and one's instruments must be designed accordingly.

In this context, theoretical expectations embodied in measuring instruments come to the fore as the media in which facts become either salient or meaningless. In Kuhn's (1977, pp. 205–206) words,

> To the extent that measurement and quantitative technique play an especially significant role in scientific discovery, they do so precisely because, by displaying significant anomaly, they tell scientists when and where to look for a new qualitative phenomenon.

Like Kuhn, Cook (1914/1979) makes repeated statements concerning the same point, for instance, that "Clearly it will be unprofitable to emphasize the value of variations unless we also suggest some standard by which those variations can be measured" (p. 400; also see pp. 431, 434–435). Cook (1914/1979, p. 431) further maintains the parallel with Kuhn, remarking on the anomalies in the orbit of Uranus that led to the discovery of Neptune, a parallel also noted in theory-informed Rasch measurement (Burdick & Stenner, 1996; Linacre, 1997).

Rasch (Bluecher, 1968, p. 124), writing before Kuhn, similarly noted that "Once a law has been established within a certain field then the law itself may serve as a tool for deciding whether or not added stimuli and/or objects belong to the original group." Rasch (1972/2010, p. 1254) speaks of modelling and instruments as delimiting the field of validity; i.e., finding those agents and objects with the properties that belong within the specified frame of reference. A basic goal of measurement then ought to be to support the identification of anomalous and unique patterns, and the deployment of methods for translating those patterns into opportunities for individual, community, or organizational growth.

4 Democracy

An ethical and political theme emerges here in the identification of anomalies. A common but mistaken assumption is that measurement excludes test or survey items, and examinees or respondents, associated with inconsistently ordered responses. An alternative approach allows the exceptional observations to inform the interpretation and application of the measure (Bohlig et al., 1998). Tools facilitating use of all data, qualitative and quantitative, fitting the measurement model and not fitting it, are available. The KidMap (Masters, 1994), for instance, is a graphical illustration of correct and incorrect test answers, or agreeable and disagreeable survey responses. Strong theoretical reasons and empirical support for the overall validity of the measurement process and results (Stenner et al., 2013; Wilson, 2005) sets the stage for integrating local departures from general expectations (Fisher Jr. & Stenner, 2018).

Individual variability is less a threat to objectivity than an opportunity for learning. Extending Cook's point, Wonk (2002), recounting Bluecher (1968), notes that Greek temple pillars

> are unique, curved, each one slightly different. They are harmonized in a united effort. They are a democracy. Whereas, the temples of the older, Eastern empires are supported by columns that are simply straight sticks, interchangeable. The phalanx of individual citizens was stronger than the massed army of slaves [which enabled 9,000 Greek citizen soldiers to defeat 50,000 Persian mercenaries and slaves at the Battle of Marathon in the fifth century BCE].

In that same spirit, should we not today expect to incorporate universal standards for reading education outcomes into the new electronic universe of digital learning environments? Could doing so aid in extending universal human rights to literacy? In the same way that each pillar in the Parthenon is unique but bears its full share of the roof's weight, each citizen in a democracy is a distinct individual but shares full responsibility for the governance of the community. So, too, each reading student should be recognizable for her or his separate value while still having access to valid information on where he or she stands relative to everyone else and relative to basic expectations for success as an adult.

5 Conclusion

Cook (1914/1979) emphasizes that:

> ...there are many more interesting instances of deliberate and delicate divergence from mathematical accuracy, and it is this divergence which gives the Parthenon its living beauty. For the essential principle of life and growth is constant variation from the rigid type.

What might we take from Cook's sense of wonder and awe before architectural and natural expressions of unique individuality that nonetheless conform to universal patterns? How might divergences from mathematical perfection be seen to bring out the living beauty of growth in reading? Can constant variation from the rigid type in the varying starting points, velocities, and decelerations of growth in reading be understood as an expression of the essential principle of life and growth? If so, it may have the potential to aid in correcting the larger problems in education potentially glossed over by inordinate focus on matters of increased text complexity (Gamson et al., 2013).

References

Bluecher, H. (1968, October XIII). A dialogue with students. In *Blücher archive lecture transcripts*. Bard College Senior Symposium. http://www.bard.edu/-bluecher/lectures/dialogue/dialogue_pf.htm

Bohlig, M., Fisher, W. P., Masters, G. N., & Bond, T. (1998). Content validity and misfitting items. *Rasch Measurement Transactions, 12*, 60.

Burdick, H., & Stenner, A. J. (1996). Theoretical prediction of test items. *Rasch Measurement Transactions, 10*, 475.

Chaitin, G. J. (1994). Randomness and complexity in pure mathematics. *International Journal of Bifurcation and Chaos, 4*, 3–15.

Cook, T. A. (1914/1979). *The curves of life*. Dover.

Fitzgerald, J., Elmore, J., Koons, H., Hiebert, E., Bowen, K., Sanford-Moore, E. E., & Stenner, A. J. (2015). Important text characteristics for early-grades text complexity. *Journal of Educational Psychology, 107*, 4–29.

Gamson, D. A., Lu, X., & Eckert, S. A. (2013). Challenging the research base of the Common Core State Standards a historical reanalysis of text complexity. *Educational Researcher, 42*, 381–391.

Hiebert, E. H. (2012). The Common Core's staircase of text complexity: Getting the size of the first step right. *Reading Today, 29*, 26–27.

Hiebert, E. H. (2013). Supporting students' movement up the staircase of text complexity. *Reading Teacher, 66*, 459–468.

Fisher, W. P., Jr., & Stenner, A. J. (2018). Ecologizing vs modernizing in measurement and metrology. *Journal of Physics: Conference Series, 1044*, 012025.

Kuhn, T. S. (1977). *The essential tension*. University of Chicago Press.

Linacre, J. M. (1997). Is Rasch general enough? *Rasch Measurement Transactions, 11*, 555.

Masters, G. N. (1994). KIDMAP—a history. *Rasch Measurement Transactions, 8*, 366.

Pollitt, J. J. (1972). *Art and experience in classical Greece*. Cambridge University Press.

Rasch, G. (1972/2010). Retirement lecture of 9 March 1972: objectivity in social sciences: a method problem (Cecilie Kreiner, Trans.). *Rasch Measurement Transactions, 24*, 1252–1272.

Rasch, G. (1980). *Probabilistic models*. University of Chicago Press.

Stenner, A. J., Fisher, W. P., Jr., Stone, M., & Burdick, D. (2013). Causal Rasch models. *Frontiers in Psychology, 4*, 1–14.

Williamson, G. L. (2018). Exploring reading and mathematics growth through psychometric innovations applied to longitudinal data. *Cogent Education, 5*, 1464424.

Williamson, G. L., Fitzgerald, J., & Stenner, A. J. (2014). Student reading growth illuminates the common core text-complexity standard. *Elementary School Journal, 115*, 230–254.

Wilson, M. (2005). *Constructing measures*. Lawrence Erlbaum Associates.

Wonk, D. (2002, June 11). Theater review: Looking back. *Gambit Weekly, 32*. Retrieved August 21, 2022, from https://www.nola.com/gambit/events/stage_previews_reviews/article_6ba980aa-5863-59dc-ad78-648424016110.html.

Index

Printed in the United States
by Baker & Taylor Publisher Services